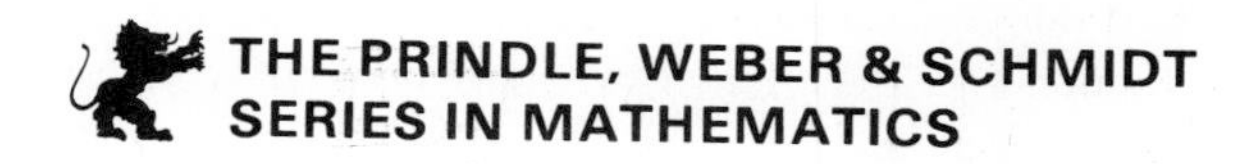

ALTHOEN/BUMCROT *Introduction to Discrete Mathematics*
BOYE/KAVANAUGH/WILLIAMS *Elementary Algebra*
BOYE/KAVANAUGH/WILLIAMS *Intermediate Algebra*
BURDEN/FAIRES *Numerical Analysis*, Fourth Edition
CASS/O'CONNOR *Fundamentals with Elements of Algebra*
CULLEN *Linear Algebra and Differential Equations*, Second Edition
DICK/PATTON *Calculus*, Volume I
DICK/PATTON *Calculus*, Volume II
DICK/PATTON *Technology in Calculus: A Sourcebook of Activities*
EVES *In Mathematical Circles*
EVES *Mathematical Circles Adieu*
EVES *Mathematical Circles Squared*
EVES *Return to Mathematical Circles*
FLETCHER/HOYLE/PATTY *Foundations of Discrete Mathematics*
FLETCHER/PATTY *Foundations of Higher Mathematics*, Second Edition
GANTNER/GANTNER *Trigonometry*
GELTNER/PETERSON *Geometry for College Students*, Second Edition
GILBERT/GILBERT *Elements of Modern Algebra*, Third Edition
GOBRAN *Beginning Algebra*, Fifth Edition
GOBRAN *Intermediate Algebra*, Fourth Edition
GORDON *Calculus and the Computer*
HALL *Algebra for College Students*
HALL *Beginning Algebra*
HALL *College Algebra with Applications*, Third Edition
HALL *Intermediate Algebra*
HARTFIEL/HOBBS *Elementary Linear Algebra*
HUMI/MILLER *Boundary Value Problems and Partial Differential Equations*
KAUFMANN *Algebra for College Students*, Fourth Edition
KAUFMANN *Algebra with Trigonometry for College Students*, Third Edition
KAUFMANN *College Algebra*, Second Edition
KAUFMANN *College Algebra and Trigonometry*, Second Edition
KAUFMANN *Elementary Algebra for College Students*, Fourth Edition
KAUFMANN *Intermediate Algebra for College Students*, Fourth Edition
KAUFMANN *Precalculus*, Second Edition
KAUFMANN *Trigonometry*
KENNEDY/GREEN *Prealgebra for College Students*
LAUFER *Discrete Mathematics and Applied Modern Algebra*
NICHOLSON *Elementary Linear Algebra with Applications*, Second Edition
PENCE *Calculus Activities for Graphic Calculators*
PENCE *Calculus Activities for the TI-81 Graphic Calculator*
PLYBON *An Introduction to Applied Numerical Analysis*
POWERS *Elementary Differential Equations*
POWERS *Elementary Differential Equations with Boundary-Value Problems*

PROGA *Arithmetic and Algebra*, Third Edition
PROGA *Basic Mathematics*, Third Edition
RICE/STRANGE *Plane Trigonometry*, Sixth Edition
SCHELIN/BANGE *Mathematical Analysis for Business and Economics*, Second Edition
STRNAD *Introductory Algebra*
SWOKOWSKI *Algebra and Trigonometry with Analytic Geometry*, Seventh Edition
SWOKOWSKI *Calculus*, Fifth Edition
SWOKOWSKI *Calculus*, Fifth Edition (Late Trigonometry Version)
SWOKOWSKI *Calculus of a Single Variable*
SWOKOWSKI *Fundamentals of College Algebra*, Seventh Edition
SWOKOWSKI *Fundamentals of College Algebra and Trigonometry*, Seventh Edition
SWOKOWSKI *Fundamentals of Trigonometry*, Seventh Edition
SWOKOWSKI *Precalculus: Functions and Graphs*, Sixth Edition
TAN *Applied Calculus*, Second Edition
TAN *Applied Finite Mathematics*, Third Edition
TAN *Calculus for the Managerial, Life, and Social Sciences*, Second Edition
TAN *College Mathematics*, Second Edition
TRIM *Applied Partial Differential Equations*
VENIT/BISHOP *Elementary Linear Algebra*, Third Edition
VENIT/BISHOP *Elementary Linear Algebra*, Alternate Second Edition
WIGGINS *Problem Solver for Finite Mathematics and Calculus*
WILLARD *Calculus and Its Applications*, Second Edition
WOOD/CAPELL *Arithmetic*
WOOD/CAPELL *Intermediate Algebra*
WOOD/CAPELL/HALL *Developmental Mathematics*, Fourth Edition
ZILL/CULLEN *Advanced Engineering Mathematics*
ZILL *A First Course in Differential Equations with Applications*, Fourth Edition
ZILL *Calculus*, Third Edition
ZILL *Differential Equations with Boundary-Value Problems*, Second Edition

BRABENEC *Introduction to Real Analysis*
EHRLICH *Fundamental Concepts of Abstract Algebra*
EVES *Foundations and Fundamental Concepts of Mathematics*, Third Edition
KEISLER *Elementary Calculus: An Infinitesimal Approach*, Second Edition
KIRKWOOD *An Introduction to Real Analysis*
RUCKLE *Modern Analysis: Measure Theory and Functional Analysis with Applications*
SIERADSKI *An Introduction to Topology and Homotopy*

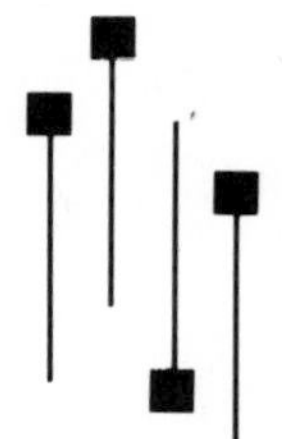

BOUNDARY VALUE PROBLEMS and PARTIAL DIFFERENTIAL EQUATIONS

M A Y E R   H U M I
Worcester Polytechnic Institute

W I L L I A M   B.   M I L L E R
Worcester Polytechnic Institute

PWS-KENT
PUBLISHING COMPANY
Boston

To my family.
M. H.
To Gerry.
W. B. M.

PWS-KENT
Publishing Company

Acquisitions Editor: Steve Quigley
Production Editor: Eve Mendelsohn Lehmann
Interior and Cover Designer: Eve Mendelsohn Lehmann
Manufacturing Coordinator: Peter Leatherwood
Artist: Santype International
Compositor: Santype International
Cover Printer: New England Book Components
Text Printer and Binder: The Maple-Vail Book Manufacturing Group

PWS-KENT Publishing Company is a division of Wadsworth, Inc.

Printed in the United States of America
91 92 93 94 95 -- 10 9 8 7 6 5 4 3 2 1

Library of Congress Cataloging-in-Publication Data

Humi, Mayer.
Boundary value problems and partial differential equations/
Mayer Humi, William B. Miller.
p. cm.
Includes index.
ISBN 0–534–92880–3
1. Boundary value problems. 2. Differential equations, Partial.
I. Miller, William, 1923– II. Title.
QA379.H86 1991
515′.353—dc20

91–21951
CIP

Preface

This book is an outgrowth of 15 years of teaching experience in a course on boundary value problems. It is intended to introduce junior and senior students to boundary value problems, with special emphasis on the modeling process that leads to partial differential equations. The book addresses the needs of engineering and science students who complete the calculus/differential equation sequence and then take a course on boundary value problems, usually after a lapse of some semesters. With this in mind, we have tried to make the derivations of formulas as explicit as possible, to allow students to follow the text with ease, and at the same time to provide them with a review of mathematical skills previously acquired. The method of presentation is intuitive and technique-oriented. In most cases, examples replace the formal theorem-proof sequence. The book does contain some asterisked sections for the more advanced students.

The book also addresses the need to acquaint students with some of the notions of modern mathematics. Thus, although we avoided the formal introduction of functional analysis, we discuss the Dirac delta function and its (weak) derivatives, Gibbs phenomena, eigenfunction expansions, and other related subjects.

Four paradigms for solving boundary value problems are included: separation of variables, closed form solutions (d'Alembert method and Green's functions), transform techniques, and numerical methods. These paradigms provide students with a rounded overview of the subject and address their practical needs in other disciplines.

The core material of the book is contained in Chapters 1–4. Instructors can pick and choose among the topics presented in the remaining chapters. Chapter 1 contains a careful derivation of several pertinent models. Chapter 2 includes an in-depth discussion of various boundary conditions. We recommend that only the first three sections be covered initially (depending on the mathematical maturity of the class). Chapter 3 covers Fourier series, and Chapter 4 discusses the method of separation of variables and the d'Alembert method for the wave equation in one space dimension.

The remaining chapters contain topics that can be covered according to the interest and maturity of the class. In Chapter 5, we introduce Bessel functions and their applications to the solution of boundary value problems in cylindrical coordinates. In Chapter 6 we do the same with Legendre polynomials and boundary value problems in spherical coordinates.

The first six chapters generally concern problems that are defined on finite intervals. In Chapter 7, students are introduced to Fourier integrals, which allow them to solve problems set on infinite or semi-infinite intervals.

One topic that has been missing from recent texts on boundary value problems is the subject of double Fourier series. Chapter 8 extends the ideas presented in Chapter 3 and demonstrates how the method of separation of variables can be extended to two-space in a time-dependent situation.

Chapter 9 is devoted to Green's functions, in order to introduce some key concepts from functional analysis that students will use in other courses such as finite elements. The presentation is geared toward giving students the practical tools needed to apply this method in various applications. Chapter 10 continues the discussion of transform methods initiated in Chapter 8, with emphasis on the Laplace transform. Chapter 11 concentrates on numerical methods, with self-contained material that assumes no background in numerical analysis. Whenever access to a mathematical routines library is available (such as IMSL, NAG, or ELLPACK) students should be encouraged to solve some practical boundary problems with the help of these packages and compare the solutions with the analytical (series) solutions.

Exercises in the book range from straightforward (necessary for building students' confidence) to challenging (for more advanced students). We have also included solutions to selected problems, but note that many exercises are formulated so that the final answer is given as part of the problem itself.

We would like to give special thanks to Mrs. C. M. Lewis, who typed and revised the manuscript several times. Thanks are also due to the staff of PWS-KENT for their patience, encouragement, and guidance throughout the process of producing this book. We are indebted to the following reviewers for their helpful comments and suggestions:

Sidney Birnbaum
California State Polytechnic University

Dennis W. Brewer
University of Arkansas

Herman Gollwitzer
Drexel University

David Green, Jr.
GMI Engineering and Management Institute

Mourad E. H. Ismail
University of South Florida

Gregory B. Passty
Southwest Texas State University

Raymond D. Terry
California Polytechnic State University

Roman Voronka
New Jersey Institute of Technology

Finally, we would like to thank our many students throughout the years who used this book in the form of lecture notes and provided us with many suggestions for its improvement.

Mayer Humi
William B. Miller

Contents

* An asterisk indicates optional sections.

CHAPTER ONE

Modeling with Partial Differential Equations

This book describes methods for solving partial differential equations. To approach this study from a practical point of view, we demonstrate, in this chapter, that these equations arise naturally as mathematical models for various physical systems. We begin by considering briefly the methodology of mathematical modeling.

SECTION 1
MATHEMATICAL MODELING

Mathematical modeling is the process whereby the evolution or the state of a real-life system is represented by a set of mathematical relations, after proper approximations and idealizations. In general this process can be divided into six steps:

1. **Objective.** In this step the real-life system to be modeled is defined. At this point it is often convenient to strip the system to its essential features so that a prototype model for it can be constructed. This reduction enables us to concentrate on the essential processes taking place in the system under consideration.
2. **Background.** In this step the pertinent laws and data about the system must be examined. In particular, if no data are available we must carry out proper experiments to obtain this information. As a result we should be able to identify the important variables that influence the evolution of the system and their relations.
3. **Approximations and idealizations.** Even when constructing a prototype model, some approximations and idealizations of reality must be made.

Thus, all mathematical models are approximations of reality to some extent. These approximations place certain limitations on the validity of the mathematical model and its correlation with the actual behavior of the system.

4. **Modeling.** At this stage the mathematical relations that govern the behavior of the system are derived. In general, we can distinguish between several types of mathematical models: probabilistic versus deterministic and continuous versus discrete. In this chapter (and in the rest of this book), we focus on deterministic and continuous physical systems that are modeled by partial differential equations. We study both the equations and the methods for solving them.
5. **Model validation.** Methods must be devised to solve the model equations and compare their predictions with the actual data about the system. If large unaccountable deviations between the model predictions and data are detected, the model must be reexamined and modified accordingly.
6. **Compounding.** At this stage the prototype model is modified to take into account some aspects of the system that were neglected earlier in order to simplify the modeling process. Also, systems with similar features might use the prototype model as a starting point for the derivation of an appropriate mathematical model.

In the rest of this chapter, we illustrate this modeling process by considering various systems that are modeled in terms of partial differential equations. In particular, we concentrate on the heat, wave, and potential equations that are important in many scientific and engineering applications. Methods to solve these equations with proper boundary conditions are discussed in the remainder of the book.

SECTION 2
THE HEAT (OR DIFFUSION) EQUATION

Objective

Build a model that describes the temperature distribution in a metal as a function of the position and time.

Discussion. As stated, this problem does not specify if the metal composition of the body is homogeneous; nor is any information given about its shape. Because such a general problem might require a complex model, we will first attempt to build a prototype model for the problem at hand and then compound it. To begin we consider the heat conduction problem in a rod of length L made of homogeneous metal with constant cross section A that is completely insulated along its lateral edges (see Figure 1). (All these assumptions are, naturally, mathematical idealizations.)

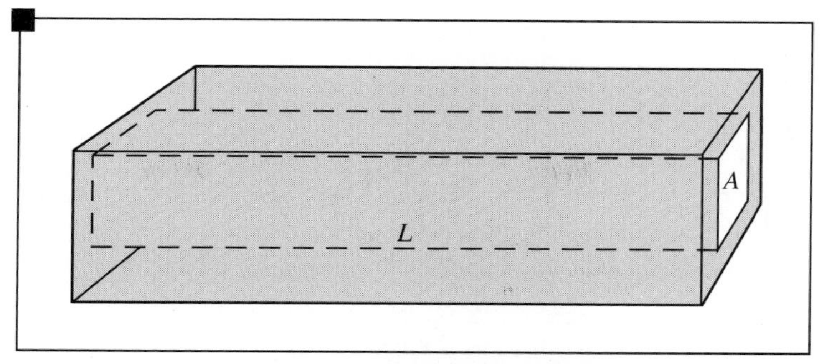

FIGURE 1 **Insulated Rod with Cross Section *A***

Background

To build an acceptable model for this problem, an understanding of the concept of the flux and the basic laws of thermodynamics (and physics) is necessary. We present here a short review of these pertinent ideas in one spatial dimension.

The Flux. Consider a flow of a certain physical quantity (such as mass, energy, heat, etc.). The flux $\mathbf{q}(\mathbf{x}, t)$ of this flow is defined as a vector in the direction of the flow [at $(\mathbf{x}, t)$] whose length is given by the amount of the quantity crossing a unit area (at $\mathbf{x}$) normal to the flow in unit time; that is,

$$|\mathbf{q}(\mathbf{x}, t)| = \lim_{\substack{\Delta S \to 0 \\ \Delta t \to 0}} \frac{\text{Quantity passing through } \Delta S \text{ in time } [t, t + \Delta t]}{\Delta S \, \Delta t}$$

where ΔS is a (small) surface area at $\mathbf{x}$ that is normal to the flow.

Thus, the approximate amount of the physical quantity passing through a surface ΔS in time Δt is given by

$$Q(\mathbf{x}, t, \Delta S, \Delta t) \approx |\mathbf{q}(\mathbf{x}, t)| \, \Delta S \, \Delta t \tag{1.1}$$

If ΔS is not normal to the flow, then it must be replaced by its projection in the direction normal to the flow.

EXAMPLE 1. Consider water flowing in a river with velocity $\mathbf{v}(\mathbf{x}, t)$. To evaluate the flux of this flow at $(\mathbf{x}, t)$, we consider a small surface element ΔS normal to $\mathbf{v}(\mathbf{x}, t)$. The amount of water flowing through ΔS in time $[t, t + \Delta t]$ is given by the quantity present at t in a tube of base ΔS and height $|\mathbf{v}| \, \Delta t$; that is,

$$Q(\mathbf{x}, t, \Delta S, \Delta t) \cong \rho |\mathbf{v}| \, \Delta S \, \Delta t$$

(where ρ is the mass density of the water). Hence,

$$\mathbf{q}(\mathbf{x}, t) = \rho \mathbf{v} \tag{1.2}$$

since the direction of the flow is given by $\mathbf{v}$ (see Figure 2). ■

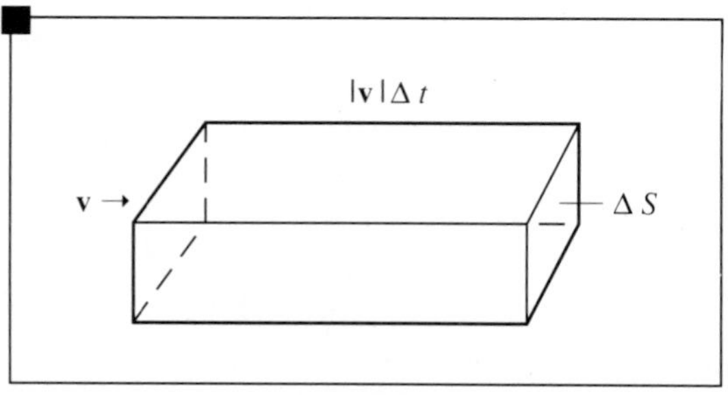

FIGURE 2 **Water Flow in a Tube**

Basic Laws of Thermodynamics. A change ΔQ in the amount of heat in a body of mass m is accompanied by a change Δu in its (equilibrium) temperature. The relationship between these changes is given by

$$\Delta Q = cm \, \Delta u \tag{1.3}$$

where c is the specific heat of the material of which the body is made; that is, the amount of heat required to raise the temperature of a body of unit mass (made of the same material) by 1 degree.

In the following discussion we assume that Q and u are normalized so that $Q = cmu$.

□ ***Remark About Units.*** In Equation (1.3) (as in any equation that relates physical quantities), a consistent set of units must be used. Thus, if the MKS system of units is used, then Q(energy) is expressed in joules, m in kilograms, u in degrees Kelvin (or Celsius), and c in joules/(kg.degK). In this book we consistently use the MKS units unless otherwise noted. □

Fourier Law of Heat Conduction. Heat is transported by diffusion in the direction opposite to the temperature gradient and at a rate proportional to it. Thus, the heat flux $\mathbf{q}(\mathbf{x}, t)$ is related to the temperature gradient by

$$\mathbf{q}(\mathbf{x}, t) = -\kappa \operatorname{grad} u(\mathbf{x}, t) = -\kappa\left(\frac{\partial u(\mathbf{x}, t)}{\partial x}, \frac{\partial u(\mathbf{x}, t)}{\partial y}, \frac{\partial u(\mathbf{x}, t)}{\partial z}\right) \tag{1.4}$$

where κ is the thermal conductivity of the material. [From Equation (1.4) we can easily infer that its units are joules/(m. sec. degK).]

Remember that the gradient of a function gives the direction in which the function increases most rapidly while in the direction opposite to it the function decreases most rapidly. Thus, a restatement of the Fourier law is that heat flows in the direction in which the temperature decreases most rapidly [and this is the reason for the minus sign in Equation (1.4)].

Principle of Energy Conservation. Because heat is a form of energy, it must be conserved. Hence, the rate of change in the amount of heat in a body must equal the rate at which heat is flowing in less the rate at which it is flowing out (we assume that no heat is generated by the body).

Approximations and Idealizations

1. Since we assumed that the material of the rod that we are considering is homogeneous, it follows that c, κ, and ρ (the material density kg/m^3) are independent of the position $\mathbf{x}$. However, for the purpose of constructing a prototype model we further assume that they are also independent of the temperature u.
2. We assume that the length of the rod remains constant in spite of the changes in its temperature.
3. We assume that the rod is perfectly insulated along its lateral surface (idealization). Hence, heat can flow only in the horizontal direction, since a vertical flow will lead to heat accumulation along the edges, which is forbidden by the Fourier law of conduction. Therefore, we infer that the temperature on a vertical cross section of the rod must be the same. Thus, the temperature u depends only on x and t; that is, $u = u(x, t)$.
4. For definiteness we assume that heat flows in the rod from left to right, which requires the left side to be warmer than the right.

Modeling

A possible approach to model the system under consideration is to use the atomic and crystal structure of the material of the rod and build a model for the heat conduction using these microscopic variables. However, because this approach will lead to a very complex set of equations, it is not useful in our context.

Thus, we present the following two methods to derive the **macroscopic** heat equation. We use the term *macroscopic* since we are using macroscopic variables such as u, c, κ, and so on to model the system.

Infinitesimal Approach. In this method we consider an infinitesimal element of the rod between x and $x + \Delta x$ and write the equation for the energy conservation in it.

Thus, since the volume of the element is $A\,\Delta x$, its mass Δm is given by $\rho A\,\Delta x$ (see Figure 3). The amount of heat in this element at time t is

$$Q(x, t, \Delta x) = c\,\Delta m u(x, t) \tag{1.5}$$

or

$$Q(x, t, \Delta x) = c\rho A u(x, t)\,\Delta x$$

The rate of change in Q is therefore given by

$$\frac{dQ}{dt} = c\rho A\,\frac{\partial u}{\partial t}\,\Delta x$$

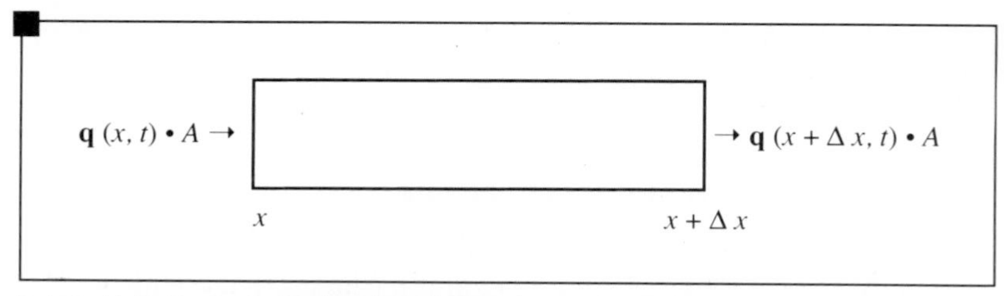

FIGURE 3 **Infinitesimal Section of the Rod**

By the principle of energy conservation, this rate of change must equal the rate at which heat is flowing in less the rate at which it is flowing out. Hence,

$$\frac{dQ}{dt} = q(x, t) \cdot A - q(x + \Delta x, t) \cdot A \tag{1.6}$$

Replacing $\frac{dQ}{dt}$ by

$$c\rho A \frac{\partial u}{\partial t} \Delta x$$

we have

$$c\rho A \frac{\partial u}{\partial t} = -A \frac{q(x + \Delta x, t) - q(x, t)}{\Delta x}$$

Letting $\Delta x \to 0$, we obtain

$$c\rho \frac{\partial u}{\partial t} = -\frac{\partial q}{\partial x}$$

But the Fourier law of heat conduction in one dimension yields $q = -\kappa(\partial u/\partial x)$ (since u is a function of x and t only!) and, therefore,

$$c\rho \frac{\partial u}{\partial t} = \kappa \frac{\partial^2 u}{\partial x^2}$$

or

$$\frac{1}{k} \frac{\partial u}{\partial t} = \frac{\partial^2 u}{\partial x^2} \tag{1.7}$$

where $k^{-1} = c\rho/\kappa$ is called the **thermal diffusivity**. Equation (1.7) is called the **heat** (or diffusion) **equation in one (space) dimension.**

Integral Approach. In this method we consider a *finite* section of the rod between a and b and use the principle of energy conservation to write an equation for the heat balance in this segment.

Since the amount of heat in an infinitesimal section of the rod between x and $x + \Delta x$ is given by Equation (1.5), the total amount of heat in the section

$[a, b]$ is given by the integral of the expression,

$$Q(t, a, b) = \int_a^b c\rho A u(x, t)\, dx$$

The rate of change in this quantity is therefore given by

$$\frac{dQ}{dt} = \int_a^b c\rho A \frac{\partial u}{\partial t}\, dx$$

By the principle of energy conservation, dQ/dt must equal the rate at which heat enters the section less the rate at which it leaves it. Thus

$$\frac{dQ}{dt} = Aq(a, t) - Aq(b, t) \tag{1.8}$$

By the fundamental theorem of calculus, this equation can be rewritten as

$$\frac{dQ}{dt} = -\int_a^b A \frac{\partial q}{\partial x}\, dx$$

from which it follows that

$$\int_a^b c\rho A \frac{\partial u}{\partial t}\, dx = -\int_a^b A \frac{\partial q}{\partial x}\, dx$$

By the Fourier law of heat conduction,

$$\int_a^b \left(c\rho A \frac{\partial u}{\partial t} - \kappa A \frac{\partial^2 u}{\partial x^2} \right) dx = 0 \tag{1.9}$$

But since a, b are arbitrary, Equation (1.9) implies that the integrand in this equation must also be zero. Hence,

$$c\rho \frac{\partial u}{\partial t} - \kappa \frac{\partial^2 u}{\partial x^2} = 0$$

which is the same equation we derived using the infinitesimal approach.

□ ***Remark.*** Although the infinitesimal and integral approaches must always (if applied correctly) yield the same result, from a conceptual modeling point of view one might be superior to the other in a given context. □

To illustrate the process of **model compounding** we now present the derivation of the heat equation in two dimensions.

EXAMPLE 2. Derive the heat equation for a thin homogeneous plate with constant cross section (height) h. Assume that the plate is perfectly insulated on the top and bottom.

Solution. Since the plate is thin and perfectly insulated,

$$u = u(x, y, t)$$

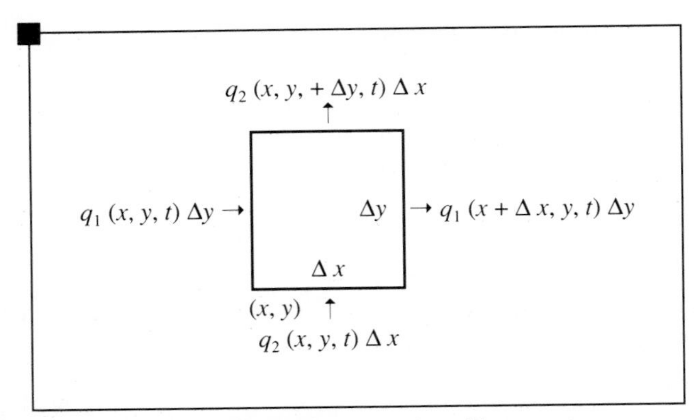

FIGURE 4 **Heat Balance in a Two-Dimensional Element**

To derive a model for u, we use the infinitesimal approach and consider a small rectangular element that is located at a point (x, y) in the plate (see Figure 4).

The amount of heat Q in this element at time t is given (approximately) by

$$Q(x, y, t, \Delta x, \Delta y) = c\rho h\, \Delta x\, \Delta y u(x, y, t)$$

Hence the rate of change in Q is

$$\frac{dQ}{dt} = c\rho h\, \Delta x\, \Delta y\, \frac{\partial u}{\partial t}(x, y, t) \tag{1.10}$$

This rate of change must equal the rate at which heat flows into the element minus the rate at which it flows out. To compute these rates we decompose $\mathbf{q}$ into

$$\mathbf{q} = q_1\mathbf{i} + q_2\mathbf{j}$$

and observe that $q_1\mathbf{i}$ is parallel to the boundary represented by the line between (x, y) and $(x + \Delta x, y)$. Hence, $q_1\mathbf{i}$ does not contribute to the flux through this boundary. Similar considerations apply to other boundaries. It follows then that

$$\frac{dQ}{dt} = q_2(x, y, t)\,\Delta x h + q_1(x, y)\,\Delta y h - q_2(x, y + \Delta y, t)\,\Delta x h$$
$$- q_1(x + \Delta x, y, t)\,\Delta y h \tag{1.11}$$

Equating Equation (1.10) with Equation (1.11) and dividing by $\Delta x\, \Delta y$ yields

$$c\rho\,\frac{\partial u}{\partial t} = -\frac{q_1(x + \Delta x, y, t) - q_1(x, y, t)}{\Delta x} - \frac{q_2(x, y + \Delta y, t) - q_2(x, y, t)}{\Delta y}$$

Letting $\Delta x, \Delta y \to 0$, we obtain

$$c\rho\,\frac{\partial u}{\partial t} = -\left(\frac{\partial q_1}{\partial x} + \frac{\partial q_2}{\partial y}\right) = -\operatorname{div}\mathbf{q}$$

To obtain an equation containing u only, we apply Equation (1.4) (in two dimensions). This leads to

$$\frac{1}{k}\frac{\partial u}{\partial t} = \nabla^2 u \tag{1.12}$$

where

$$\nabla^2 u = \frac{\partial^2 u}{\partial x^2} + \frac{\partial^2 u}{\partial y^2}$$

is the Laplace operator and $k = c\rho/\kappa$. Equation (1.12) is the **heat conduction equation in two dimensions.** ■

SECTION 2 EXERCISES

1. Generalize the prototype model to a situation in which heat is being generated in the rod at a rate of $r(x, t)$ per unit volume.

   □ ***Hint.*** The rate at which heat is being generated in any infinitesimal slice $[x, x + \Delta x]$ is given by $r(x, t)A\,\Delta x$. Use this to modify Equation (1.6) or (1.8). □

2. Generalize the prototype model to the case in which c, κ, and ρ are functions of x (nonhomogeneous rod).

3. Repeat Exercise 2 when c, κ, and ρ are functions of the temperature u.

4. Generalize the prototype model to the case $A = A(x)$ where $A(x)$ is a function that varies slowly with x so we can still assume approximately that the temperature u is a function of x and t.

5. Newton's law of cooling states that for a noninsulated rod the rate of the heat loss per unit length is proportional to $u - T_0$ where u is the rod temperature and T_0 is the temperature of the surroundings. Show that the heat equation in one dimension now takes the form

   $$c\rho\,\frac{\partial u}{\partial t} = \kappa\,\frac{\partial^2 u}{\partial x^2} - a(u - T_0)$$

   where a is a constant.

   □ ***Note.*** In this case we must assume that the rod is "thin" in order to be able to justify the approximation that temperature u is a function of x and t. □

6. A rod is made of a material that undergoes a chemical reaction as a result of which the specific heat and conductivity are changing with respect to time [i.e., $c = c(t)$ and $\kappa = \kappa(t)$]. Derive the corresponding heat equation in one dimension.

7. Derive the heat conduction equation for a three-dimensional body by considering an infinitesimal volume.

8. If the temperature inside a homogeneous and isotropic sphere is given to be a function of the radial distance r only, show that the heat conduction equation is given by

   $$\frac{1}{k}\frac{\partial u}{\partial t} = \left(\frac{\partial^2 u}{\partial r^2} + \frac{2}{r}\frac{\partial u}{\partial r}\right)$$

   □ ***Hint.*** Use infinitesimal spherical shells. □

9. Derive the heat conduction equation for an isotropic homogeneous and laterally insulated circular plate in which $u = u(r, t)$.

□ ***Hint.*** Consider infinitesimal rings. □

10. Show that if $u(x, y)$ is a solution of the heat equation, then $\partial u/\partial x$ and $\partial u/\partial t$ are also solutions of this equation.

***11.** Show that Burger's equation,

$$\frac{\partial u}{\partial t} + u\,\frac{\partial u}{\partial x} = \upsilon\,\frac{\partial^2 u}{\partial x^2}$$

can be transformed into the heat equation by introducing

$$w = -\,2\upsilon\,\frac{\partial}{\partial x}\,(\ln u)$$

SECTION 3
MODELING WAVE PHENOMENA

Objective

Construct a prototype mathematical model for the transverse vibrations of a string with fixed ends (see Figure 5).

Background

In general, an in-depth treatment of wave phenomena requires considering the elastic properties of matter and leads to a complicated set of equations. To overcome this difficulty we make the following simplifying approximations and idealizations so that a prototype model can be constructed by applying

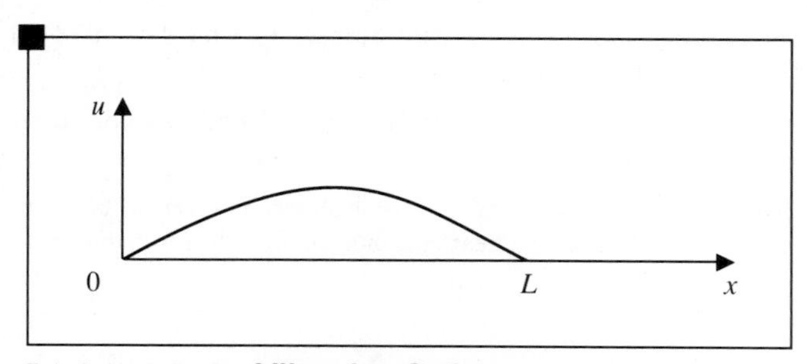

FIGURE 5 **Vibrating String**

only Newton's second law $\mathbf{F} = m\mathbf{a}$ (i.e., the force equals the mass multiplied by the acceleration) to the system under study.

Approximations and Idealizations

1. The string is rigidly attached at its endpoints.
2. The string vibrates in one plane.
3. No external forces act on the string (prototype model).
4. The string does not suffer from damping forces (prototype model).
5. The string is homogeneous. In particular, this implies that the linear density ρ and the mass per unit length m of the string are constant.
6. The deflection of the string from equilibrium and its slope are always small. Consequently, we are able to make the following two approximations:
   a. The magnitude of the tension force $\mathbf{T}(x, t)$ in the string is constant; that is, $|\mathbf{T}(x, t)| = T$.
   b. The string is rigid longitudinally; that is, a point on the string moves only in the vertical direction.
7. The tension force in the string is always tangential to it. This is usually expressed by saying that the string is assumed to be *perfectly flexible*.

Modeling

Consider a small segment of the string between x and $x + \Delta x$ as shown in Figure 6. Before we can apply Newton's second law to the motion of this segment, we must make the following observations:

1. By approximation 6b, the segment is not moving in the horizontal direction.
2. The mass of the segment is given by $\rho\ \Delta s$. However, since we are considering only small deflections, $|u| \ll 1$, it follows that $\Delta s \approx \Delta x$.
3. The acceleration of the segment in the vertical direction is given by $(\partial^2 u/\partial t^2)$.

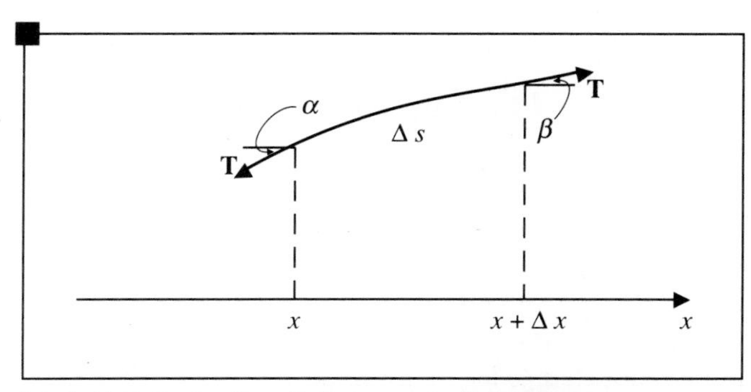

FIGURE 6 **Small Segment of the String**

4. The sum of the vertical forces acting on the segment is $T \sin\beta - T\sin\alpha$. (See Figure 6.)

Combining all these observations we infer from Newton's second law that

$$\rho\,\Delta x\,\frac{\partial^2 u}{\partial t^2} = T(\sin\beta - \sin\alpha) \tag{1.13}$$

By approximation 6, however, the slope of the string is always small, hence

$$\sin\alpha \cong \tan\alpha = \frac{\partial u}{\partial x}(x, t) \tag{1.14}$$

$$\sin\beta \cong \tan\beta = \frac{\partial u}{\partial x}(x + \Delta x, t) \tag{1.15}$$

Substituting Equations (1.13) and (1.14) in Equation (1.15) and dividing by Δx, we obtain

$$\rho\,\frac{\partial^2 u}{\partial t^2} = T\,\frac{\dfrac{\partial u}{\partial x}(x + \Delta x, t) - \dfrac{\partial u}{\partial x}(x, t)}{\Delta x}$$

Letting $\Delta x \to 0$, we finally obtain

$$\frac{1}{c^2}\frac{\partial^2 u}{\partial t^2} = \frac{\partial^2 u}{\partial x^2} \tag{1.16}$$

where $c^2 = T/\rho$.

Equation (1.16) is called the **wave equation in one dimension**.

□ ***Remark.*** From Figure 6 we can infer that the sum of the horizontal forces acting on the string segment is $T(\cos\beta - \cos\alpha) \neq 0$. Hence the segment must have an acceleration in the horizontal direction, which contradicts approximation 6b. However, since α, β are small, we can argue that $T(\cos\beta - \cos\alpha)$ is negligible. □

Compounding

EXAMPLE 3. Derive a model equation for very small vibrations of a vertically suspended chain whose length is L and whose mass density per unit length ρ is constant. ■

Approximations

1. Since the amplitude u of the vibrations is small, we assume that a point on the chain does not change its x-coordinate (see Figure 7).
2. The tension $T(x, t)$ in the chain cannot be assumed to be constant in the

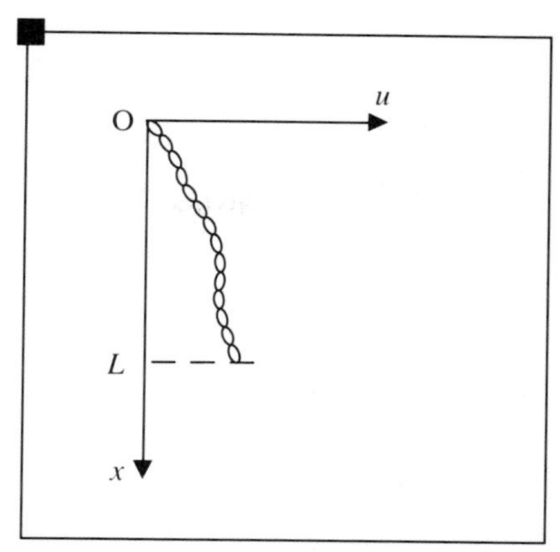

FIGURE 7 **Vibrating Chain**

context of this problem. In fact, in the equilibrium (vertical) position of the chain,

$$T(x) = \rho g(L - x) \tag{1.17}$$

In the following model we assume that Equation (1.17) gives an acceptable approximation for the tension in the vibrating chain when $|u| \ll 1$ and $\left|\frac{\partial u}{\partial t}\right| \ll 1$.

3. Other approximations and idealizations of the prototype model remain intact.

Modeling

To construct a mathematical model we once again consider a small section of chain between $[x, x + \Delta x]$. Applying Newton's second law in the horizontal direction to such a section (see Figure 8), we obtain

$$\rho \, \Delta x \, \frac{\partial^2 u}{\partial t^2} = T(x + \Delta x) \sin \beta - T(x) \sin \alpha$$

But since α, β are small, we can once again use the approximations given by Equations (1.14) and (1.15). Hence,

$$\rho \, \frac{\partial^2 u}{\partial t^2} = \frac{1}{\Delta x} \left[T(x + \Delta x) \frac{\partial u}{\partial x} (x + \Delta x, t) - T(x) \frac{\partial u}{\partial x} (x, t) \right]$$

Letting $\Delta x \to 0$, it follows then that

$$\rho \, \frac{\partial^2 u}{\partial t^2} = \frac{\partial}{\partial x} \left[T(x) \frac{\partial u}{\partial x} (x, t) \right]$$

Substituting Equation (1.17) for $T(x)$, we finally obtain

$$\frac{\partial^2 u}{\partial t^2} = g \, \frac{\partial}{\partial x} \left[(L - x) \frac{\partial u}{\partial x} \right] \tag{1.18}$$

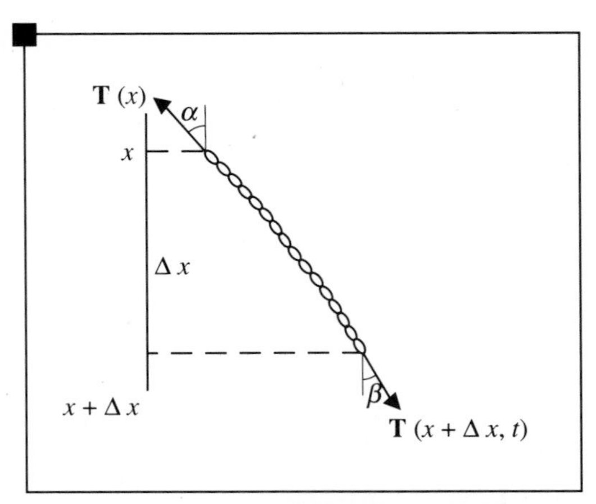

FIGURE 8 **Small Section of the Vibrating Chain**

Other examples of physical systems describing wave phenomena will be discussed in Sections 4 and 5.

SECTION 3 EXERCISES

12. Derive a model equation for the vibration of the string if a vertical external force $F(x, t)$ per unit length is acting on it. Especially consider the case of the gravitational force where $F = \rho g$.

13. Derive a model equation for the vibration of the string when its motion is subject to an elastic restraint and a damping force. (The restraint can be thought of as a force of ku per unit length acting to return the string to its equilibrium position. The damping force is given by $b(\partial u/\partial t)$ per unit length acting to oppose its motion.)

14. Generalize the model equation for the vibration of the string to the case where $\rho = \rho(x)$.

15. Longitudinal elastic waves. Consider an elastic homogeneous rod with a constant cross section placed along the x-axis. If we apply longitudinal stresses to the rod, then a particle whose rest position is x will find itself at $x + u(x, t)$, where we are assuming that the displacement $u(x, t) = u(x, t)$. It is known that the local stress, T (force/unit area), in an elastic bar satisfies $T = Eu_x$ where E is called the *elastic modulus*. Derive a model equation for $u(x, t)$ by considering a small section of the rod and the stress difference between its ends (see Figure 9).

□ ***Hint.*** Let ρ be the mass density of the bar and A its cross section. The mass of the small section is $\rho A\ \Delta x$, and its acceleration is $\partial^2 u/\partial t^2$. The stress difference across the section is $EA[u_x(x + \Delta x, t) - u_x(x, t)]$. Note that $c = \sqrt{\rho/E}$ is called the *speed of sound in the medium.* □

16. Derive a model equation for the vibrations of a membrane whose edges are fixed. Make similar assumptions as for the vibrating string.

□ ***Hint.*** Consider a small rectangular area and apply the same analysis as for the string in the x and y directions. □

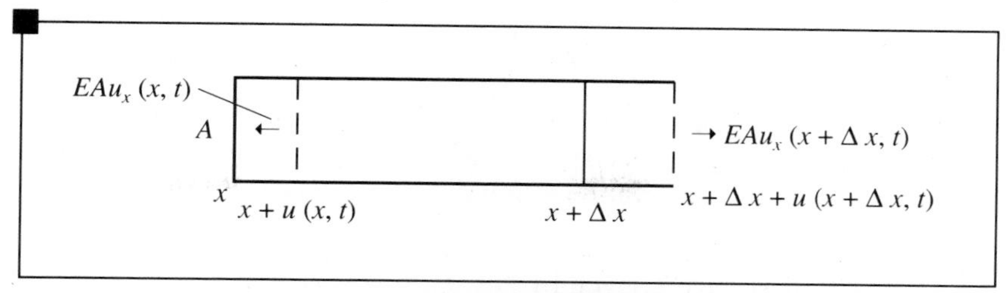

FIGURE 9 **Force Diagram on a Small Element of an Elastic Rod**

17. The pressure p and the mass flow rate u of a fluid flowing in a long pipe are related approximately by the equations

$$\frac{\partial p}{\partial t} = -c\frac{\partial u}{\partial x}$$

$$\frac{\partial u}{\partial t} = -\frac{\partial p}{\partial x}$$

where c is the compressibility of the fluid. Show that both p and u satisfy the wave equation in one dimension.

18. Show that

$$u(x, t) = f(x - ct) + g(x + ct)$$

is a solution of the wave equation in one dimension when f and g are any "smooth functions."

19. Derive a model equation for the small vibrations of a string that is rotating around the x-axis at a constant angular velocity ω. Assume that at each moment all the points of the string are in one plane.

20. A mass m is attached to the end of a suspended chain of length L and linear density ρ. Derive a model equation for the small vibrations of this system.

21. The results of Exercise 15 can be applied to the air vibrations in an organ pipe. However, since it is not "easy" to follow the position of air molecules, it is more natural to derive an equation for the pressure in the pipe. If it is known that the pressure is proportional to $\partial u/\partial x$, show that it must satisfy the wave equation.

SECTION 4*
SHALLOW WATER WAVES

Objective

Derive a prototype model equation that describes the phenomena of slow waves in shallow water (i.e., waves in a pool or on the seashore).

Background

An in-depth treatment of wave phenomena in fluids requires a knowledge of fluid mechanics. We simplify the derivation of this problem by applying Newton's second law and the following elementary facts.

1. The **hydrostatic pressure** p (force/unit area) in a fluid at a point of depth D below its surface is given by

   $$p = \rho g D$$

   where ρ is the mass density of the fluid and g the acceleration of gravity.
2. The **principle of mass conservation** states that the rate of change of mass in a given volume equals the rate at which mass is entering the volume less the rate at which it is leaving it.

Approximations and Idealizations

1. Water is incompressible and hence has a constant density ρ independent of the pressure.
2. The motion of the fluid under consideration is two dimensional; that is, each fluid particle is constrained to move in two dimensions, x and y (see Figure 10).
3. Because we are considering only slow waves, water acceleration is small and we can approximate the pressure at a point (x, y) beneath the water surface by the hydrostatic pressure

   $$p(x, y, t) = \rho g[h(x, t) - y]$$

   where $h(x, t)$ is the wave height.

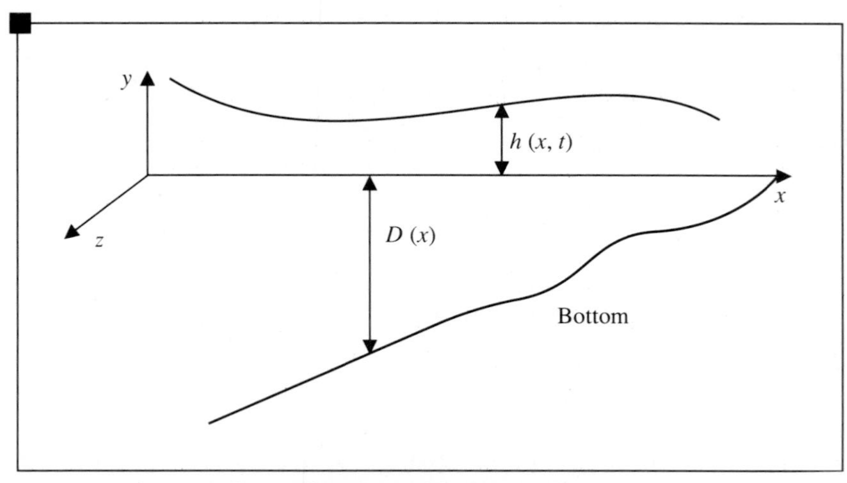

FIGURE 10 **Cross Section of Shallow Wave**

4. The force per unit volume in the fluid is given by the negative of the pressure gradient. In particular, the force per unit volume in the x direction is

$$F = -p_x = -\rho g h_x(x, t)$$

where $p_x = \partial p/\partial x$, and so on.

5. Since the force F is independent of y, it is reasonable to assume that the x component of a particle velocity u is a function of x and t only; that is, $u = u(x, t)$.

Modeling

Consider a small volume ΔV of water. Its mass is $\rho\, \Delta V$ and its acceleration is du/dt. The force acting on it in the x direction is $-\rho g h_x(x, t)\, \Delta V$. Hence by Newton's second law,

$$\rho\, \Delta V \frac{du}{dt} = -\rho g h_x(x, t)\, \Delta V \tag{1.19}$$

But $u = u[x(t), t]$, therefore,

$$\frac{du}{dt} = \frac{\partial u}{\partial t}\frac{dt}{dt} + \frac{\partial u}{\partial x}\frac{dx}{dt} = \frac{\partial u}{\partial t} + \frac{\partial u}{\partial x} u$$

Thus we infer from Equation (1.19) that

$$u_t + u_x u = -g h_x \tag{1.20}$$

Because Equation (1.20) contains two unknown quantities, we need another independent equation that relates h and u in order to be able to solve our model. Such an equation can be obtained by using the principle of mass conservation.

To apply this principle consider the column of water between x and $x + \Delta x$ and the horizontal length Δz in the z direction (see Figure 11).

The difference between the amounts of water in this volume at $t + \Delta t$ and t is

$$\Delta z \cdot \Delta x \cdot \rho[h(x, t + \Delta t) - h(x, t)]$$

This quantity must equal the difference between the amount of water entering this volume in time Δt less the amount leaving it. Hence,

$$\begin{aligned} &\Delta z \cdot \Delta t \cdot \rho\{[D(x) + h(x, t)]u(x, t) \\ &\qquad - [D(x + \Delta x) + h(x + \Delta x, t)]u(x + \Delta x, t)\} \\ &= \Delta z \cdot \Delta x \cdot \rho[h(x, t + \Delta t) - h(x, t)] \end{aligned} \tag{1.21}$$

Dividing Equation (1.21) by Δx, Δt and letting them approach zero, we obtain in the limit

$$h_t = -[u(h + D)]_x \tag{1.22}$$

Equations (1.20) and (1.22) constitute a system of coupled partial differential equations describing shallow water waves.

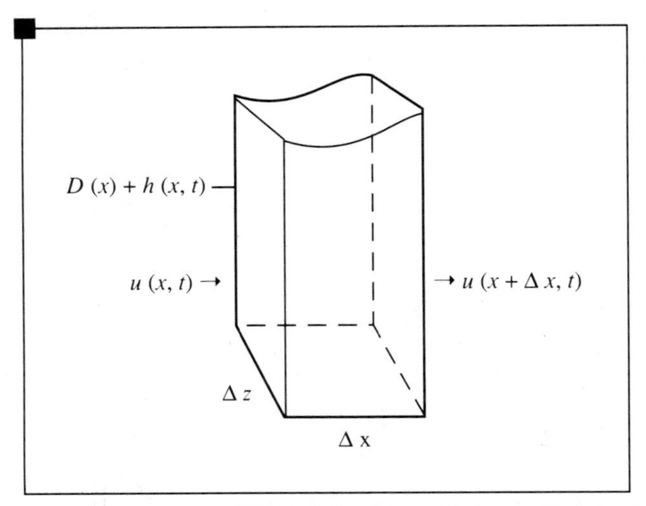

FIGURE 1 1 **Water Flow in a Volume Element**

SECTION 4 EXERCISES

22. If D is a constant and u, h and their derivatives are small (so that nonlinear terms in these quantities can be neglected), show that $h(x, t)$ must satisfy the wave equation

$$h_{tt} = (gD)h_{xx}$$

23. Under the same conditions as in Exercise 22, what is the equation that is satisfied by $u(x, t)$?

24. Repeat Exercises 22 and 23 when D is a linear function of x; that is,

$$D = ax + b \qquad a > 0, b < 0$$

25. Modify Equation (1.20) to include the action of a damping force (in the x direction) that is proportional to u.

SECTION 5* UNIFORM TRANSMISSION LINE

Objective

Derive a prototype model equation for the voltage and current in a long, uniform two-wire transmission line.

Background

(For a detailed introduction to electric circuits, see the appendix to this chapter.) The most common forms of transmission lines are coaxial and two-wire types. The coaxial transmission line consists of two concentric circular cylinders of metal. The two-wire types consist of two parallel wires, one of which is used as ground. We consider here the two-wire line.

The passage of an electric current through a cable always involves a leakage, which leads to a loss of electric energy. For short distances this loss can usually be ignored. However, over long distances, which are found in transmission lines, these losses must be taken into account.

Modeling

Since the transmission line is uniform, we assume that the resistance (R), capacitance (C), inductance (L), and leakage (G) *per unit length* of the transmission line are constant.

To derive the required model equations, we consider a small section of the line between x and $x + \Delta x$ and apply Kirchoff's laws to a circuit that is equivalent to it (see Figure 12). Thus $R\,\Delta x$, $L\,\Delta x$, $C\,\Delta x$, and $G\,\Delta x$ are, respectively, the resistance, inductance, capacitance, and conductance of the section. The quantity $G\,\Delta x$ (where G is expressed in mhos or siemens) is a "virtual" resistance so that the power lost through it to the ground is equal to that due to leakage.

Applying Kirchoff's second law between A, D, we obtain

$$e(x, t) - e(x + \Delta x, t) = R\,\Delta x\, i(x, t) + L\,\Delta x\,\frac{\partial i(x, t)}{\partial t}$$

Similarly applying Kirchoff's first law at the node B, we have

$$i(x, t) - i(x + \Delta x, t) = C\,\Delta x\,\frac{\partial e(x + \Delta x, t)}{\partial t} + G\,\Delta x\, e(x + \Delta x, t)$$

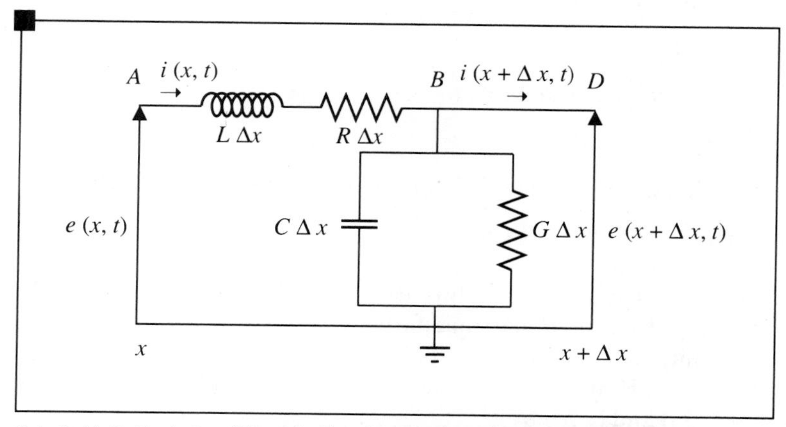

FIGURE 12 **Circuit Equivalent to a Small Section of the Wire**

Dividing these equations by Δx and letting Δx go to zero, we arrive at two differential equations:

$$\frac{\partial e}{\partial x}(x, t) = -Ri(x, t) - L\frac{\partial i(x, t)}{\partial t} \tag{1.23}$$

$$\frac{\partial i}{\partial x}(x, t) = -Ge(x, t) - C\frac{\partial e(x, t)}{\partial t} \tag{1.24}$$

Finally, we can find an equation for $e(x, t)$ only by differentiating Equations (1.23) and (1.24) with respect to x and t, respectively, to eliminate $i(x, t)$. We obtain

$$e_{xx} = LCe_{tt} + (LG + RC)e_t + RGe \tag{1.25}$$

Similarly, we can show that $i(x, t)$ satisfies

$$i_{xx} = LCi_{tt} + (LG + RC)i_t + RGi \tag{1.26}$$

Equations (1.25) and (1.26) are known as the **telegraph equations**.

Special Cases

High-Frequency Limit. To analyze qualitatively this limit, consider the case where

$$e(x, t) = A(x)\sin(\omega t + \varphi_1) \tag{1.27}$$

$$i(x, t) = B(x)\sin(\omega t + \varphi_2) \tag{1.28}$$

and $\omega \gg 1$. Under these assumptions the second term in the right-hand side of Equation (1.23), whose "effective coefficient" is $L\omega$ (the impedance), is much larger than the first term, whose effective coefficient is R. Hence Equation (1.23) can be approximated in this limit by

$$e_x = -Li_t \tag{1.29}$$

Similarly, Equation (1.24) reduces to

$$i_x = Ce_t \tag{1.30}$$

Combining Equations (1.29) and (1.30) we obtain the wave equation

$$e_{xx} = LCe_{tt}, \qquad i_{xx} = LCi_{tt}$$

Low-Frequency Limit. In this case, i and e change very slowly with time. Therefore, from a similar qualitative analysis as performed for the high frequency limit, with $\omega \ll 1$, we obtain that the effective coefficient of the second term ($L\omega$) is much smaller than R. Consequently, we can approximate Equations (1.23) and (1.24) by

$$e_x = -Ri, \qquad i_x = -Ge \tag{1.31}$$

and, therefore,

$$e_{xx} = RGe, \qquad i_{xx} = RGi \tag{1.32}$$

These are ordinary differential equations for i and e.

Submarine Cable. Earlier this century telecommunication signals between the United States and Europe were transmitted by cables that were laid down on the ocean floor. For these cables, $G \cong 0$ because of their extreme insulation. Moreover, the signal frequency ω is low. Under these circumstances we infer from Equations (1.27) and (1.28) that the impedance $L\omega$ is much smaller than R. Hence, Equation (1.23) can be approximated by

$$e_x = -Ri$$

Furthermore, since we have assumed $G = 0$, Equation (1.24) simplifies to

$$i_x = -C\frac{\partial e}{\partial t}$$

Combining these two equations, we can approximate Equations (1.23) and (1.24) as

$$e_{xx} = RCe_t, \qquad i_{xx} = RCi_t \tag{1.33}$$

which shows that e and i satisfy the one-dimensional diffusion equation.

SECTION 5 EXERCISES

26. Compare Equations (1.25) and (1.26) with the model equation for a spring mass system with friction and identify the physical meaning of each term in these equations.

27. Derive a differential equation for the voltage $e(t)$ in the circuit in Figure 13 if the current $i(t)$ is known.

28. The microscopic form of Maxwell's equations is

$$\nabla \cdot \mathbf{E} = 4\pi\rho \tag{1.34}$$

$$\nabla \cdot \mathbf{H} = 0 \tag{1.35}$$

$$\nabla \times \mathbf{H} = \frac{4\pi}{c}\mathbf{J} + \frac{1}{c}\frac{\partial \mathbf{H}}{\partial t} \tag{1.36}$$

$$\nabla \times \mathbf{E} + \frac{1}{c}\frac{\partial \mathbf{H}}{\partial t} = 0 \tag{1.37}$$

where $\mathbf{E}$ and $\mathbf{H}$ are, respectively, the electric and magnetic fields, $\mathbf{J}$ is the electric current, and ρ is the charge density. From Equation (1.35) we infer the existence of a vector potential $\mathbf{A}(x)$ so that

$$\mathbf{H} = \nabla x \mathbf{A}$$

Substituting this into Equation (1.37) we obtain

$$\nabla x\left(\mathbf{E} + \frac{1}{c}\frac{\partial \mathbf{A}}{\partial t}\right) = 0$$

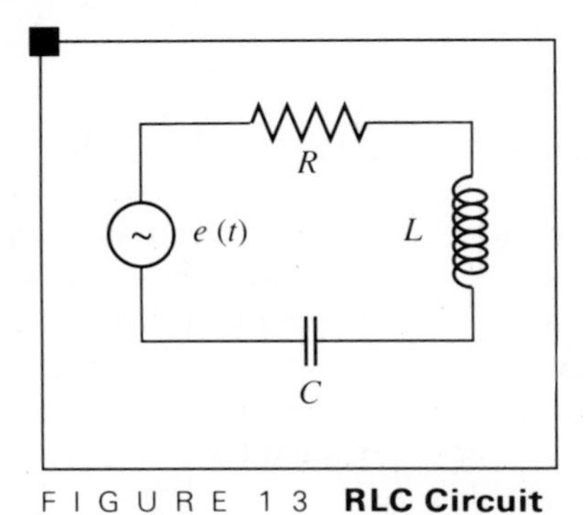

FIGURE 13 **RLC Circuit**

Hence, there exists a scalar potential function $\Phi(x)$ so that

$$\mathbf{E} + \frac{1}{c}\frac{\partial \mathbf{A}}{\partial t} = -\nabla\Phi$$

Show that if $\mathbf{A}$, Φ are chosen to satisfy the (gauge) condition

$$\nabla \mathbf{A} + \frac{1}{c}\frac{\partial \Phi}{\partial t} = 0$$

then $\mathbf{A}$ and Φ must satisfy the following inhomogeneous wave equations:

$$\nabla^2\Phi - \frac{1}{c^2}\frac{\partial^2\Phi}{\partial t^2} = -4\pi\rho$$

$$\nabla^2\mathbf{A} - \frac{1}{c^2}\frac{\partial^2\mathbf{A}}{\partial t^2} = -\frac{4\pi}{c}\mathbf{J}$$

where

$$\nabla^2 f = \frac{\partial^2 f}{\partial x^2} + \frac{\partial^2 f}{\partial y^2} + \frac{\partial^2 f}{\partial z^2}$$

SECTION 6
THE POTENTIAL (OR LAPLACE) EQUATION

Objective

Derive model equations to compute the gravitational field of a material body.

□ ***Remark.*** Although we restrict our discussion to the gravitational field, the static electric field would warrant similar treatment. □

Background

Newton's law of gravitation states that a point mass M attracts another point mass m by a force

$$\mathbf{F} = -\frac{GMm}{r^2}\,\mathbf{e}_r \tag{1.38}$$

where G is the gravitational constant, r is the distance between the two masses, and $\mathbf{e}_r$ is a unit vector along $\mathbf{r}$ (pointing outward from M; i.e., in the direction in which $\mathbf{r}$ increases) (see Figure 14).

Since Equation (1.38) can be rewritten as

$$\mathbf{F} = \left(-\frac{GM}{r^2}\,\mathbf{e}_r\right)m$$

we introduce the gravitational field generated by the mass M as

$$\mathbf{F} = -\frac{GM}{r^2}\,\mathbf{e}_r$$

Thus, the gravitational field is the force that would have acted on a test particle of unit mass at a point due to the presence of the mass M.

The gravitational field admits a potential, which states that there exists a scalar function Φ so that

$$\mathbf{F} = \nabla\Phi$$

In fact, for the gravitational field of a point mass M, the potential function Φ is

$$\Phi(r) = \frac{GM}{r}$$

□ ***Remark.*** Following the standard engineering convention (see, e.g., Standard Mathematical Tables published by CRC), spherical coordinates (see Figure 15) are defined as

$$x = r\sin\theta\cos\varphi$$

$$y = r\sin\theta\sin\varphi$$

$$z = r\cos\theta$$

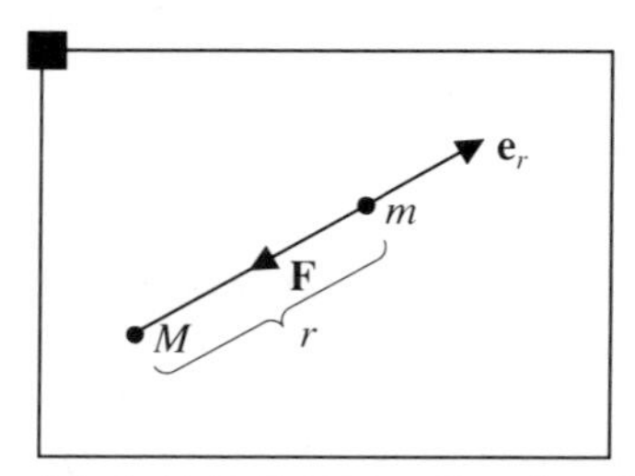

FIGURE 14 **Gravitational Force Between Two Masses M, m**

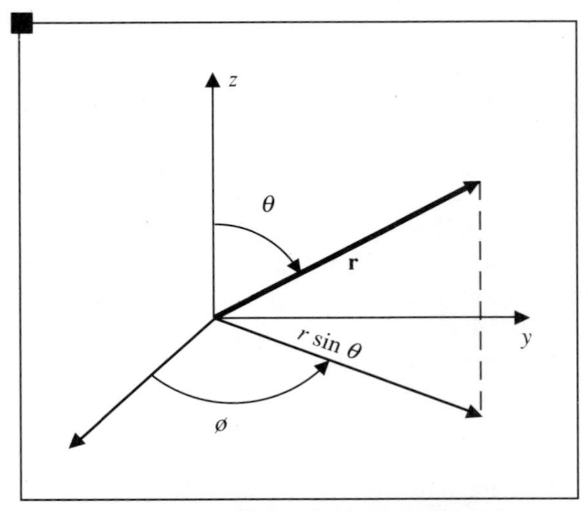

FIGURE 15 **Spherical Coordinates**

and the expression of the gradient operator is

$$\nabla f = \frac{\partial f}{\partial r}\,\mathbf{e}_r + \frac{1}{r}\frac{\partial f}{\partial \theta}\,\mathbf{e}_\theta + \frac{1}{r\sin\theta}\frac{\partial f}{\partial \varphi}\,\mathbf{e}_\varphi$$

In our case, Φ is a functionof r only. □

The superposition principle states that the gravitational field at a point due to two point masses M_1, M_2 is equal to the (vector) sum of their gravitational fields, which can be written as

$$\mathbf{F}_{\text{total}} = \mathbf{F}_1 + \mathbf{F}_2 = -\frac{GM_1}{r_1^2}\,\mathbf{e}_{r_1} - \frac{GM_2}{r_2^2}\,\mathbf{e}_{r_2}$$

where r_1, r_2 are the distances from M_1, M_2 to the point under consideration (see Figure 16).

As a corollary we observe that if Φ_1, Φ_2 are the potential functions for the gravitational field of the masses M_1, M_2, respectively, then

$$\mathbf{F}_{\text{total}} = \nabla\Phi_1 + \nabla\Phi_2 = \nabla(\Phi_1 + \Phi_2)$$

where

$$\Phi_i = \frac{GM_i}{r_i} \qquad i = 1, 2$$

Thus, the potential function of the total gravitational field is given by the (scalar) sum of the individual potential functions.

Similarly, if we are given a finite number of point masses M_i, $i = 1, \ldots, n$ with gravitational fields $\mathbf{F}_i$ and potential functions Φ_i, then

$$\mathbf{F}_{\text{total}} = \sum_{i=1}^{n} \mathbf{F}_i = \sum_{i=1}^{n} \nabla\Phi_i = \nabla\left(\sum_{i=1}^{n} \Phi_i\right)$$

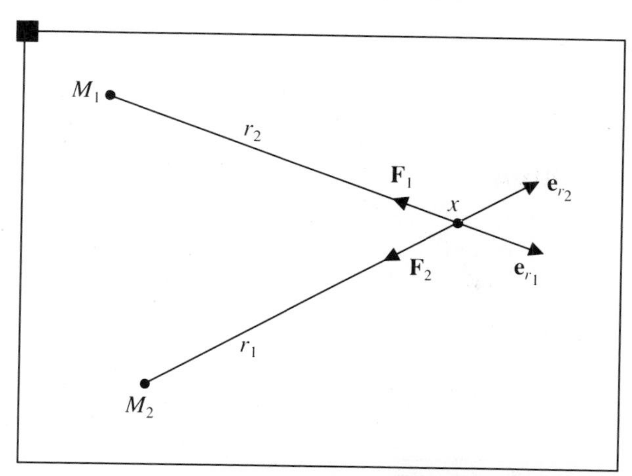

FIGURE 16 **Superposition Principle for the Gravitational Field**

□ ***Remark.*** If the total potential function Φ of a gravitational field is known, then the gravitational field itself is given as $\mathbf{F} = \nabla\Phi$. This is one of the reasons for the introduction of the potential function, since Φ is simply a (scalar) sum of the individual potentials and hence easier to compute than the (vector) sum of the gravitational fields. □

Idealizations

1. We assume that the concept of a point mass is valid. As a matter of fact, we note that due to the discrete nature of matter the notion of a (mathematical) point particle with mass m has no physical meaning.
2. We assume that the field generated by a point particle does not act on itself; otherwise, various contradictions will creep in.

Modeling

To compute the gravitational field due to a continuous mass distribution with mass density $\rho(\mathbf{x})$ in a volume V, we divide the volume into small cells of volume ΔV_i. If we consider each of these cells as a point particle of mass $\rho(\mathbf{x}_i') \, \Delta V_i$ (where $\mathbf{x}_i' \in \Delta V_i$), then the gravitational field due to it at a point $\mathbf{x} = (x, y, z)$ is

$$\Delta \mathbf{F}_i = \frac{-G\rho(\mathbf{x}_i') \, \Delta V_i}{r_i^2} \mathbf{e}_{r_i}$$

where

$$r_i = |\mathbf{x} - \mathbf{x}_i'| = \sqrt{(x - x_i')^2 + (y - y_i')^2 + (z - z_i')^2}$$

(see Figure 17).

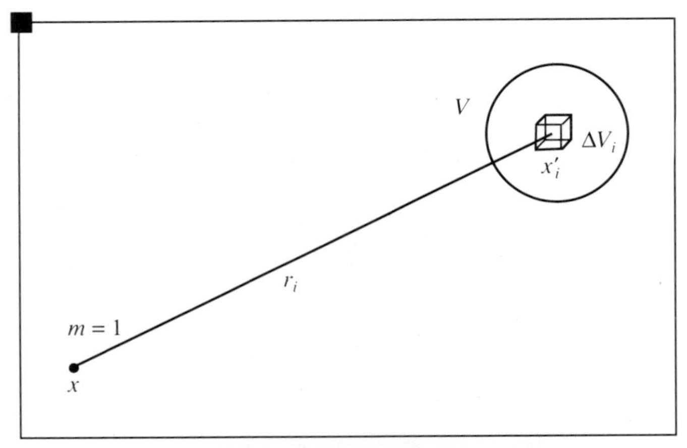

FIGURE 1 7 **Gravitational Potential at x Due to ΔV_i**

The gravitational potential associated with this field is

$$\Delta\Phi_i = \frac{G\rho(\mathbf{x}'_i)\,\Delta V_i}{r_i}$$

Hence, the potential due to the whole mass is given approximately by

$$\Phi = \sum_i \Delta\Phi_i = \sum_i \frac{G\rho(\mathbf{x}'_i)\,\Delta V_i}{r_i} \tag{1.39}$$

By letting $\Delta V_i \to 0$, the sum in Equation (1.39) will be replaced by the volume integral over V,

$$\begin{aligned} \Phi(x, y, z) &= \int_V \frac{G\rho(\mathbf{x}')\,dV}{r} \\ &= \int_V \frac{G\rho(x', y', z')\,dx'\,dy'\,dz'}{\sqrt{(x-x')^2 + (y-y')^2 + (z-z')^2}} \end{aligned} \tag{1.40}$$

Notice that the integral is over the volume of the body whose coordinates are denoted by x', y', z', whereas the coordinates of the point where the potential is being computed are denoted by x, y, z.

To find the differential equation satisfied by $\Phi(x, y, z)$, we compute

$$\nabla\Phi = \left(\frac{\partial\Phi}{\partial x}, \frac{\partial\Phi}{\partial y}, \frac{\partial\Phi}{\partial z}\right)$$

and $\nabla \cdot (\nabla\Phi) = \nabla^2\Phi$ to show that

$$\nabla^2\Phi = \frac{\partial^2\Phi}{\partial x^2} + \frac{\partial^2\Phi}{\partial y^2} + \frac{\partial^2\Phi}{\partial z^2} = 0$$

This equation is called the **potential or Laplace equation** (in three dimensions).

Compounding

In various applications we must consider other equations that are closely related to the Laplace equation in three dimensions. Three examples of such equations follow:

1. The Laplace equation in n dimensions is

$$\nabla^2 u = \sum_{i=1}^{n} \frac{\partial^2 u}{\partial x_i^2} = 0$$

2. The Poisson equation is given by

$$\nabla^2 u = f(\mathbf{x})$$

   It can be shown that the gravitational field inside a body satisfies such an equation with $f(\mathbf{x}) = 4\pi\rho(\mathbf{x})$. (See Exercise 34).

3. The Helmholtz equation is

$$\nabla^2 u + k^2 u = 0$$

   where k is a constant.

EXAMPLE 4. Compute the gravitational field of a solid sphere of radius a and a constant mass density ρ.

Solution. As noted earlier, in all such problems we first compute the gravitational potential Φ and then evaluate the gravitational field F as $\Delta\Phi$.

In this problem, because the sphere is isotropic the gravitational potential must be the same for all points whose distance from the center of the sphere is the same. It follows that if we let the center of the sphere coincide with the origin of the coordinate system, then all points with $r =$ constant must have the same potential. It is enough, therefore, to evaluate Φ at $(0, 0, z)$. To do so we infer from Equation (1.40) that

$$\Phi(0, 0, z) = G\rho \int_V \frac{dx'\, dy'\, dz'}{\sqrt{(x')^2 + (y')^2 + (z - z')^2}}$$

To compute this integral we introduce spherical coordinates on x', y', z'. This transformation is given by

$$x' = r \sin\theta \cos\varphi$$
$$y' = r \sin\theta \sin\varphi$$
$$z' = r \cos\theta$$

and

$$dV = r^2 \sin\theta \, d\theta \, d\varphi \, dr$$

If we write the volume integral for $\Phi(0, 0, z)$ as an iterated integral, we see that

$$\Phi(0, 0, z) = G\rho \int_0^a \int_0^\pi \int_0^{2\pi} \frac{r^2 \sin\theta \, d\varphi \, d\theta \, dr}{\sqrt{r^2 + z^2 - 2rz\cos\theta}}$$

$$= 2\pi G\rho \int_0^a \int_0^\pi \frac{r^2 \sin\theta \, d\theta \, dr}{\sqrt{r^2 + z^2 - 2rz\cos\theta}}$$

To compute the integral over θ, we introduce the substitution

$$w = r^2 + z^2 - 2rz\cos\theta$$

Remembering that z is a constant in this computation, we obtain

$$\Phi(0, 0, z) = \frac{\pi G\rho}{z} \int_0^a r\,dr \int_{(r-z)^2}^{(r+z)^2} w^{-1/2}\,dw$$

$$= \frac{2\pi G\rho}{z} \int_0^a r\,dr \cdot w^{1/2}\Big|_{(r-z)^2}^{(r+z)^2}$$

But

$$w^{1/2}\Big|_{(r-z)^2}^{(r+z)^2} = (r + z) - |r - z| = 2r$$

since $r \le a \le z$, and it follows that

$$\Phi(0, 0, z) = \frac{4\pi a^3 \rho G}{3z} = \frac{GM}{z}$$

where M is the total mass of the body. For a general point whose distance from the origin is R, we obtain

$$\Phi(R) = \frac{GM}{R} \tag{1.41}$$

Equation (1.41) implies that the potential of a solid sphere of mass M is equivalent to that of a point particle with the same mass situated at its center. ■

SECTION 6 EXERCISES

29. Compute the gravitational field of a spherical shell with constant density ρ, inner radius R_1, and outer radius R_2.

□ ***Hint.*** Consider two cases: one for a point outside the shell and another for a point inside the cavity. □

30. Derive an expression for the gravitational potential of a thin metal wire bent to form a circle of radius a if its mass density per unit length is ρ.

31. Derive a model equation for the gravitational field generated by a flat circular ring $a < r < b$ whose mass density per unit area ρ is constant. Consider points in the ring plane only.

32. Compute the expression for the two-dimensional Laplace equation in polar coordinates.

33. Compute the expression for the Laplace equation in three dimensions in cylindrical and spherical coordinates.

34. Derive the Poisson equation as follows:
(a) Show that the gravitational field for a point of mass m satisfies

$$\int_S \mathbf{F} \cdot d\mathbf{S} = 4\pi m$$

where $\mathbf{S}$ is an (arbitrary) sphere around m.
(b) Then deduce that the gravitational field of a continuous mass distribution satisfies

$$\int_S \mathbf{F} \cdot d\mathbf{S} = 4\pi \int_V \rho \, dV$$

(c) Use the fact that $\mathbf{F} = \nabla\varphi$ and the divergence theorem to obtain the Poisson equation.

35. Show that $\ln(x^2 + y^2)$ and $1/\sqrt{x^2 + y^2 + z^2}$ are solutions of the Laplace equation in two and three dimensions, respectively.

SECTION 7*
THE CONTINUITY EQUATION

Objective

Derive a model equation for the traffic flow on a highway without exits and with one entrance and one lane (see Figure 18).

Discussion. One possible approach to modeling the traffic flow is to describe each car as a finite element on the highway and then write a *discrete* model, which describes the motion of each such car. However, if there are many cars on the highway, this approach is not practical and it is better to construct a *continuous* model, which treats these cars as "smeared out" quantities. We construct such a continuous model in this section.

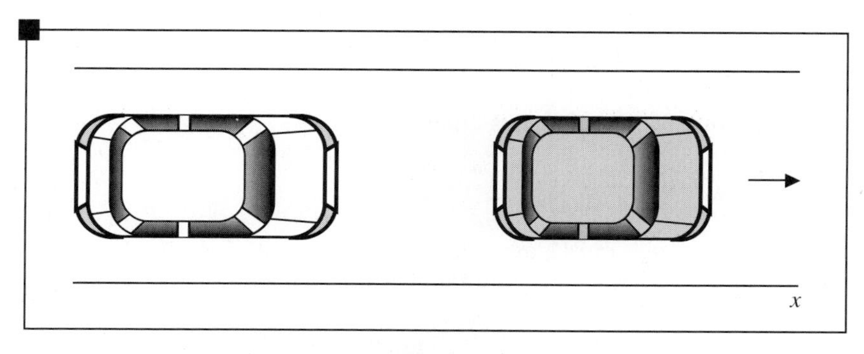

FIGURE 18 **One-Lane Highway**

Approximations and Idealizations

1. We assume that the highway is infinite.
2. We define the car density $\rho(x, t)$ as

$$\rho(x, t) \cong \frac{\text{Number of cars on the interval } [x, x + \Delta x] \text{ at time } t}{\Delta x}$$

   where Δx must be large compared to a car length. [Otherwise, $\rho(x, t) = 1$ if there is a car at x in time t, or $\rho(x, t) = 0$ if there is none.] In fact, the same approximation is made whenever we define the mass density of a "continuous" body made of discrete atoms and molecules. Hence, the equations we derive in this section apply also to fluid flow in a pipe.
3. We assume that there are no accidents on the highway (or that their number is negligible). Hence, we can formulate the principle of "car conservation" (which is equivalent to that of mass conservation) as follows:

   The rate at which the number of cars on the segment [a, b] *is changing equals the rate at which they enter less the rate at which they are leaving it.*
4. We define the concept of car flux $q(x, t)$ in the same way that we defined this concept in Section 2. However, in this context it is natural to define the flux per lane rather than per unit area. Equivalently, this can be considered as letting the unit length be equal to the width of the lane. Moreover, note that

$$q(x, t) = \rho(x, t)u(x, t) \tag{1.42}$$

   where $u(x, t)$ is the car's speed at x at time t, and

$$q = \frac{\text{Cars}}{\text{Lane} \cdot \text{Time}}$$

Modeling

To derive a model equation for the traffic flow, consider a finite section of the road between a and b. The number n of cars in this segment at time t is

$$n(t, a, b) = \int_a^b \rho(x, t)\, dt$$

Hence, the rate of change in this quantity is

$$\frac{dn}{dt} = \int_a^b \frac{\partial \rho(x, t)}{\partial t}\, dt \tag{1.43}$$

This rate of change must equal the flux of cars entering at a less the flux of cars leaving at b. (Remember that the flux was defined per lane!) Therefore,

$$\frac{dn}{dt} = q(a, t) - q(b, t) = -\int_a^b \frac{\partial q}{\partial x}(x, t)\, dx \tag{1.44}$$

Thus, we infer from Equations (1.43) and (1.44) that

$$\int_a^b \left(\frac{\partial \rho}{\partial t} + \frac{\partial q}{\partial x}\right) dx = 0$$

And since a, b are arbitrary, it follows that

$$\frac{\partial \rho}{\partial t} + \frac{\partial q}{\partial x} = 0$$

Using Equation (1.42) to substitute for $q(x, t)$, we finally obtain

$$\frac{\partial \rho}{\partial t} + \frac{\partial(\rho u)}{\partial x} = 0$$

This equation is the **continuity equation in one dimension**. Notice, however, that this equation contains two unknown quantities, ρ and u. Therefore, to solve it we must either be able to express $u = u(\rho)$ or find an additional equation that relates these two quantities.

Compounding

To derive the version of the continuity equation in three dimensions, we consider a fluid flow with mass density $\rho(\mathbf{x}, t)$. Let V be a volume contained in the flow. The mass of the fluid in V at time t is given by

$$m(t, V) = \int_V \rho(\mathbf{x}, t)\, d\mathbf{x}$$

Hence, the rate of change of mass in V is

$$\frac{dm}{dt} = \int_V \frac{\partial \rho}{\partial t}(\mathbf{x}, t)\, d\mathbf{x}$$

Now let S denote the boundary of V and $\mathbf{n}(\mathbf{x})$ the unit *outward* normal to S. The total mass flow rate of the fluid across S in the outward direction is

$$\int_S \mathbf{q} \cdot \mathbf{n}\, dS = \int_S (\rho \mathbf{u}) \cdot \mathbf{n}\, dS$$

The mass conservation principle implies, however, that the rate of change of mass in V must equal the rate at which the mass is crossing S in the *inward* direction. Therefore,

$$\int_V \frac{\partial \rho}{\partial t}(\mathbf{x}, t)\, dV = -\int_S \rho \mathbf{u} \cdot \mathbf{n}\, dS \qquad \textbf{(1.45)}$$

To convert the right-hand side of Equation (1.45) into a volume integral, we now invoke the divergence theorem, which states that for any smooth vector field $\mathbf{F}$ in V,

$$\int_S \mathbf{F} \cdot \mathbf{n}\, dS = \int_V \operatorname{div} \mathbf{F}\, dV$$

This yields

$$\int_V \left[\frac{\partial \rho}{\partial t} + \operatorname{div}(\rho \mathbf{u}) \right] dV = 0 \tag{1.46}$$

Since V is arbitrary we infer that the integrand in Equation (1.46) must be zero, or,

$$\frac{\partial \rho}{\partial t} + \operatorname{div}(\rho \mathbf{u}) = 0 \tag{1.47}$$

Equation (1.47), which is a first-order partial differential equation, is called the **continuity equation in three dimensions.**

SECTION 7 EXERCISES

36. From your experience guess the general form of the relationship between u and ρ in a one-lane highway.

37. Derive a model equation for an infinite one-lane highway where cars are entering and leaving the highway at constant rates α, β per mile, respectively. Generalize to the case where $\alpha = \alpha(x, t)$, $\beta = \beta(x, t)$.

38. Compound the model in Exercise 37 to include accidents at a rate $a(x, t)$ per mile.

39. Consider fluid flow in a long cylindrical pipe with constant cross section A whose axis lies along the x-axis. Let $\rho(x, t)$ be the density of the fluid and $q(x, t)$ be its flux. If the walls of the pipe are made of porous material that allows the fluid in the pipe to leak out at a rate L per unit length, show that

$$\frac{\partial \rho}{\partial t} + \frac{\partial q}{\partial x} + \frac{L}{A} = 0$$

40. Derive model equations for the car densities $\rho_1(x, t)$, $\rho_2(x, t)$ in a two-lane infinite highway with no entries or exits where cars are moving from lane 1 to lane 2 at a rate of $a(\rho_2)$ per mile and at a rate $b(\rho_1)$ per mile from lane 2 to lane 1.

41. Compound the model of Exercise 40 to include entries and exits.

APPENDIX
RLC CIRCUITS

In this appendix we discuss electrical circuits that contain resistors, capacitors, and inductances only. We do not consider circuits with nonlinear resistors or logic circuits.

Objective

Build a model that will predict the electric current at any point of an electric circuit.

Background

An electric circuit usually contains four basic components:

1. Resistances are denoted by R_k and depicted as
2. Capacitors are denoted by C_k and depicted as
3. Inductances are denoted L_k and depicted as
4. External power sources are denoted by e_k and depicted as
   a. for direct current sources (batteries, etc.).
   b. for alternating current sources.

Each of these components in a circuit diagram is considered "pure" (e.g., a resistor has zero capacity and inductance, etc.).

Three basic physical quantities in an electric circuit are of interest:

1. The electric current, $i(x, t)$
2. The electric potential (or voltage), $V(x, t)$. This potential is measured with respect to a fixed reference point called **ground** (since it is usually the earth's potential that is used for this purpose).
3. The electric charge $Q(x, t)$

The units (in the MKS system) of the quantities introduced above are as follows:

R—ohms (denoted by Ω)

L—henries

C—farads

Q—coulombs

i—amperes

V—volts

□ ***Remark.*** Sometimes we use a quantity called the conductance G instead of the resistance R where $G = 1/R$. The unit of conductance is called **mho** (*ohm* spelled backward) or **siemens**. □

The relationship between the components and the physical quantities in an electric circuit is as follows:

1. The voltage drop across a resistance is related to the current passing through it by

$$V = Ri \tag{A.1}$$

2. The potential drop through an inductance L is given by

$$V = L\frac{di}{dt} \tag{A.2}$$

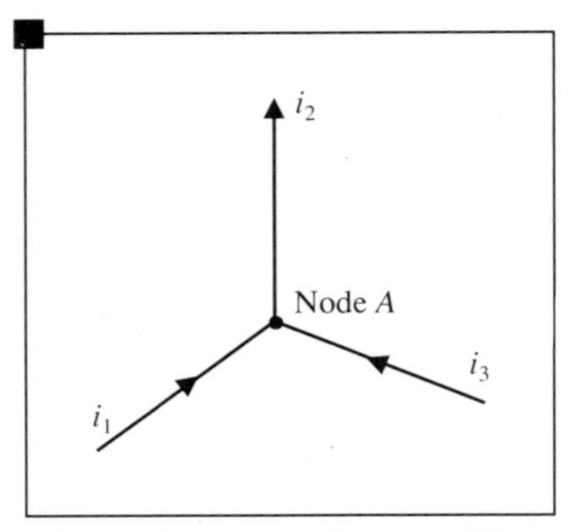

FIGURE 19 **Current Flow at a Node**

3. The total electric charge Q on a capacitor is given by

$$Q = CV \tag{A.3}$$

Hence, the virtual current i in the capacitor is given by

$$i = \frac{dQ}{dt} = C\frac{dV}{dt} \tag{A.4}$$

Kirchhoff's laws form the basis for the analysis of all electric circuits. To introduce these laws we define the following terms:

1. A **node** in an electric circuit is a point where three or more wires are joined together (see Figure 19).
2. A **loop** in an electric circuit is a sequence of circuit elements that start and end at the same point.

First Kirchhoff Law. The algebraic sum of all the currents at a node is 0. In Figure 19 we have at the node A,

$$i_1 - i_2 + i_3 = 0$$

By convention, currents coming to the node are considered positive and those leaving it are negative.

Second Kirchhoff Law. The algebraic sum of the voltage drops around a loop in an electric circuit is equal to the algebraic sum of the external voltage sources in the loop. We assume that the resistance, capacity, or inductance of a given electrical component is independent of the environmental factors (such as temperature, humidity, etc.) and the previous history of the circuit. Cables connecting circuit components have zero resistance, capacity, and inductance.

With this background we can in principle analyze any given circuit.

EXAMPLE 5. A simple RLC circuit is illustrated in Figure 20. To write the equation governing the flow of current i in the circuit, we first note that because there are no nodes in this circuit only Kirchhoff's second law applies.

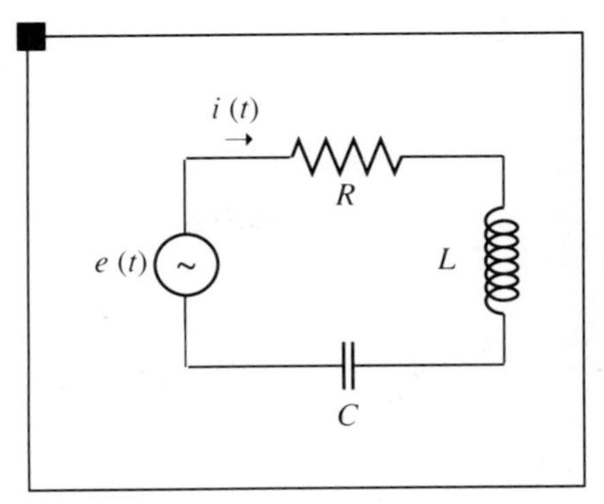

FIGURE 20 **RLC Circuit**

Hence,

$$e(t) = Ri + \frac{Q}{C} + L\,\frac{di}{dt} \tag{A.5}$$

If $e(t)$ is differentiable, then by differentiating Equation (A.4) with respect to t and using Equation (A.4), we obtain

$$\frac{de}{dt} = R\,\frac{di}{dt} + \frac{1}{C}\,i + L\,\frac{d^2 i}{dt^2} \tag{A.6}$$

If de/dt is known, then Equation (A.6) constitutes a second-order (inhomogeneous) differential equation with constant coefficients for the current i in the circuit. If e is not differentiable, we can substitute $i = Q/c$ in Equation (A.5) to obtain

$$E(t) = \frac{Q}{C} + R\,\frac{dQ}{dt} + L\,\frac{d^2 Q}{dt^2} \tag{A.7}$$

■

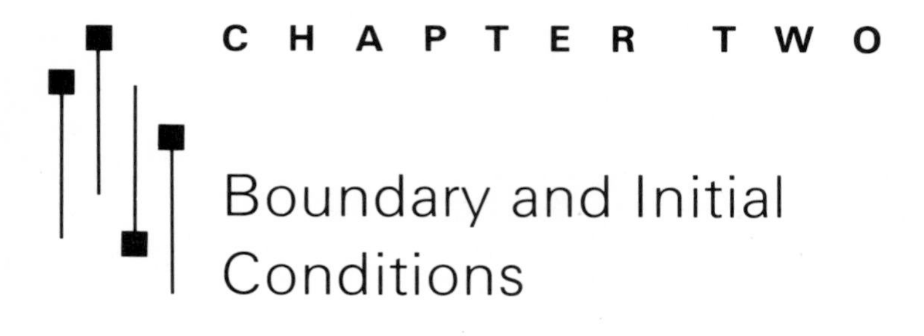

CHAPTER TWO

Boundary and Initial Conditions

SECTION 1
CLASSIFICATION OF PARTIAL DIFFERENTIAL EQUATIONS

Many physical laws express a linear relationship between cause and effect. For example, doubling the strength of the cause (or source) may result in a doubling of the effect produced by the source. Because the mathematical formulation of these laws leads to linear differential equations, the study of these equations and their solutions is integral in many applications. This chapter begins with a brief review of terminology.

A **partial differential equation (PDE)** is a relation of the form

$$F(\mathbf{x}, u, u_{x_1}, \ldots, u_{x_n}, u_{x_1x_2}, \ldots, u_{x_i \ldots x_k}) = 0 \tag{2.1}$$

where

$$u_{x_i} = \frac{\partial u}{\partial x_i}, \qquad u_{x_ix_j} = \frac{\partial^2 u}{\partial x_i \, \partial x_j}$$

and so on, and F is some general function.

In Equation (2.1), $\mathbf{x} = (x_1, \ldots, x_n)$ are the independent variables that range over some domain of R^n. The function u is the dependent (unknown) variable. It is assumed that all the derivatives of $u = u(x_1, \ldots, x_n)$ that appear in Equation (2.1) exist.

Definition 1. The **order** of the partial differential equation (2.1) is the order of the highest derivative that appears in this equation.

EXAMPLE 1

(a) The Korteweg–DeVries (K–dV) equation,

$$u_t - 6uu_x + u_{xxx} = 0$$

is a third-order partial differential equation.

(b) Burger's equation,

$$u_t + uu_x = \nu u_{xx}$$

is a second-order equation.

(c) The order of the equation

$$u_{xxx} + xu_{xy} + yu^2 = x + y$$

is three. ■

Definition 2. A nth-order partial differential equation is called **linear** if F is a polynomial of the **first degree** in u and its derivatives.

From Definition 2 we infer that the general form of such second-order linear partial differential equations is

$$a(x_1, \ldots, x_n)u + \sum_i b_i(x_1, \ldots, x_n)\frac{\partial u}{\partial x_i} + \sum_{i,j} c_{ij}(x_1, \ldots, x_n)\frac{\partial^2 u}{\partial x_i\, \partial x_j} + \cdots = f(x_1, \ldots, x_n) \tag{2.2}$$

This differential equation is said to be **homogeneous** if $f(x_1, \ldots, x_n) = 0$; otherwise it is said to be **inhomogeneous**.

EXAMPLE 2

(a) The partial differential equation

$$u_{xx} + 2xu_y = \sin x$$

is a linear second-order inhomogeneous equation. If we replace $\sin x$ by zero on the right-hand side, then the equation becomes homogeneous.

(b) The partial differential equation

$$u_x + uu_y = f(x, y)$$

is a nonlinear equation since the term uu_y is a monomial of the second degree.

(c) The heat, wave, and Laplace equations that were introduced in Chapter 1 are all linear second-order equations.

(d) The K–dV and Burger's equations as shown in Example 1 are nonlinear partial differential equations. ■

THEOREM 1. If u_1, u_2 are solutions of a linear homogeneous partial differential equation [i.e., an equation of the form in Equation (2.2) with $f = 0$], then $w = c_1 u_1 + c_2 u_2$, where c_1, c_2 are constants, is also a solution of the equation.

Proof. To prove this theorem we first note that because the differentiation operator is linear we can write

$$\frac{\partial^m}{\partial x^m}(c_1 u_1 + c_2 u_2) = c_1 \frac{\partial^m u_1}{\partial x^m} + c_2 \frac{\partial^m u_2}{\partial x^m} \qquad m = 1, 2, \ldots$$

Hence if we substitute w in Equation (2.2), we obtain

$$c_1\left[au_1 + \sum_i b_i \cdot (u_1)_{x_i} + \cdots\right] + c_2\left[au_2 + \sum_i b_i \cdot (u_2)_{x_i} + \cdots\right]$$
$$= c_1 \cdot 0 + c_2 \cdot 0 = 0 \quad \blacksquare$$

EXAMPLE 3. If u_1, u_2 are solutions of the heat equation in one (spatial) dimension, then

$$\frac{1}{k}\frac{\partial u_i}{\partial t} = \frac{\partial^2 u_i}{\partial x^2} \qquad i = 1, 2$$

and $c_1 u_1 + c_2 u_2$ is also a solution of this equation. This follows from the fact that

$$\frac{1}{k}\frac{\partial}{\partial t}(c_1 u_1 + c_2 u_2) = c_1\left(\frac{1}{k}\frac{\partial u_1}{\partial t}\right) + c_2\left(\frac{1}{k}\frac{\partial u_2}{\partial t}\right)$$
$$= c_1 \frac{\partial^2 u_1}{\partial x^2} + c_2 \frac{\partial^2 u_2}{\partial x^2} = \frac{\partial^2}{\partial x^2}(c_1 u_1 + c_2 u_2) \quad \blacksquare$$

This property of linear homogeneous equations (as expressed by Theorem 1) is called the **superposition principle** and plays a central role in solving these equations. Observe, however, that although this theorem is stated for two solutions only, it is true for any finite linear combination of solutions. Furthermore, under proper restrictions it is also true for an infinite number of solutions. If u_i, $i = 1, 2, \ldots$ are solutions of a homogeneous linear partial differential equation, then

$$w = \sum_{i=1}^{\infty} c_i u_i \tag{2.3}$$

is also a solution of this equation. Throughout this book, we assume that the conditions needed for Equation (2.3) to be true are satisfied.

For linear inhomogeneous equations we have Theorem 2.

THEOREM 2. If u_1, u_2 are solutions of a linear inhomogeneous equation [i.e., an equation of the form in Equation (2.2) with $f \not\equiv 0$], then $u_1 - u_2$ is a solution of the corresponding homogeneous equation.

Proof. By assumption we have

$$au_1 + \sum_{i=1}^{n} b_i \cdot (u_1)_{x_i} + \cdots = f(x) \tag{2.4}$$

$$au_2 + \sum_{i=1}^{n} b_i \cdot (u_2)_{x_i} + \cdots = f(x) \tag{2.5}$$

Subtracting Equation (2.5) from Equation (2.4), we obtain

$$a(u_1 - u_2) + \sum_{i=1}^{n} b_i \cdot (u_1 - u_2)_{x_i} + \cdots = 0$$

which is the desired result. ■

We deduce from Theorem 2 that the general solution of an inhomogeneous linear partial differential equation is given by the sum of one particular solution of the inhomogeneous equation plus the general solution of the homogeneous equation.

Linear second-order partial differential equations in two (or more) variables are further classified locally ($\equiv$ pointwise) into three classes: elliptic, hyperbolic, and parabolic. Definition 3 shows how we can classify the general linear second-order partial differential equation in two variables x, y, which can be written as

$$a_{11}(x, y)\frac{\partial^2 u}{\partial x^2} + 2a_{12}(x, y)\frac{\partial^2 u}{\partial x \partial y} + a_{22}(x, y)\frac{\partial^2 u}{\partial y^2} + \cdots = f(x, y) \tag{2.6}$$

Definition 3. We say that Equation (2.6) is elliptic, hyperbolic, or parabolic at (x, y) if

$$\Delta(x, y) = a_{12}^2(x, y) - a_{11}(x, y)a_{22}(x, y)$$

is negative, positive, or zero, respectively, at this point.

Obviously, if the coefficients a_{11}, a_{12}, a_{22} are constants, then the equation belongs to one of these classes for all points (x, y).

EXAMPLE 4

(a) The heat equation

$$\frac{1}{k}\frac{\partial u}{\partial t} = \frac{\partial^2 u}{\partial x^2}$$

is parabolic.

(b) The wave equation

$$\frac{1}{c^2}\frac{\partial^2 u}{\partial t^2} = \frac{\partial^2 u}{\partial x^2}$$

is hyperbolic.

(c) The potential equation

$$\nabla^2 u = \frac{\partial^2 u}{\partial x^2} + \frac{\partial^2 u}{\partial y^2} = 0$$

is elliptic. ■

The importance of this classification stems from the fact that the solutions of equations in the same class have many properties in common. Furthermore, methods used to investigate a particular equation in a given class can be extended in general to other equations in the same class.

The motivation for the names elliptic, hyperbolic, and parabolic comes from analytic geometry. Thus, if we replace formally in Equation (2.6) $\partial u/\partial x$ by ξ, $\partial u/\partial y$ by η, and identify $\partial^2 u/\partial x^2$ with ξ^2, $\partial^2 u/\partial x \partial y$ with $\xi\eta$, and so on, then (for fixed x, y) we obtain the quadratic curve

$$a_{11}(x, y)\xi^2 + 2a_{12}(x, y)\xi\eta + a_{22}(x, y)\eta^2 + \cdots = f(x, y)$$

This curve is an ellipse, hyperbola, or parabola if Δ is negative, positive, or zero, respectively.

SECTION 1 EXERCISES

1. Determine which of the following equations is linear

(a) $\dfrac{\partial^2 u}{\partial x^2} + x^2 \dfrac{\partial u}{\partial y} = x^2 + y^2$

(b) $y^2 \dfrac{\partial^2 u}{\partial x^2} + u \dfrac{\partial u}{\partial x} x^2 \dfrac{\partial^2 u}{\partial y^2} = 0$

(c) $\left(\dfrac{\partial u}{\partial x}\right)^2 + \dfrac{\partial^2 u}{\partial y^2} = 0$

2. Label the following partial differential equations as either elliptic, parabolic, or hyperbolic.

(a) $\dfrac{\partial^2 u}{\partial x^2} - 2 \dfrac{\partial^2 u}{\partial x \partial y} + \dfrac{\partial^2 u}{\partial y^2} = 0$

(b) $\dfrac{\partial^2 u}{\partial x^2} + \dfrac{\partial^2 u}{\partial y^2} + \dfrac{\partial u}{\partial x} = 0$

(c) $\dfrac{\partial^2 u}{\partial x^2} - \dfrac{\partial^2 u}{\partial y^2} + 2 \dfrac{\partial u}{\partial x} = u$

(d) $\dfrac{\partial^2 u}{\partial x^2} + \dfrac{\partial u}{\partial x} + 2 \dfrac{\partial u}{\partial y} = 0$

3. Find a particular solution for $\nabla^2 u = 1$.

□ ***Hint.*** Try $u = u(x)$. □

4. Find a particular solution for $\nabla^2 u = 1/r^2$, where $r^2 = x^2 + y^2$

□ ***Hint.*** In polar coordinates,

$$\nabla^2 u = \frac{1}{r}\frac{\partial}{\partial r}\left(r\frac{\partial u}{\partial r}\right) + \frac{1}{r^2}\frac{\partial^2 u}{\partial \theta^2}$$

Use this equation to find a particular solution to this problem in the form $u = u(r)$. □

5. Show that if $v = v(x)$ is a solution of

$$\frac{d^2 v}{dx^2} + g(x) = 0$$

and $w = w(x, t)$ is a solution of

$$\frac{1}{k}\frac{\partial w}{\partial t} = \frac{\partial^2 w}{\partial x^2}$$

then $u = w + v$ is a solution of

$$\frac{1}{k}\frac{\partial u}{\partial t} = \frac{\partial^2 u}{\partial x^2} + g(x)$$

□ ***Note.*** This method of decomposing the solution of a "complicated" equation into a sum of two functions, each of which satisfies a simpler differential equation, is a variation of the superposition principle. It is important in many applications. □

6. Show how to reduce the solution of

$$\frac{1}{c^2}\frac{\partial^2 u}{\partial t^2} = \frac{\partial^2 u}{\partial x^2} + q(w)$$

to that of the usual wave equation in one spatial dimension [without $q(w)$].

□ ***Hint.*** Write $u = w(x, t) + v(x)$ where $v(x)$ satisfies $v'' + q = 0$. □

7. Explain why the superposition principle does not hold for Burger's equation.

8. Verify directly the superposition principle for the wave and Laplace equations.

SECTION 2
BOUNDARY AND INITIAL CONDITIONS

It is well known from the theory of ordinary differential equations that after finding a fundamental set of solutions $u_1, \ldots, u_n$ to a given linear equation, it is necessary to use the boundary conditions to select the solution that corresponds to the physical state of the system under consideration. For ordinary differential equations, however, it is easy to solve this problem since the adjustment of the constants $c_1, \ldots, c_n$ in the general solution

$$y = c_1 u_1 + \cdots + c_n u_n \qquad \textbf{(2.7)}$$

to the boundary conditions is reducible to a finite set of algebraic equations that can be solved with little effort.

The corresponding problem for partial differential equations is far more complicated, since, in general, the number of independent solutions for such equations is infinite. Moreover, the boundaries of the regions of the independent variables over which we desire to solve the equations are not discrete points as in the one-dimensional case but are continuous curves or surfaces. Thus, the complete formulation of a physical system in terms of partial differential equations requires careful attention not only to the equations that govern the system but also to the correct formulation of the boundary conditions. Furthermore, we observe that most differential equations that we encounter in applications are mathematical expressions of physical laws (e.g., the heat equation is an expression of the law of energy conservation). Therefore we must specify, in addition to the boundary conditions (BC), the initial conditions (IC) of the system in order to obtain a unique solution. Adjusting the arbitrary constants in the general solution to these boundary and initial conditions is usually at least as difficult as solving the partial differential equation itself. Hence, it is natural to refer to these problems (including the solution of the partial differential equation) as boundary value problems (BVP).

To study the boundary conditions systematically, recall that for an nth-order ordinary differential equation the boundary conditions are imposed at the endpoints of the interval of the independent variable over which the solution of the equation is desired. Furthermore, these conditions consist of relations between the unknown function and its derivatives up to order $n - 1$.

From a physical point of view, it is usually clear then that when we consider a partial differential equation of order n in, for example, x, y, we have to specify first the region D in R^2 over which the solution of this equation is desired. The appropriate boundary conditions will involve then the value of the unknown function u and its partial derivatives up to order $n - 1$ on the boundary ∂D of D.

By a solution to a boundary value problem on an open region D, we mean a function u that satisfies the differential equation on D and is continuous on $D \cup \partial D$, and satisfies the specified boundary conditions on ∂D.

Definition 4. The boundary conditions are linear if they express a linear relationship between u and its partial derivatives (up to the appropriate order) on ∂D. (In other words, a boundary condition is linear if it is expressed as a linear equation between u and its derivatives on ∂D.)

In particular, for second-order partial differential equations ($n = 2$), linear boundary conditions can take one of the following three forms:

1. The boundary conditions specify the values of the unknown function u on the boundary. This type of boundary condition is called the **Dirichlet condition.**
2. The boundary conditions specify the derivative of u in the direction normal to the boundary, which is written as $\partial u/\partial \mathbf{n}$. This type of boundary condition is called the **Neumann condition.**

3. The boundary conditions specify a linear relationship between u and its normal derivative on the boundary. These are referred to as **mixed boundary conditions** or **Robin's boundary conditions**. The general form of such a boundary condition is

$$\left[\alpha u + \beta \frac{\partial u}{\partial \mathbf{n}}\right]\bigg|_{\partial D} = f(\mathbf{x})\bigg|_{\partial D} \qquad \alpha, \beta \text{ constants}$$

□ ***Remarks.***

1. The normal derivative on the boundary $\partial u/\partial \mathbf{n}$ is defined as

$$\frac{\partial u}{\partial \mathbf{n}} = \text{grad} u \cdot \mathbf{n} = \left(\frac{\partial u}{\partial x_1}, \ldots, \frac{\partial u}{\partial x_n}\right) \cdot \mathbf{n}$$

where $\mathbf{n}$ is the outward normal to ∂D.

2. We observe that a mixed boundary value problem can have a Dirichlet condition on one part of the boundary, a Neumann condition on another part, and a Robin condition on still another part of the boundary. □

EXAMPLE 5. If $u(x, t)$ is the displacement of a vibrating string and its ends are fixed at $x = 0$ and $x = L$, then the conditions $u(0, t) = 0$ and $u(L, t) = 0$ are Dirichlet conditions. ■

EXAMPLE 6. Suppose $u(x, t)$ is the temperature in a rod of length a. If the rod is perfectly insulated at $x = 0$ and $x = a$, the heat flux at these points is zero. From the Fourier law of heat conduction, it follows that the appropriate boundary conditions are $(\partial u/\partial x)(0, t) = 0$ and $(\partial u/\partial x)(a, t) = 0$. These are Neumann boundary conditions. ■

EXAMPLE 7. Suppose in Example 6 that poor insulation is used at the ends of the rod. Then the boundary conditions might take the form

$$u(0, t) + \frac{\partial u}{\partial x}(0, t) = 0 \quad \text{and} \quad u(a, t) + \frac{\partial u}{\partial x}(a, t) = u_0$$

This is an example of Robin's condition. ■

In a way similar to the superposition principle for the solutions of linear partial differential equations, we have the superposition principle for linear boundary conditions in Theorem 3.

THEOREM 3. If u_1, u_2 are solutions of a linear homogeneous partial differential equation with linear boundary conditions

$$\left[\alpha u_1(\mathbf{x}) + \beta \frac{\partial u_1(\mathbf{x})}{\partial n}\right]\bigg|_{\partial D} = f(\mathbf{x})\bigg|_{\partial D} \tag{2.8}$$

$$\left[\alpha u_2(\mathbf{x}) + \beta \frac{\partial u_2(\mathbf{x})}{\partial \mathbf{n}}\right]\bigg|_{\partial D} = g(\mathbf{x})\bigg|_{\partial D} \tag{2.9}$$

where α, β are constants, then $w = u_1 + u_2$ is a solution of the partial differential equation that satisfies the boundary conditions

$$\left[\alpha w(\mathbf{x}) + \beta \frac{\partial w(\mathbf{x})}{\partial \mathbf{n}}\right]\Bigg|_{\partial D} = (f + g)(\mathbf{x})\Bigg|_{\partial D} \tag{2.10}$$

■

Theorem 3 is particularly useful in applications in which the boundary conditions are complex.

EXAMPLE 8. Consider the solution of the Laplace equation $\nabla^2 u = 0$ in the rectangle shown in Figure 1 with the linear boundary conditions

$$u(x, 0) = f_1(x), \qquad u(x, b) = f_2(x) \tag{2.11}$$

and

$$u(0, y) = g_1(y), \qquad u(a, y) = g_2(y) \tag{2.12}$$

■

We will see in Chapter 4 that in order to solve this boundary value problem using the method of separation of variables, we need (homogeneous) boundary conditions that are zero on a pair of opposite sides of the rectangle. Therefore, we split this problem into two parts: (see Example 9 in Chapter 4).

$$\begin{array}{ll} \nabla^2 u_1 = 0 & \nabla^2 u_2 = 0 \\ u_1(x, 0) = f_1(x) & u_2(x, 0) = 0, \quad u_2(x, b) = 0 \\ u_1(x, b) = f_2(x) & u_2(0, y) = g_1(y) \\ u_1(0, y) = 0, \quad u_1(a, y) = 0 & u_2(a, y) = g_2(y) \end{array} \tag{2.13}$$

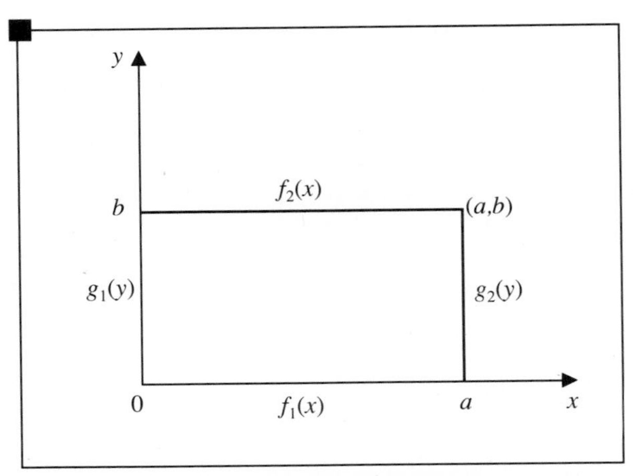

FIGURE 1 **Boundary Conditions on a Plate**

Obviously, if we can solve for u_1, u_2, then $u_1 + u_2$ is a solution of the Laplace equation, which satisfies all the boundary conditions [Equations (2.11) and (2.12)]. However, Neumann boundary conditions usually do not specify a unique solution of a BVP as is demonstrated by Example 9.

EXAMPLE 9. Consider the solution of $\nabla^2 u = 0$ on the rectangle in Figure 1 with the Neumann boundary conditions

$$\frac{\partial u}{\partial y}(x, 0) = f_1(x), \qquad \frac{\partial u}{\partial y}(x, b) = f_2(x) \tag{2.14}$$

and

$$\frac{\partial u}{\partial x}(0, y) = g_1(y), \qquad \frac{\partial u}{\partial x}(a, y) = g_2(y) \tag{2.15}$$

Then it is obvious that if u is a solution of this BVP, then $w = u + c$, where c is a constant, is also a solution of the Laplace equation with the same boundary conditions. Thus, Neumann boundary conditions determine the solution of this boundary value problem only up to a constant. ■

We now consider the appropriate initial conditions that have to be joined with the boundary conditions in order to ensure a unique solution of a given partial differential equation. As we observed in this section, these initial conditions are determined by the physical law that is expressed by the differential equation.

EXAMPLE 10. The heat equation is an expression of the law of energy conservation. Hence, an appropriate initial condition is the heat (or temperature) distribution in the body at some time $t = t_0$. Thus, a unique solution of this equation is obtained only when we specify the appropriate boundary conditions as well as the initial temperature distribution in the body.

A typical BVP for the heat conduction in a homogeneous rod with constant cross section that is insulated laterally is given by

$$\frac{1}{k}\frac{\partial u}{\partial t} = \frac{\partial^2 u}{\partial x^2} \qquad 0 < x < a, \ 0 < t \tag{2.16}$$

$u(0, t) = T_0$, $u(a, t) = T_1$, $0 < t$ (Dirichlet boundary conditions), and $u(x, 0) = f(x)$, $0 < x < a$ (initial condition).

To motivate the need for the initial condition from a more mathematical point of view, we observe that the region over which we want to solve Equation (2.16) is represented in R^2 by an "infinite slot" (see Figure 2).

To formulate a "well-posed" boundary value problem, one with a unique solution, on this region D, we need to specify u (and possibly its derivatives) on ∂D. It is seen then that the boundary conditions at the end of the rod, $u(0, t) = T_0$, $u(a, t) = T_1$ specify these conditions on the horizontal sides of the slot. The initial condition $u(x, 0) = f(x)$ specifies u on the third part of ∂D, which is on the x-axis.

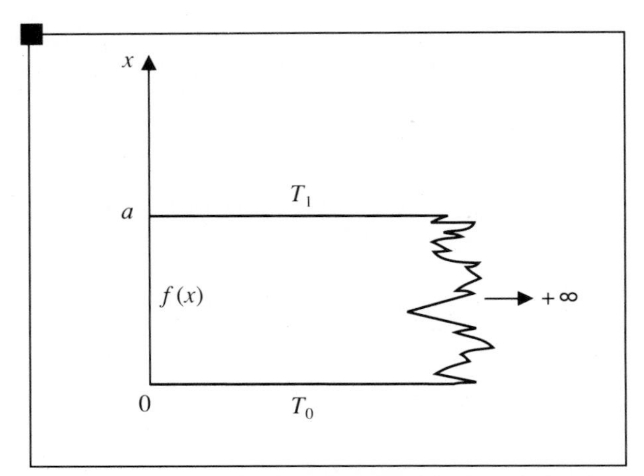

FIGURE 2 **Heat Conduction on a Rod of Infinite Length**

From another point of view, we note that for a fixed t, Equation (2.16) "reduces" to a second-order ordinary differential equation. Thus we need two boundary conditions on u for each t. Similarly, if we let x be fixed, this equation reduces to a first-order ordinary differential equation and we infer that only one boundary condition on u for each x is needed. This boundary condition is supplied by the initial condition $u(x, 0) = f(x)$.

At this juncture you may wonder whether some additional condition must be specified at the "boundary" $t = \infty$. The previous discussion shows that no such condition is needed in our context. However, the problem does have an implicit (physical) constraint that requires $u(x, t)$ to be bounded for all t. (Infinite temperature is not an acceptable solution.) Such implicit constraints and boundary conditions are discussed in Section 3. ■

EXAMPLE 11. The wave equation for the motion of a string,

$$\frac{1}{c^2}\frac{\partial^2 u}{\partial t^2} = \frac{\partial^2 u}{\partial x^2} \qquad 0 < x < a,\ 0 < t \tag{2.17}$$

is an expression of Newton's second law. When we use this law to solve for the displacement of a point mass, we need the initial position and velocity of the particle. It follows then that to obtain a unique solution of Equation (2.17), we need the boundary conditions on the string at its ends and initial conditions regarding the position and velocity of the string at time 0.

$$u(x, 0) = f(x), \qquad \frac{\partial u}{\partial t}(x, 0) = g(x) \qquad 0 < x < a \tag{2.18}$$

■

EXAMPLE 12. Write the partial differential equation and boundary and initial conditions for a string whose ends are fixed at $x = 0$ and $x = a$ and whose displacement and velocity at $t = 0$ are $f(x)$ and $g(x)$, respectively (see Figure 3).

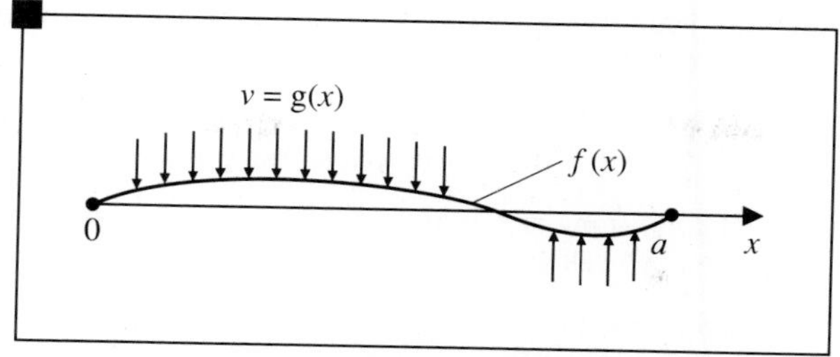

FIGURE 3 **Vibrating String**

Solution

$$\frac{1}{c^2}\frac{\partial^2 u}{\partial t^2} = \frac{\partial^2 u}{\partial x^2} \qquad 0 < x < a,\ t > 0$$

$$u(0, t) = 0 \qquad t > 0$$

$$u(a, t) = 0 \qquad t > 0$$

$$u(x, 0) = f(x) \qquad 0 < x < a$$

$$\frac{\partial u}{\partial t}(x, 0) = g(x) \qquad 0 < x < a$$

■

These initial conditions can be justified more formally, as in Example 11. Note, however, that no initial conditions have to be imposed on the Laplace equation $\nabla^2 u = 0$ since it describes a static distribution of sources.

SECTION 2 EXERCISES

9. Show how to decompose the following boundary value problem into two boundary value problems, each with homogeneous boundary conditions $\nabla^2 u = 0$ on the rectangle in Figure 1.

$$\frac{\partial u}{\partial y}(x, 0) = f_1(x), \qquad \frac{\partial u}{\partial y}(x, b) = f_2(x)$$

$$u(0, y) = g_1(y), \qquad u(a, y) = g_2(y)$$

10. Classify the boundary value problems that appear in Exercise 9.

11. If

$$\frac{\partial u}{\partial t} = -\frac{\partial v}{\partial x}, \qquad \frac{\partial v}{\partial t} = -\frac{\partial u}{\partial x} \qquad 0 \le x \le a,\ 0 < t$$

and

$$u(x, 0) = \sin x, \qquad u(a, t) = 0, \qquad v(x, 0) = -\cos x, \qquad v(0, t) = 0$$

(a) show that u satisfies the wave equation in one spatial dimension.

(b) find the boundary and initial conditions that have to be imposed on u to obtain a unique solution of the wave equation in part (a).

12. Discuss from a practical point of view the nature of the boundary conditions that have to be imposed on the solution of $\nabla^2 u = 0$ in the quarter plane $0 < x$, $0 < y$ at infinity.

13. Explain formally why the initial conditions in Equation (2.18) with the boundary conditions $u(0, t) = u(a, t) = 0$ represent a "well-posed" problem for Equation (2.17).

14. Give formal arguments to explain why the boundary conditions in Equations (2.11) and (2.12) ensure a unique solution for $\nabla^2 u = 0$.

15. Show that any two solutions of the Neumann problem for the Laplace equation on a domain D can differ only by a constant.

SECTION 3
IMPLICIT BOUNDARY CONDITIONS

Although in many cases the use of boundary and initial conditions will lead to a unique solution of the boundary value problem, in other instances some implicit restrictions have to be taken into account. Moreover, there are problems in which some of the boundary conditions might not be apparent and must be deduced through a careful analysis of the problem.

The first such practical restriction occurs whenever a partial differential equation models a real-life system. The corresponding solution must be bounded. Thus, any solution of the partial differential equation that becomes unbounded over part of the region D of interest must normally be eliminated from further consideration. Moreover, when we consider the solution of a partial differential equation over an infinite domain, this boundedness condition on the solution at infinity usually serves as the boundary condition on the solution at this point.

□ ***Remark.*** Solutions that are singular only at some points of D are useful in certain contexts, such as Green's functions (see Chapter 9). □

EXAMPLE 13. The heat equation in one dimension,

$$\frac{1}{k}\frac{\partial u}{\partial t} = \frac{\partial^2 u}{\partial x^2} \tag{2.19}$$

over $x \in (-\infty, \infty)$, $0 < t$ admits the following solutions:

$$u_0(x, t) = Ax + B \tag{2.20}$$

$$u_\lambda(x, t) = \cosh \lambda x \cdot e^{k\lambda^2 t} \qquad 0 < \lambda \text{ real} \tag{2.21}$$

This equation has other solutions. However, it is obvious that for any fixed t,

$$\lim_{x \to \pm\infty} u_\lambda(x, t) = \infty \qquad 0 \le \lambda$$

which implies that these are not acceptable solutions from a physical point of view. ■

EXAMPLE 14. Formulate an explicit boundary value problem for Equation (2.19) over $x \in (-\infty, \infty)$, $0 < t$.

Solution. Since u must be bounded at infinity in both x and t, the following restrictions must apply:

$$\lim_{x \to \pm\infty} u(x, t) < \infty, \qquad \lim_{t \to \infty} u(x, t) < \infty$$

The appropriate initial condition is given by

$$u(x, 0) = f(x) \qquad -\infty < x < \infty$$

where $f(x)$ is bounded for all x [e.g., $f(x) = e^{-x^2}$]. ■

Boundedness conditions are also important when we consider bounded regions.

EXAMPLE 15. Laplace equation

$$\nabla^2 u = \frac{1}{r}\frac{\partial}{\partial r}\left(r\frac{\partial u}{\partial r}\right) + \frac{1}{r^2}\frac{\partial^2 u}{\partial \theta^2} = 0 \tag{2.22}$$

on the disk $0 \le r \le a$ admits the solutions

$$u_n(r, \theta) = \frac{1}{r^n}(A_n \sin n\theta + B_n \cos n\theta) \tag{2.23}$$

and

$$w_n(r, \theta) = r^n(C_n \sin n\theta + D_n \cos n\theta) \qquad n = 1, 2, \ldots \tag{2.24}$$

We observe that the solutions $u_n(r, \theta)$ have a singularity (i.e., become unbounded) at $r = 0$. Therefore they are not acceptable solutions on this domain. The solutions $w_n(r, \theta)$ are acceptable since they are bounded. If, on the other hand, we change the domain of interest and consider the points outside the disk; i.e., $a < r$, then the solutions $w_n(r, \theta)$ are unbounded and must be discarded as unacceptable solutions on this domain, whereas $u_n(r, \theta)$ are bounded. ■

The Laplace equation on the disk is interesting since its complete formulation as a boundary value problem illustrates the need for careful examination of implicit boundary conditions.

EXAMPLE 16. Formulate the complete boundary value problem for the Laplace equation on the disk $0 \le r \le a$ in R^2.

Solution. As a first step toward the solution of our problem, we must choose a unique range for θ so that every point on the disk will be represented by a unique pair of (r, θ). We make the standard choice of $-\pi < \theta \le \pi$.

To begin with it is obvious that the region in question has a boundary $r = a$, and hence a Dirichlet-type boundary condition will be of the form

$$u(a, \theta) = f(\theta) \tag{2.25}$$

with $u(r, \theta)$ being bounded for all points on the disk. Now although this might seem to be the desired complete formulation of the boundary value problem, we observe that points (r, θ) with $\theta = \pi - \varepsilon$ and $\theta = -\pi + \varepsilon$, $0 < \varepsilon \ll 1$ are represented by values of θ "far apart" but are still close to each other on the disk. Thus, we must ensure that u and its derivative with respect to θ are continuous at the fictitious boundary $\theta = \pm\pi$. We do so by writing

$$u(r, \pi) = u(r, -\pi), \qquad \frac{\partial u}{\partial \theta}(r, \pi) = \frac{\partial u}{\partial \theta}(r, -\pi) \qquad 0 \le r \le a \tag{2.26}$$

The conditions in Equations (2.25) and (2.26), along with the boundedness condition that was discussed in Example 15, form a complete formulation of the desired boundary value problem with Dirichlet boundary conditions. ■

□ ***Remark.*** Since $\theta = -\pi$ is outside the domain of θ, $u(r, -\pi)$ and $(\partial u/\partial\theta)(r, -\pi)$ should be interpreted as $\lim_{\theta \to -\pi^+} u(r, \theta)$ and $\lim_{\theta \to -\pi^+} (\partial u/\partial \theta)(r, \theta)$, respectively. □

Initial Value Problems

In order to specify a unique solution of a second-order ordinary differential equation, we can impose either boundary or initial conditions. In the latter case the value of the unknown function and its derivative must be specified at some point. Similarly, for second-order partial differential equations we can consider problems where the unknown function u is specified on a surface S together with its first-order derivative normal to S. Such problems are called **initial value problems.** It is possible to show that if the coefficients of the differential equation, the initial conditions, and S are "smooth" enough, then such initial conditions lead to a unique solution of the equation in a region D containing S. This statement is called the **Cauchy–Kovalevsky theorem**. In this book we will not consider such problems. However, we present here an intriguing example due to Hadamard.

■ EXAMPLE 17. Consider the initial value problem

$$\nabla^2 u = 0 \qquad \text{in } R^2$$

$$u(x, 0) = 0, \qquad \frac{\partial u}{\partial y}(x, 0) = \frac{1}{k}\sin kx \tag{2.27}$$

where $k > 0$ is an integer. It is easy to verify that the solution of this initial value problem is given by

$$u(x, y) = \frac{1}{k^2}\sin kx \cdot \sinh ky \tag{2.28}$$

What is unusual about this solution is that since

$$\left|\frac{\sin kx}{k}\right| < \frac{1}{k}$$

it follows that when k is large the initial conditions [Equation (2.27)] are close to $u(x, 0) = 0$, $(\partial u/\partial y)(x, 0) = 0$ whose unique solution is $u = 0$. On the other hand, the solution [Equation (2.28)] becomes unbounded as $k \to \infty$ even for small values of $|y|$. This shows that small changes in the initial conditions on u in this problem lead to large changes in the solution. Stated differently, the solution of Laplace equations does not depend continuously on the initial data, at least in this case. ■

SECTION 3 EXERCISES

16. Formulate the Neumann boundary value problem for the Laplace equation on the disk $0 \le r \le a$.

17. Verify that Equations (2.20) and (2.21) are solutions of Equation (2.19).

18. Verify that Equations (2.23) and (2.24) are solutions of Equation (2.22).

19. Verify that Equation (2.28) is the solution of the initial value problem in Equation (2.27).

20. Formulate the Dirichlet boundary value problem for the wave equation in one dimension for an infinite string. Determine if this problem can be considered an initial value problem.

21. Formulate the Dirichlet boundary value problem for the Laplace equation inside the sphere $0 \le r \le a$ in three dimensions. (Use spherical coordinates.)

22. Repeat Exercise 21 for a cylinder of radius r and height h. (Use cylindrical coordinates.)

23. Explain why the implicit boundary condition

$$\frac{\partial u}{\partial r}(r, \pi) = \frac{\partial u}{\partial r}(r, -\pi)$$

is not needed in Example 16.

SECTION 4
FURTHER EXAMPLES OF BOUNDARY VALUE PROBLEMS

In this section we present various examples of physical systems and their formulation as boundary value problems. We will focus on problems that are modeled by the heat and wave equation in one dimension.

EXAMPLE 18. Formulate the boundary value problem that governs the heat conduction in a homogeneous rod of constant cross section A and length a

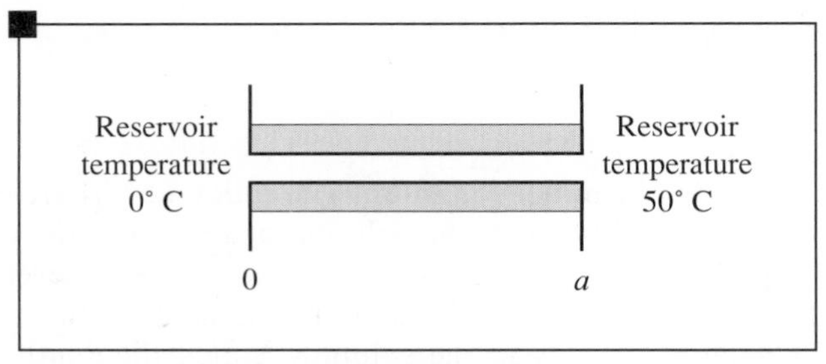

FIGURE 4 **Heat Flow in a Rod**

that is insulated laterally if its left and right ends are in contact with heat reservoirs with temperatures 0° C and 50° C, respectively (see Figure 4).

Solution. Obviously, the partial differential equation that governs heat conduction is

$$\frac{1}{k}\frac{\partial u}{\partial t} = \frac{\partial^2 u}{\partial x^2} \qquad 0 < x < a,\ 0 < t \tag{2.29}$$

Since the left and right sides of the rod are in contact with heat reservoirs whose temperatures remain constant in spite of the heat conducted by the rod, we have

$$u(0, t) = 0, \qquad u(a, t) = 50 \tag{2.30}$$

which are Dirichlet boundary conditions. To complete the formulation of this boundary value problem, we need only the initial condition of the rod,

$$u(x, 0) = f(x) \qquad 0 < x < a \tag{2.31}$$

■

EXAMPLE 19. Reconsider Example 18 if the two ends of the rod are perfectly insulated.

Solution. When the end of the rod is perfectly insulated, the heat flux there is zero. Hence, from Fourier's law of heat conduction, we have

$$q(a, t) = -kA\frac{\partial u}{\partial x}(a, t) = 0$$

$$q(0, t) = -kA\frac{\partial u}{\partial x}(0, t) = 0$$

Since k, A are not zero, these equations imply the Neumann boundary conditions

$$\frac{\partial u}{\partial x}(0, t) = \frac{\partial u}{\partial x}(a, t) = 0 \qquad t > 0 \tag{2.32}$$

Equation (2.32) together with Equations (2.29) and (2.31) gives the complete formulation of this boundary value problem. ■

EXAMPLE 20. Reconsider the differential equation in Equation (2.29) if the flux at $x = a$ is given by Newton's cooling law, which states that the heat flux is proportional to the temperature difference between the end of the rod and the ambient temperature. Furthermore, assume that the initial temperature in the rod is

$$\frac{50(a - x)}{a}$$

Solution. From Newton's cooling law we have at $x = a$,

$$q(a, t) = \alpha[u(a, t) - u_a]$$

where α is a constant and u_a is the ambient temperature. Hence, from Fourier's law of heat conduction, we conclude that

$$kA \frac{\partial u}{\partial x}(a, t) + \alpha u(a, t) = \alpha u_a$$

which is a Robin's-type boundary condition. To complete the formulation of this boundary value problem, we need to observe only that

$$u(0, t) = 50, \qquad u(x, 0) = \frac{50(a - x)}{a} \qquad 0 < x < a, t > 0$$

■

EXAMPLE 21. Formulate the boundary value problem for a vibrating string of length a that is rigidly fixed at its ends.

Solution. The equation that governs the vibrations is the wave equation in one dimension:

$$\frac{1}{c^2}\frac{\partial^2 u}{\partial t^2} = \frac{\partial^2 u}{\partial x^2} \qquad 0 < x < a, 0 < t \tag{2.33}$$

Since the ends of the string are rigidly fixed, we infer that

$$u(0, t) = u(a, t) = 0 \qquad t > 0 \tag{2.34}$$

To complete the formulation we need the initial position and velocity of the string. [Remember that Equation (2.33) is equivalent to Newton's second law.]

$$u(x, 0) = f(x), \qquad \frac{\partial u}{\partial x}(x, 0) = g(x) \qquad 0 < x < a \tag{2.35}$$

Equations (2.33)–(2.35) give the complete formulation for the Dirichlet boundary value problem of this system.

■

EXAMPLE 22. Reconsider the vibrating string if its ends are fastened to air bearings that are free to move on a rod at right angles to the x-axis (see Figure 5).

Solution. Because the bearings cannot exert a vertical component of force, the tension in the string instantly tends to pull the bearings up or down. The bearings then slide in such a way as to keep the wire perpendicular to the

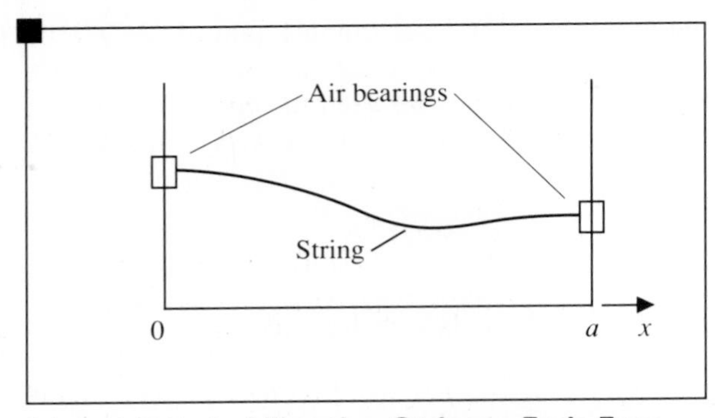

FIGURE 5 **Vibrating String—Ends Free**

rod. Therefore, the slope of the string at the bearings is always zero and

$$\frac{\partial u}{\partial x}(0, t) = 0, \qquad \frac{\partial u}{\partial x}(a, t) = 0 \qquad t > 0 \tag{2.36}$$

which is a Neumann's boundary condition. Equations (2.33), (2.35), and (2.36) give the required formulation of this boundary value problem. ■

EXAMPLE 23. Reconsider Example 21 if the two ends of the string are attached to springs that can move only in the vertical direction, as shown in Figure 6.

Solution. At each end of the string there are two forces to consider. One is the vertical force due to the tension in the string, and the other is the force exerted by the springs. At any time $t > 0$ we assume that these two forces are equal in magnitude but opposite in direction. The vertical force F due to the

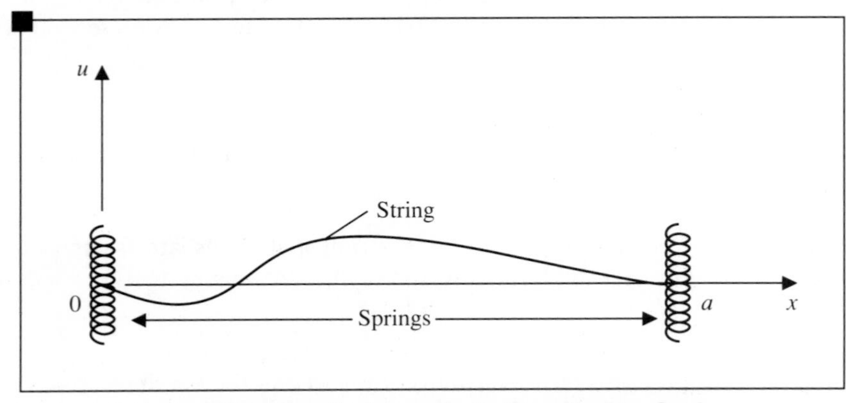

FIGURE 6 **Vibrating String—Ends Attached to Springs**

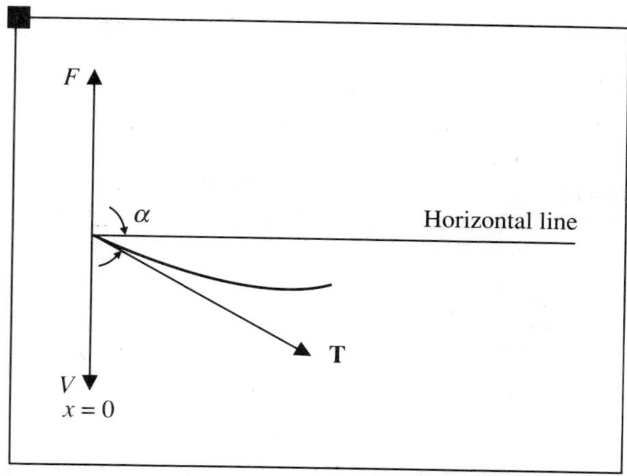

FIGURE 7 **Vector Diagram—Left End Attached to Spring**

spring is given by Hooke's law,

$$F = -ku(0, t)$$

where k is the spring constant. Similarly, the vertical force V due to the tension T in the string is

$$V = T \sin \alpha$$

(see Figure 7). However, for small vibrations, α is assumed to be very near zero, and we can write

$$\sin \alpha \cong \tan \alpha = \frac{\partial u}{\partial x}(0, t)$$

Since $F + V = 0$, we conclude that

$$u(0, t) - \frac{T}{k}\frac{\partial u}{\partial x}(0, t) = 0 \qquad t > 0 \tag{2.37}$$

Similarly, at $x = a$ we have

$$u(a, t) - \frac{T}{k}\frac{\partial u}{\partial x}(a, t) = 0 \qquad t > 0 \tag{2.38}$$

These are Robin's boundary conditions. Equations (2.33), (2.35), (2.37), and (2.38) give the required formulation of this boundary value problem. ■

SECTION 4 EXERCISES

24. In formal style, write the differential equation and boundary and initial conditions for finding the temperature $u(x, t)$ in a laterally insulated rod of length L if the following

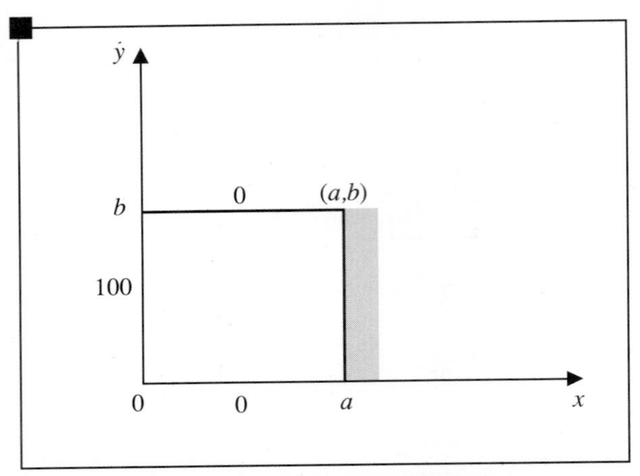

FIGURE 8 **Boundary Conditions for Exercise 26**

conditions hold:

(a) The temperature at $x = L$ is 100.
(b) The end at $x = 0$ is perfectly insulated.
(c) The initial temperature in the rod is x^2.

25. Let $u(x, t)$ be the temperature in a laterally insulated rod of length L. Write the differential equation and boundary and initial conditions if the

(a) left-hand end has a temperature of $0.01e^t$.
(b) right-hand end has a temperature of 100.
(c) initial temperature of the rod is $\cos x$.

26. The steady-state temperature of a rectangular plate a by b insulated laterally is given by $u(x, y)$ (see Figure 8). In formal style, write the differential equation and boundary conditions if the

(a) temperature along the y-axis is 100.
(b) temperature along the x-axis is 0.
(c) edge at $x = a$ is perfectly insulated.
(d) temperature along the edge $y = b$ is 0.

27. Along with the one-dimensional equation for finding the temperature $u(x, t)$, write the following boundary and initial conditions:

(a) The temperature at $x = 0$ is 50°.
(b) At $x = L$ the temperature equals the rate of change of temperature with respect to x.
(c) The initial temperature is x^2.

28. In formal style, write the differential equation and boundary conditions for the steady-state temperature $u(x, y)$ in a rectangular plate insulated laterally that measures a by b where a is along the x-axis if the

(a) temperature along edge $y = b$ is $\sin \pi x/a$.
(b) temperature along the x-axis is zero.
(c) edge along the y-axis is perfectly insulated.
(d) temperature along edge $x = a$ is zero.

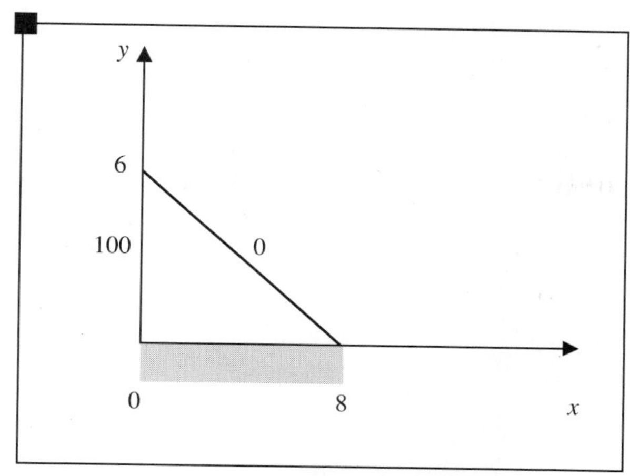

FIGURE 9 **Boundary Conditions for Exercise 29**

29. Given the triangular plate insulated laterally as shown in Figure 9, write the differential equation and boundary conditions for the steady-state temperature $u(x, y)$ if the

(a) temperature along the y-axis is 100.
(b) edge along the x-axis is perfectly insulated.
(c) "slanted edge" is kept at zero temperature.

30. In formal style, write the differential equation and boundary and initial conditions for the solution $y(x, t)$ of a vibrating string if

(a) it is fastened to a bearing free to move in the y direction at $x = 0$.
(b) the end at $x = L$ is fixed.
(c) the initial displacement is $\sin \pi x/L$.
(d) the initial velocity is 1.

31. Write the boundary conditions governing the temperature in a rod of length L under the following conditions:

	$x = 0$	$x = L$
(a)	Temperature $= 0$	Temperature $= 0$
(b)	Temperature $= 50$	Temperature $= 0$
(c)	Temperature $= \cos 10t$	Temperature $= 50$
(d)	Perfectly insulated	Temperature $= 0$
(e)	Perfectly insulated	Perfectly insulated
(f)	Temperature + rate of change of temperature with respect to $x = 0$	Heat flow across plane at $x = L$ is 10
(g)	Temperature equals heat flow across plane at $x = 0$	Perfectly insulated

32. Write the BVP for the temperature $u(x, t)$ in a laterally insulated rod in which no heat is generated. The temperature at $x = 0$ is zero, and the end $x = L$ is perfectly insulated. The initial temperature is $100 - x$.

33. A plate of radius c, insulated laterally, has the shape of one quarter of a circular disk. Assuming no heat is generated inside, write the boundary value problem for the steady-state temperature $u(r, \theta)$ if the temperature on the circumference is zero, the temperature on one edge is $50°$ K, and the other edge is perfectly insulated.

34. A rectangular plate a by b in the x by y directions is insulated laterally. Assume the flow of heat is in a steady-state condition. Write the boundary value problem if the following boundary conditions hold:

$x = 0$	Temperature $= 0$
$x = a$	Temperature $= 0$
$y = 0$	Temperature $= 100$
$y = b$	Perfectly insulated

35. A circular plate of radius 10 centimeters is insulated laterally. Assume the flow of heat is steady state. Write the boundary value problem if the upper semicircle is at temperature 100 and the lower semicircle is at temperature zero.

36. Write the boundary and initial conditions for a vibrating string of length L under the following conditions:

	$x = 0$	$x = L$	Initial Displacement	Initial Velocity
(a)	Fixed	Fixed	0	$g(x)$
(b)	Fixed	Fixed	$\sin (x - L)$	x^2
(c)	Free*	Fixed	$f(x)$	0
(d)	Free	Free	$\cos x$	0
(e)	Displacement $= \sin t$	Fixed	$H(x)$	$2h(x)$
(f)	Displacement $= e^{-t}$	displacement $=$ slope	0	0
(g)	Displacement $= -$ slope	Fixed	$f(x)$	$g(x)$

37. A submarine cable of length 6000 kilometers stretches from Boston to London. Write the differential equation and boundary and initial conditions for finding the voltage $e(x, t)$ if

(a) the voltage at $x = 0$ is $100 \sin t$.
(b) $x = 6000$ is a dead short.
(c) the initial voltage is 0.

38. A high-frequency line connects a transmitter to its antenna. Write the differential equation and boundary and initial conditions of the current $i(x, t)$ if the

(a) current at $x = 0$ is 0.
(b) rate of change of the current with respect to x at $x = a$ is $2 \cos 100t$.
(c) initial current is $\sin \pi x/a$.
(d) rate of change of the current initially is 0.

39. The pressure $p(x, t)$ on an organ pipe satisfies the wave equation. Write the differential equation and boundary and initial conditions. Assume that air pressure in the organ loft is p_0 and that the

(a) pipe is open at $x = 0$.
(b) pipe is closed at $x = 9.75$ meters.
(c) initial pressure $= 2p_0$.
(d) rate of change of pressure initially is zero.

---

* The string is attached to a free-moving air bearing.

40. Write the differential equation and boundary and initial conditions for finding the displacement $u(x, y, t)$ of a circular rubber diaphragm of radius 4 centimeters, center at origin, if the

(a) circumference is fixed.
(b) initial displacement is xy.
(c) initial velocity is zero.

41. Write the boundary and initial conditions for the voltage $e(x, t)$ on a high-frequency transmission line of length 80 meters under the following conditions:

	$x = 0$	$x = 80$	Initial Voltage	Rate of Change of Voltage
(a)	Short	Short	100	$g(x)$
(b)	$50v$	Short	0	$\sin x$
(c)	Open	Open	$f(x)$	$g(x)$
(d)	Short	R_1	0	0

42. Write the boundary and initial conditions for the current $e(x, t)$ on a high-frequency transmission line of length 80 meters under the following conditions.

	$x = 0$	$x = 80$	Initial Current	Rate of Change of Current
(a)	Open	Open	$f(x)$	$g(x)$
(b)	$3a$	Open	100	0
(c)	R_1	Short	0	0
(d)	Short	$\cos t$	$\sin x$	$\sin x$

43. Write boundary and initial conditions for voltage on a 6000-kilometer submarine cable if the voltage at the sending end is zero and the receiving end is open. The initial voltage is 1 volt.

44. Write the boundary and initial conditions for finding the longitudinal displacement $\psi(x, t)$ of a bar of length L if the initial displacement and velocity are zero and the right-hand end is fixed. The left-hand end is moved by an electric motor so that the plane is displaced by $(L/1000) \cos 10t$ meters.

45. Write the boundary and initial conditions for the acoustic pressure $p(x, t)$ in an organ pipe 10 meters long under the following conditions:

	$x = 0$	$x = 10$	Initial Pressure	Rate of Change of Pressure
(a)	Open	Open	5000 N/m^2	0
(b)	Open	Closed	0	50 N/m^2-sec.
(c)	Closed	Open	2500	100
(d)	Closed	Closed	$f(x)$	0
(e)	1000 N/m^2*	Open	0	0
(f)	200*	Closed	0	0

46. Write the differential equation and boundary and initial conditions for finding the temperature $u(x, y, z, t)$ in a brick. Assume that the brick has the dimensions and physical quantities shown in Figure 10 and that the initial temperature is $u(\mathbf{x}, 0) = xy \sin z$.

* Acoustic pressure.

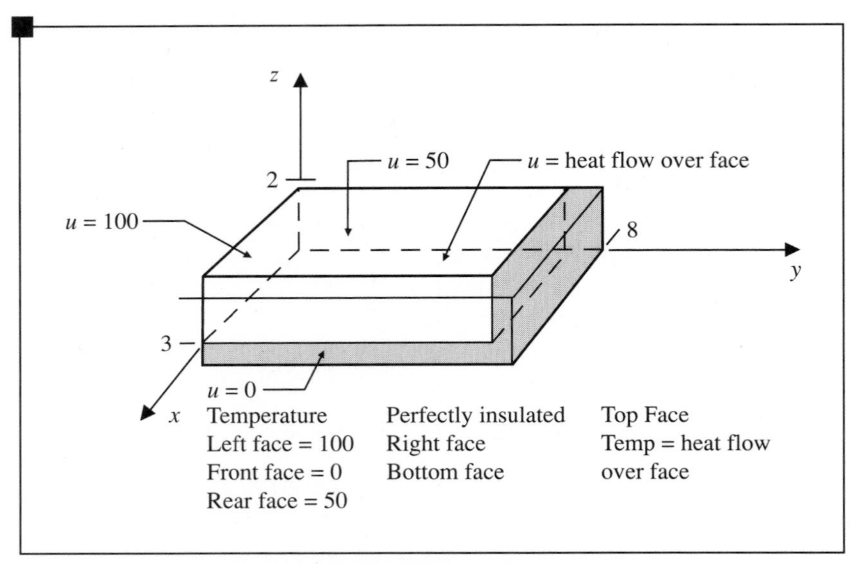

FIGURE 10 **Boundary Conditions for Exercise 46**

SECTION 5*
UNIQUENESS FOR BOUNDARY VALUE PROBLEMS

Given a boundary value problem for the heat, wave, or potential equations, the question naturally arises whether the solution of any such problem is unique. In general, it is possible to appeal to physical intuition to convince ourselves that this must be true. However, the uniqueness of the solution can be shown to hold using rigorous arguments.

In this section we demonstrate the uniqueness of the solution for several boundary value problems. Similar proofs exist for other boundary value problems that were described in previous sections.

EXAMPLE 24. Let D be a finite domain in R^2 and ∂D be its boundary. Show that the solution of the Dirichlet problem

$$\nabla^2 u = 0 \qquad \text{in } D \tag{2.39}$$

and

$$u\bigg|_{\partial D} = f \tag{2.40}$$

is unique.

Solution. To prove the uniqueness of the solution of this problem; we first note that the solutions of the Laplace equation are referred to as **harmonic functions**. One of the basic properties of these functions is given by the maximum–minimum value theorem.

THEOREM 4. If u is continuous on $D \cup \partial D$ and harmonic in D (i.e., $\nabla^2 u = 0$ in D), then the maximum and minimum values of u are obtained on ∂D.

Let now u_1, u_2 be two solutions of the BVPs in Equations (2.39) and (2.40). It is easy to see that $u_2 - u_1$ is harmonic and $u_1 - u_2|_{\partial D} = 0$. Hence the maximum and minimum values of $u_1 - u_2$ on D are zero, which says $u_1 = u_2$ on D. ■

We prove a similar result for the Dirichlet problem for the wave equation

$$\frac{1}{c^2}\frac{\partial^2 u}{\partial t^2} = \frac{\partial^2 u}{\partial x^2} \qquad 0 \le x \le a \tag{2.41}$$

$$u(0, t) = u(a, t) = 0 \qquad t > 0 \tag{2.42}$$

$$u(x, 0) = f(x), \qquad \frac{\partial u}{\partial t}(x, 0) = g(x) \qquad 0 < x < a \tag{2.43}$$

We derive first the "energy equation" for Equation (2.41). To this end we multiply this equation by u_t and integrate over $[0, a]$, from which we obtain

$$\frac{1}{c^2}\int_0^a u_{tt}u_t\,dx = \int_0^a u_{xx}u_t\,dx \tag{2.44}$$

Rewriting $u_{tt}u_t$ as

$$\frac{1}{2}\frac{\partial}{\partial t}(u_t^2)$$

and integrating the right-hand side of Equation (2.44) by parts leads to

$$\begin{aligned}\frac{1}{2}\frac{d}{dt}\frac{1}{c^2}\int_0^a u_t^2\,dx &= u_x u_t\Big|_0^a - \int_0^a u_x u_{xt}\,dx \\ &= u_x u_t\Big|_0^a - \int_0^a \frac{\partial}{\partial t}\left(\frac{1}{2}u_x^2\right)dx \\ &= u_x u_t\Big|_0^a - \frac{d}{dt}\int_0^a \frac{1}{2}u_x^2\,dx\end{aligned} \tag{2.45}$$

Hence,

$$\frac{1}{2}\frac{d}{dt}\int_0^a \left[\frac{1}{c^2}u_t^2 + u_x^2\right]dx = u_x u_t\Big|_0^a \tag{2.46}$$

The expression

$$E(u, t) = \frac{1}{2}\int_0^a \left[\frac{1}{c^2}u_t^2 + u_x^2\right]dx \tag{2.47}$$

is called the **energy integral** of u on $[0, a]$, and Equation (2.46) is called the **energy equation** for Equation (2.41). ■

EXAMPLE 25. The solution of the boundary value problem in Equations (2.41)–(2.43) is unique.

Solution. If u_1, u_2 are two distinct solutions of the boundary value problem in Equations (2.41)–(2.43), then $w = u_1 - u_2$ is a solution of Equation (2.41) with the following boundary and initial conditions:

$$w(0, t) = w(a, t) = 0 \qquad t > 0 \tag{2.48}$$

$$w(x, 0) = \frac{\partial w}{\partial t}(x, 0) = 0 \qquad 0 < x < a \tag{2.49}$$

Applying the energy equation to w, we infer from Equation (2.49) that

$$\frac{d}{dt}\int_0^a \left[\frac{1}{c^2} w_t^2 + w_x^2\right] dx = 0 \tag{2.50}$$

which tells us that $E(w, t)$ is a constant. To compute this constant we observe that at $t = 0$, $w_t(x, 0) = 0$ and

$$w_x(x, 0) = \left.\frac{\partial w(x, t)}{\partial x}\right|_{t=0} = \frac{\partial w(x, 0)}{\partial x} = 0$$

Hence, for all $t \geq 0$,

$$E(w, t) = E(w, 0) = 0 \tag{2.51}$$

But the integrand in Equation (2.47) is nonnegative. Therefore, we deduce from Equation (2.51) that

$$w_x(x, t) = w_t(x, t) = 0 \qquad t \geq 0, 0 \leq x \leq a$$

which implies that $w(x, t)$ is a constant. However, $w(x, 0) = 0$. Therefore, if $w(x, t)$ is a constant, it must follow that $w(x, t) = 0$ and $u_1 = u_2$. ■

The uniqueness of the solution for the heat equation with Dirichlet boundary conditions,

$$\frac{1}{k}\frac{\partial u}{\partial t} = \frac{\partial^2 u}{\partial x^2} \qquad 0 \leq x \leq a, 0 < t \tag{2.52}$$

$$u(0, t) = T_0, \qquad u(a, t) = T_1 \tag{2.53}$$

$$u(x, 0) = f(x) \tag{2.54}$$

follows from the existence of a maximum principle for Equation (2.52).

THEOREM 5. (Maximum principle for the heat equation.) The solution of Equation (2.52) on $0 \leq x \leq a$ and for $0 < t < T$ attains its maximum and minimum on the partial boundary B in the (x, t) plane, which can be described as

$$B = \{(x, 0), (0, t), (a, t) \,|\, 0 \leq x \leq a, 0 \leq t \leq T\}$$ ■

EXAMPLE 26. The solution of Equations (2.52)–(2.54) is unique. In fact, if u_1, u_2 are two solutions of this boundary value problem, then $w = u_1 - u_2$ satisfies the following boundary and initial conditions:

$$w(0, t) = w(a, t) = w(x, 0) = 0$$

or $w = 0$ on B. By the maximum principle this implies that $w = 0$ for all x, t under consideration. ■

At this stage you might wonder whether all the boundary value problems discussed earlier in this chapter have solutions at all (the existence problem). We will not answer this question directly here. However, we will show by construction in Chapter 4 that in general (e.g., when the boundary conditions are "smooth"), the answer to this question is affirmative. Nevertheless, a solution to some boundary value problems exists only if some "compatibility condition" is satisfied, as is demonstrated by Example 27.

EXAMPLE 27. Show that the Neumann problem,

$$\nabla^2 u = 0 \qquad \text{on } D \tag{2.55}$$

$$\frac{\partial u}{\partial \mathbf{n}} = f \qquad \text{on } \partial D \tag{2.56}$$

where D is a finite domain in R^2, has a solution only if the compatibility condition

$$\oint_{\partial D} f\, ds = 0 \tag{2.57}$$

is satisfied.

Solution. To prove Equation (2.57) we use Green's first identity, which states that for any two (smooth) functions u, v in R^2 and a finite domain D,

$$\iint_D (v\nabla^2 u + \nabla v \cdot \nabla u)\, dA = \oint_{\partial D} v \frac{\partial u}{\partial \mathbf{n}}\, ds \tag{2.58}$$

If we let $v = 1$ and u be the solution of Equations (2.55) and (2.56), then it is obvious from Equation (2.58) that Equation (2.57) must be true for a solution of this boundary value problem to exist. ■

SECTION 5 EXERCISES

47. Find the generalization of the energy equation for the inhomogeneous wave equation

$$\frac{1}{c^2}\frac{\partial^2 u}{\partial t^2} = \frac{\partial^2 u}{\partial x^2} + r(x) \tag{2.59}$$

48. Use the results of Exercise 47 to show the uniqueness of the solution for the Dirichlet problem for Equation (2.59).

49. Show that if $\nabla^2 u = 0$ on a domain D and $(\partial u/\partial \mathbf{n}) = 0$ on ∂D, then u is a constant on D.

□ ***Hint.*** Use Equation (2.58) with $u = v$. □

50. Use the result of Exercise 49 to show that the solution of the Neumann problem for the Laplace equation on a domain D is unique up to a constant.

51. Prove the following energy equation for the heat equation [Equation (2.52)]:

$$\frac{1}{k}\frac{d}{dt}\int_0^a \left[\frac{1}{2}u^2\right] dx = u_x u\Big|_0^a - \int_0^a u_x^2\, dx$$

□ ***Hint.*** Multiply Equation (2.52) by u and integrate over $[0, a]$. □

52. Use the result of Exercise 51 to prove the uniqueness of the solution of the Cauchy problem for Equation (2.52).

53. Show that the solution of the Poisson equation $\nabla^2 u = f$ with the boundary condition in Equation (2.40) is unique.

CHAPTER THREE

Fourier Series

SECTION 1
INTRODUCTION

A representation of a function in terms of an infinite series has many useful applications. You are probably familiar with the Taylor series expansion of a function $f(x)$ around $x = a$:

$$f(x) = \sum_{n=0}^{\infty} \frac{f^{(n)}(a)}{n!} (x - a)^n \tag{3.1}$$

Such an expansion of $\ln(1 + x)$ around $x = 0$ leads to

$$\ln(1 + x) = \sum_{n=0}^{\infty} (-1)^n \frac{x^{n+1}}{n + 1} \tag{3.2}$$

This expression allows us to evaluate $\ln 2$ by the basic arithmetic operations. However, the expansion in Equation (3.1) requires $f(x)$ to be infinitely differentiable (or at least differentiable up to some order if we use a truncated expansion). Furthermore, there are cases, although somewhat "pathological" from a practical point of view, in which the expansion fails to converge to $f(x)$.

An example of such a case is the Taylor expansion of $f(x) = e^{-1/x^2}$ around 0. Even when the series in Equation (3.1) converges to f, there are many cases in which the rate of convergence is slow; that is, we have to take a large number of terms in the expansion in order to obtain a reasonable approximation to $f(x)$. This is particularly true for functions that remain bounded as $x \to \infty$. In fact, since $(x - a)^n \to \infty$ as $x \to \infty$ $(n > 0)$, we infer that for such functions the convergence of Equation (3.1) to f for large x must depend on cancellations between the various terms.

In applications to boundary value problems, we frequently encounter functions that are only piecewise continuous, and it follows that the expansion in Equation (3.1) is not appropriate for these functions. Furthermore, in many applications these functions are periodic [i.e., there exists $p > 0$ so that $f(x) = f(x + p)$]. Thus the Taylor expansion, even if it exists, converges slowly to $f(x)$.

To overcome these difficulties we introduce in this chapter expansions of the form

$$\sum_{n=0}^{\infty} a_n \varphi_n(x)$$

where $\varphi_n(x)$ are suitably chosen functions. These expansions have two merits:

1. The computation of a_n requires the evaluation of an integral involving $f(x)$ and $\varphi_n(x)$. Thus, $f(x)$ is required to be integrable but not differentiable. In particular, piecewise continuous functions are integrable.
2. The functions $\varphi_n(x)$ can be chosen so that all of them have the same period as f. This is done in an attempt to ensure a rapid convergence to f. In fact, since the sine and cosine functions are known to be periodic, it is natural to attempt expansion of periodic functions in terms of these known functions. These expansions are called Fourier series.

SECTION 2
PERIODIC FUNCTIONS

In investigating the solution of boundary value problems, we often deal with functions that are periodic, that is, repeat their pattern over and over again. This idea can be made more precise by stating the following definition.

Definition 1. If for all x a function $f(x)$ is defined and $f(x) = f(x + p)$ where $p \neq 0$, then the function f is said to be a **periodic function** with period p.

□ ***Note.*** If we are dealing with a function that is not defined at certain points, the function is still called periodic provided that the points at which f is not defined are p units apart. An example of such a function is shown in Figure 1. □

■ **EXAMPLE 1.** The function $f(x) = \sin bx$ is periodic with period $p = 2\pi/b$ since

$$f(x) = \sin bx = \sin (bx + 2\pi) = \sin b\left(x + \frac{2\pi}{b}\right) = f\left(x + \frac{2\pi}{b}\right) \quad ■$$

■ **EXAMPLE 2.** Define $f(x)$ as follows:

$$f(x) = \begin{cases} -1 & 2n \leq x < 2n + 1 \\ 1 & 2n + 1 \leq x < 2n + 2 \end{cases} \quad n \text{ is an integer and } -\infty < n < \infty.$$

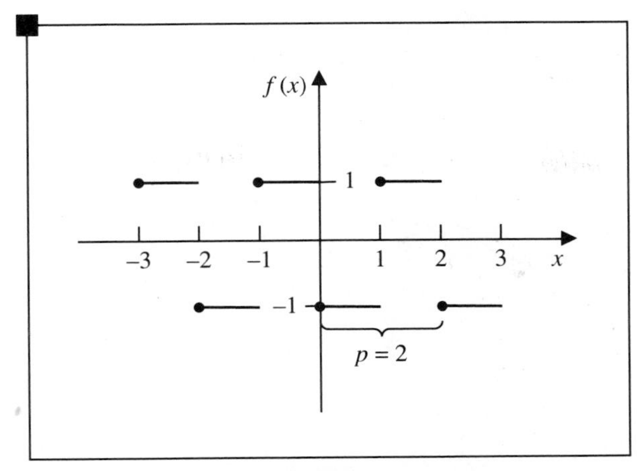

FIGURE 1 **Square Wave**

Plotting this function, we see in Figure 1 that it takes the form of a square wave with period 2. ■

Upon investigating Figure 1 more carefully, we notice that there are repeating patterns other than the one selected. Clearly, we see that p can equal 4, 6, and so on. It is easily shown that this is a property inherent in periodic functions.

THEOREM 1. If f is a function with period p, then $2p, 3p, \ldots, np, -p, -2p, \ldots, -np$ (n is an integer not equal to zero) are also periods of f; that is, $f(x) = f(x + np)$.

Proof. We can prove this result by mathematical induction for $n > 0$.

Step 1. By hypothesis, $f(x) = f(x + p)$, and therefore the statement is true for $n = 1$.

Step 2. Assume the statement is true for $n = k$; that is, $f(x) = f(x + kp)$. But $f(x) = f(x + kp) = f(x + kp + p) = f[x + (k + 1)p]$ and, therefore the statement is true for $n = k + 1$.

By mathematical induction,

$$f(x) = f(x + np) \qquad \text{for } n = 0, 1, 2, \ldots$$

The proof for $n < 0$ is similar. ■

Although a periodic function has an infinite number of periods, it is customary to call the smallest positive period, if it exists, the **primitive period** (or sometimes just **period**).

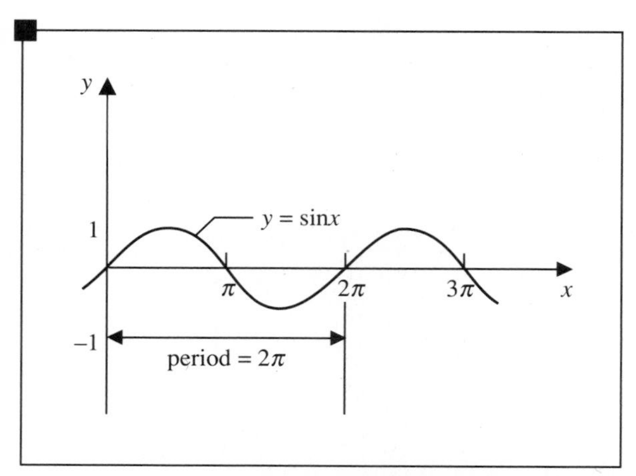

FIGURE 2 **Example of Periodic Function**

EXAMPLE 3. We usually state the period of sin x or cos x as 2π since 2π is the smallest positive period in either case (see Figure 2). ■

SECTION 3
INTRODUCTION TO FOURIER SERIES: ORTHOGONALITY AND NORMALITY

J. B. J. Fourier (1768–1830) is given credit for discovering a new type of series quite different from a Taylor series. Its terms are sine and cosine functions instead of powers of x. We will show how these two periodic functions can conveniently represent other periodic functions.

Consider the series

$$\frac{a_0}{2} + \sum_{n=1}^{\infty} \left\{ a_n \cos \frac{n\pi x}{L} + b_n \sin \frac{n\pi x}{L} \right\}$$

where a_0, a_n, and b_n are constants. For any values of a_0, a_n, and b_n, this series is called a **trigonometric** series. If, however, the coefficients a_0, a_n, and b_n are selected according to a particular rule (which is derived below), the coefficients are called **Fourier coefficients** and the series is known as a **Fourier series**. Thus, Fourier series are trigonometric series, but not all trigonometric series are Fourier series.

In this chapter we investigate the following problem: Given a function f that is periodic with period $2L$, is it possible to find values for a_0, a_n, and b_n so that f can be represented by the series

$$\frac{a_0}{2} + \sum_{n=1}^{\infty} \left\{ a_n \cos \frac{n\pi x}{L} + b_n \sin \frac{n\pi x}{L} \right\}$$

Notice that all these functions have period $2L$; by using these functions we hope for a quick convergence to f.

In this section we investigate a special property held by the sequence of functions

$$\left\{1, \sin\frac{n\pi x}{L}, \cos\frac{n\pi x}{L}\right\} \qquad n = 1, 2, \ldots$$

known as **orthogonality**.

Definition 2. If $f(x)$ and $g(x)$ are two functions that are not identically zero and integrable over the interval (a, b), and $w(x) > 0$ is also integrable over (a, b), then if

$$\int_a^b f(x)g(x)w(x)\,dx = 0$$

we say the function $f(x)$ is **orthogonal** to $g(x)$ with respect to the weighting function $w(x)$ over the interval $[a, b]$.

EXAMPLE 4. Let $f(x) = x$, $g(x) = 5x^3/2 - 3x/2$. The integral

$$\int_{-1}^{1} x\left(\frac{5x^3}{2} - \frac{3x}{2}\right) dx = \left.\frac{x^5}{2} - \frac{x^3}{2}\right|_{-1}^{1} = 0$$

Therefore, $f(x) = x$ is orthogonal to $g(x) = 5x^3/2 - 3x/2$ over the interval $[-1, 1]$. Notice in this example that the weighting function is the function identically equal to 1. ■

EXAMPLE 5. Show that the Laguerre polynomials $L_0 = 1$ and $L_1 = 1 - x$ are orthogonal over the interval $[0, \infty)$ with respect to the weighting function e^{-x}.

Applying these functions to the integral in Definition 2, we have

$$\int_0^\infty e^{-x}(1)(1 - x)\,dx = \lim_{b\to\infty} be^{-b} = 0$$

where the limit is evaluated using L'Hospitals' rule. Therefore, 1 is orthogonal to $(1 - x)$ with respect to the weighting function e^{-x} over $[0, \infty)$. ■

We have stated the general definition of orthogonality, but when dealing with Fourier series the value of $w = 1$. Therefore, we will omit w from our discussion.

Definition 3. If $\{\varphi_n(x)\}$ is a set of functions such that $\varphi_n(x)$ is orthogonal to $\varphi_m(x)$ when $m \neq n$, then $\{\varphi_n(x)\}$ is called a **set of orthogonal functions**.

We will show that the set of functions

$$\left\{1, \sin\frac{n\pi x}{L}, \cos\frac{n\pi x}{L}\right\} \qquad n = 1, 2, \ldots$$

is an orthogonal set over the interval $[-L, L]$. Occasionally it is convenient to represent 1 as $\cos(0\pi x/L)$.

THEOREM 2. The set of functions $\{1, \sin(n\pi x/L), \cos(n\pi x/L)\}$ form an orthogonal set over $[-L, L]$; that is,

$$\int_{-L}^{L} \sin\frac{m\pi x}{L}\cos\frac{n\pi x}{L}\,dx = 0 \qquad m, n = 0, 1, 2, \ldots$$

$$\int_{-L}^{L} \sin\frac{m\pi x}{L}\sin\frac{n\pi x}{L}\,dx = 0 \qquad m, n = 1, 2, \ldots, m \neq n$$

$$\int_{-L}^{L} \cos\frac{m\pi x}{L}\cos\frac{n\pi x}{L}\,dx = 0 \qquad m, n = 0, 1, 2, \ldots, m \neq n$$

Proof. To justify the result for the first integral shown with $m, n = 0, 1, 2, \ldots$, it is necessary to look at two cases.

Case 1. For $m \neq n$,

$$\begin{aligned}\int_{-L}^{L} \sin\frac{m\pi x}{L}\cos\frac{n\pi x}{L}\,dx &= \frac{1}{2}\int_{-L}^{L}\sin\frac{(m+n)\pi x}{L}\,dx + \frac{1}{2}\int_{-L}^{L}\sin\frac{(m-n)\pi x}{L}\,dx \\ &= -\frac{L}{2(m+n)\pi}\cos\frac{(m+n)\pi x}{L}\bigg|_{-L}^{L} - \frac{L}{2(m-n)\pi}\cos\frac{(m-n)\pi x}{L}\bigg|_{-L}^{L} \\ &= -\frac{L}{2(m+n)\pi}[\cos(m+n)\pi - \cos(m+n)(-\pi)] \\ &\quad - \frac{L}{2(m-n)\pi}[\cos(m-n)\pi - \cos(m-n)(-\pi)] \\ &= 0\end{aligned}$$

Case 2. If $m = n \neq 0$, we use the following argument. The integral

$$\int_{-L}^{L} \sin\frac{n\pi x}{L}\cos\frac{n\pi x}{L}\,dx = \frac{L}{n\pi}\sin^2\frac{n\pi x}{L}\bigg|_{-L}^{L} = 0$$

Similarly, it can be shown that

$$\left.\begin{aligned}\int_{-L}^{L} \sin\frac{m\pi x}{L}\sin\frac{n\pi x}{L}\,dx = 0 \\ \int_{-L}^{L} \cos\frac{m\pi x}{L}\cos\frac{n\pi x}{L}\,dx = 0\end{aligned}\right\} \quad m \neq n$$

■

It is obvious that the integral of the square of a continuous function f times a positive weighting function over the interval $[a, b]$ is nonzero unless $f(x) \equiv 0$ over (a, b). This idea leads to Theorem 3.

THEOREM 3. The values of the integrals of the square of a function chosen from the set $\{1, \cos(n\pi x/L), \sin(n\pi x/L)\}$ are as follows:

$$\int_{-L}^{L} 1 \cdot 1\, dx = 2L$$

$$\left.\begin{aligned} \int_{-L}^{L} \sin^2 \frac{n\pi x}{L}\, dx = L \\ \int_{-L}^{L} \cos^2 \frac{n\pi x}{L}\, dx = L \end{aligned}\right\} \quad n = 1, 2, \ldots$$

Proof. The first integral in the theorem can be written as

$$\int_{-L}^{L} 1 \cdot 1\, dx = x \Big|_{-L}^{L} = 2L$$

The third integral yields

$$\int_{-L}^{L} \cos^2 \frac{n\pi x}{L}\, dx = \frac{x}{2} + \frac{L}{4n\pi} \sin \frac{2n\pi x}{L} \Big|_{-L}^{L} = L$$

■

The calculation of

$$\int_{-L}^{L} \sin^2 \frac{n\pi x}{L}\, dx$$

is similar.

The integral $(\int_a^b f^2(x)\, dx)^{1/2}$ is called the **norm** of f over $[a, b]$ and is denoted $\| f \|$. If the value of the integral happens to equal 1, the given function is said to be **normalized** (or **normal**) over $[a, b]$. We see from Theorem 3 that none of the functions 1, $\sin(n\pi x/L)$, $\cos(n\pi x/L)$ are normal over the interval $[-L, L]$.

If a function f is not normal, that is, $\int_a^b f^2(x)\, dx = k \neq 1$, a constant multiple of f that is normal can always be found. It is easily shown that $f/\| f \|$ is a normal function over $[a, b]$.

EXAMPLE 6. Normalize the functions 1 and $\sin(n\pi x/L)$ over the interval $[-L, L]$.

Since the integral

$$\int_{-L}^{L} 1^2\, dx = 2L$$

the function $f(x) = 1$ is normalized by multiplying $f(x)$ by $1/\sqrt{2L}$. The function $\psi(x) = 1/\sqrt{2L}$ is normal over the interval $[-L, L]$ because

$$\int_{-L}^{L} \left(\frac{1}{\sqrt{2L}}\right)^2 dx = \frac{2L}{2L} = 1$$

In a similar way the function

$$\psi_n(x) = \frac{1}{\sqrt{L}} \sin \frac{n\pi x}{L}$$

is normal over $[-L, L]$. ■

THEOREM 4. If $\{\varphi_n(x)\}$ is an orthogonal set over the interval $[a, b]$, the set $\{\psi_n(x)\}$ where

$$\psi_n(x) = \frac{\varphi_n(x)}{\|\varphi_n(x)\|}$$

is both orthogonal and normal over $[a, b]$. When a set possesses both properties, it is said to be an **orthonormal** set.

Proof. The integral

$$\int_a^b \psi_m(x)\psi_n(x)\,dx = \int_a^b \frac{\varphi_m(x)\varphi_n(x)}{\|\varphi_m\|\,\|\varphi_n\|}\,dx$$
$$= \frac{1}{\|\varphi_m(x)\|\,\|\varphi_n(x)\|} \int_a^b \varphi_m(x)\varphi_n(x)\,dx = 0 \qquad m \neq n$$

Therefore, the set remains orthogonal after modification. Next the integral

$$\int_a^b \psi_n^2(x)\,dx = \int_a^b \frac{\varphi_n^2(x)}{\|\varphi_n(x)\|^2}\,dx$$
$$= \frac{1}{\|\varphi_n(x)\|^2} \int_a^b \varphi_n^2(x)\,dx = \frac{\|\varphi_n^2(x)\|}{\|\varphi_n^2(x)\|} = 1$$

This fact tells us that each function $\psi_n(x)$ is normalized. The set $\{\psi_n(x)\}$ is an **orthonormal set.** ■

Orthonormal sets are used in mathematical analysis to make the work simpler. In our study of Fourier series, all we will need is the property of orthogonality.

SECTIONS 2 AND 3 EXERCISES

1. Determine whether the functions listed below, defined on $-\infty < x < \infty$, are periodic. If they are periodic, state (1) a period and (2) the primitive period (if it exists).
   (a) $\cos x$
   (b) $\tan 3x$
   (c) x^2
   (d) $[x] - x$ where [ ] is the greatest integer function
   (e) $\sin x \cos x$
   (f) $2 \sin x - 3 \cos x$
   (g) 5
   (h) $\sinh x$
   (i) $f(x) = (x - 2n)^2$, $2n - 1 \leq x \leq 2n + 1$, $n = \ldots -2, -1, 0, 1, 2, \ldots$

2. Are the following pairs of functions orthogonal over the interval indicated? [Take the weighting function $w(x) = 1$.]
   (a) 1 and x, $[-2, 2]$
   (b) 1 and x, $[0, 2]$
   (c) $\sin x$ and $\sin 2x$, $[0, \pi]$
   (d) $\sin x$ and $\cos x$, $[0, \pi]$
   (e) $\sinh x$ and $\cosh x$, $[-1, 1]$
   (f) $P_2(x) = (1/2)(3x^2 - 1)$ and $P_3(x) = (1/2)(5x^3 - 3x)$, $[-1, 1]$

3. Given the functions $bx + 2$ and $x - 1$, find b so that the two functions are orthogonal over $[0, 2]$.

4. Prove $\sin(n\pi x/L)$ and $\sin(m\pi x/L)$ are orthogonal over $[-L, L]$, $n \neq m$.

5. Prove $\cos(n\pi x/L)$ and $\cos(m\pi x/L)$ are orthogonal over $[-L, L]$, $n \neq m$.

6. Prove $\int_{-\pi}^{\pi} \sin^2 x\, dx = \pi$.

7. Are the following pairs of functions orthogonal with respect to the given weighting function over the interval indicated?
   (a) $f_1 = e^x, f_2 = e^{-x}$; $w(x) = \sin x$, $[0, \pi]$
   (b) $T_0 = 1$, $T_1 = x$; $w(x) = 1/\sqrt{1 - x^2}$, $[-1, 1]$
   (c) $L_0 = 1$, $L_2 = (1/2)[x^2 - 4x + 2]$; $w(x) = e^{-x}$, $[0, \infty)$
   (d) $U_0 = 1$, $U_1 = 2x$; $w(x) = \sqrt{1 - x^2}$, $[-1, 1]$

8. Normalize the function given below over the interval indicated.
   (a) x; $w(x) = 1$, $[0, 2]$
   (b) $x^2 + 1$; $w(x) = 1$, $[0, 1]$
   (c) $\sin n\pi x$; $w(x) = 1$, $[0, \pi]$

SECTION 4
DETERMINING FOURIER COEFFICIENTS

Let $f(x)$ be a periodic function with period $2L$. Under what conditions can the function $f(x)$ be represented by a series of the form

$$\frac{a_0}{2} + \sum_{n=1}^{\infty} \left\{ a_n \cos \frac{n\pi x}{L} + b_n \sin \frac{n\pi x}{L} \right\}$$

At this point we will use a tilde ($\sim$) instead of an equal sign ($=$) and write

$$f(x) \sim \frac{a_0}{2} + \sum_{n=1}^{\infty} \left\{ a_n \cos \frac{n\pi x}{L} + b_n \sin \frac{n\pi x}{L} \right\} \tag{3.3}$$

since, first, we do not know if the series converges and, second, if it converges, if it converges to the function $f(x)$.

The argument given in this section for finding the constants a_0, a_n, and b_n is predicated on the fact that the series converges to $f(x)$ and the operations used are permissible. Later on in Theorem 13 we will state a set of conditions under which a certain class of functions $f(x)$ possesses such a series representation.

To find a_0 we multiply both sides of Equation (3.3) by 1 and integrate both sides of the equation from $-L$ to L. On the right-hand side, we are assuming that integration of an infinite sum can be replaced by the sum (infinite) of individual integrals. After carrying out these operations, we have

$$\int_{-L}^{L} f(x)\,dx = \frac{a_0}{2}\int_{-L}^{L} dx + \sum_{n=1}^{\infty}\left\{a_n \int_{-L}^{L} \cos\frac{n\pi x}{L}\,dx + b_n \int_{-L}^{L} \sin\frac{n\pi x}{L}\,dx\right\} \tag{3.4}$$

But from Theorems 2 and 3 we can write Equation (3.4) as

$$\int_{-L}^{L} f(x)\,dx = 2L\frac{a_0}{2} + \sum_{n=1}^{\infty}\{a_n\cdot 0 + b_n\cdot 0\} \tag{3.5}$$

Solving Equation (3.5) we find

$$a_0 = \frac{1}{L}\int_{-L}^{L} f(x)\,dx$$

To find a_m, multiply both sides of Equation (3.3) by $\cos(m\pi x/L)$. Assuming first that the product of an infinite sum equals the sum of products and then that the integral of an infinite sum equals the sum of integrals, we can write

$$\int_{-L}^{L} f(x)\cos\frac{m\pi x}{L}\,dx = \frac{a_0}{2}\int_{-L}^{L}\cos\frac{m\pi x}{L}\,dx + \sum_{n=1}^{\infty}\left\{a_n\int_{-L}^{L}\cos\frac{n\pi x}{L}\cos\frac{m\pi x}{L}\,dx + b_n\int_{-L}^{L}\sin\frac{n\pi x}{L}\cos\frac{m\pi x}{L}\,dx\right\}$$

Then, applying Theorems 2 and 3 again, we can write

$$\int_{-L}^{L} f(x)\cos\frac{m\pi x}{L}\,dx = 0 + a_m\int_{-L}^{L}\cos^2\frac{m\pi x}{L}\,dx + \sum_{n=1}^{\infty} b_n\cdot 0 = a_m L$$

Solving for a_m we have

$$a_m = \frac{1}{L}\int_{-L}^{L} f(x)\cos\frac{m\pi x}{L}\,dx$$

Since m is a dummy variable, we can write

$$a_n = \frac{1}{L}\int_{-L}^{L} f(x)\cos\frac{n\pi x}{L}\,dx \qquad n = 1, 2, \ldots$$

Similarly, we can show that

$$b_n = \frac{1}{L}\int_{-L}^{L} f(x)\sin\frac{n\pi x}{L}\,dx \qquad n = 1, 2, \ldots$$

From these results we can form the following definition of a Fourier series.

Definition 4. Let f be a function whose period is $2L$. The **Fourier series** of f is the series

$$\frac{a_0}{2} + \sum_{n=1}^{\infty} \left\{a_n \cos\frac{n\pi x}{L} + b_n \sin\frac{n\pi x}{L}\right\}$$

provided the coefficients

$$a_0 = \frac{1}{L}\int_{-L}^{L} f(x)\,dx$$

$$a_n = \frac{1}{L}\int_{-L}^{L} f(x)\cos\frac{n\pi x}{L}\,dx$$

$$b_n = \frac{1}{L}\int_{-L}^{L} f(x)\sin\frac{n\pi x}{L}\,dx$$

all exist. These coefficients are called **Fourier coefficients.**

The existence of a Fourier series for f does not automatically guarantee that the series actually converges to $f(x)$. However, a well-known theorem (see Theorem 13) states that if f is continuous at x_0, then the Fourier series actually converges to $f(x_0)$. On the other hand, when f has a jump discontinuity at x_0, then the Fourier series converges to

$$\frac{f(x_0^+) + f(x_0^-)}{2}$$

where $f(x_0^+) = \lim_{x\to x_0} f(x)$, $x > x_0$ and $f(x_0^-) = \lim_{x\to x_0} f(x)$, $x < x_0$. For convenience we will assume from here on that we are dealing with functions that satisfy Theorem 13. Therefore, we will drop the tilde ($\sim$) and use the equal sign ($=$).

EXAMPLE 7. Find the Fourier coefficients and Fourier series for the function

$$f(x) = \begin{cases} -3 & -5 + 10n < x < 10n \\ 3 & 10n < x < 5 + 10n \end{cases} \qquad n = 0, \pm 1, \pm 2, \ldots$$

which is shown in Figure 3.

The first thing to do when calculating Fourier coefficients is to carefully ascertain the smallest value of L. Postponing this determination often results in incorrect answers.

It is easily seen from Figure 3 that the period is $2L = 10$, and therefore $L = 5$. To find the values of a_0 and a_n, we write

$$a_0 = \frac{1}{5}\int_{-5}^{0} (-3)\,dx + \frac{1}{5}\int_{0}^{5} 3\,dx = 0$$

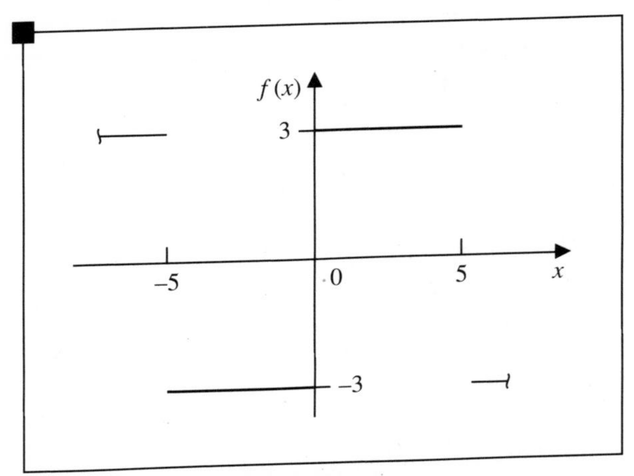

FIGURE 3 **Square Wave**

and

$$a_n = \frac{1}{5}\int_{-5}^{0}(-3)\cos\frac{n\pi x}{5}\,dx + \frac{1}{5}\int_{0}^{5} 3\cos\frac{n\pi x}{5}\,dx = 0$$

The value of b_n is given by

$$\begin{aligned} b_n &= \frac{1}{5}\int_{-5}^{0}(-3)\sin\frac{n\pi x}{5}\,dx + \frac{1}{5}\int_{0}^{5} 3\sin\frac{n\pi x}{5}\,dx \\ &= \frac{3}{n\pi}\cos\frac{n\pi x}{5}\Big|_{-5}^{0} - \frac{3}{n\pi}\cos\frac{n\pi x}{5}\Big|_{0}^{5} \\ &= \frac{3}{n\pi}(1-\cos n\pi) - \frac{3}{n\pi}(\cos n\pi - 1) \\ &= \frac{6}{\pi}(1-\cos n\pi) \end{aligned}$$

The Fourier series can be written as

$$f(x) = \frac{6}{\pi}\sum_{n=1}^{\infty}\frac{1-\cos n\pi}{n}\sin\frac{n\pi x}{5}$$

We will see in the discussion following Table 2 that this answer can be simplified. ■

EXAMPLE 8. Find the Fourier series for the function

$$f(x) = \begin{cases} 0 & (2n-1)\pi < x \le 2n\pi \\ x - 2n\pi & 2n\pi \le x < (2n+1)\pi \end{cases} \qquad n = 0, \pm 1, \pm 2, \dots$$

which is shown in Figure 4.

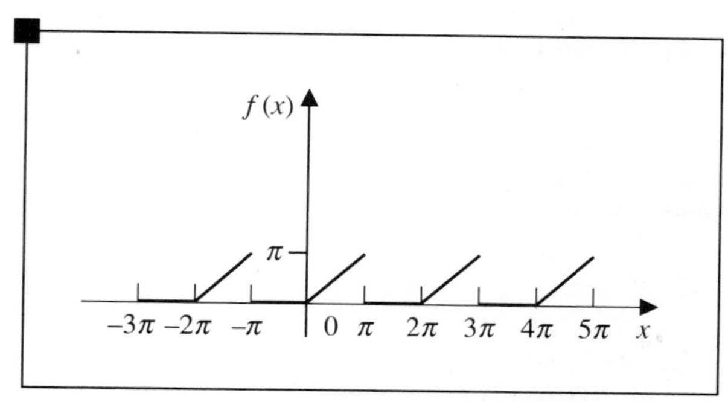

FIGURE 4 **Function in Example 8**

As stated in Example 7, the first thing we do is find the value of L. This is

$$2L = 2\pi$$

or

$$L = \pi$$

Then

$$\begin{aligned} a_0 &= \frac{1}{\pi}\int_{-\pi}^{\pi} f(x)\,dx = \frac{1}{\pi}\int_{-\pi}^{0} 0\,dx + \frac{1}{\pi}\int_{0}^{\pi} x\,dx \\ &= \left(\frac{1}{\pi}\right)\frac{x^2}{2}\bigg|_0^{\pi} = \frac{\pi}{2} \end{aligned}$$

Next we see that the coefficient

$$\begin{aligned} a_n &= \frac{1}{\pi}\int_0^{\pi} x\cos nx\,dx = \frac{1}{n^2\pi}\left[\cos nx + nx\sin nx\right]\bigg|_0^{\pi} \\ &= \frac{1}{n^2\pi}\left[\cos n\pi - 1\right] \end{aligned}$$

and the coefficient

$$b_n = \frac{1}{\pi}\int_0^{\pi} x\sin nx\,dx = -\frac{1}{n}\cos n\pi$$

The Fourier series is given by

$$f(x) = \frac{\pi}{4} + \sum_{n=1}^{\infty}\left\{\frac{(\cos n\pi - 1)}{n^2\pi}\cos nx - \frac{1}{n}\cos n\pi\sin nx\right\} \tag{3.6}$$

■

Although the answer given by Equation (3.6) is correct, as in Example 7, it can be written in a more compact form.

Tables 1 and 2 make the work of calculating Fourier coefficients a great deal easier.

TABLE 1 Integrals for Fourier Series

$$\int \sin nx\, dx = -\frac{1}{n}\cos nx$$

$$\int x \sin nx\, dx = \frac{1}{n^2}[\sin nx - nx\cos nx]$$

$$\int x^2 \sin nx\, dx = \frac{1}{n^3}[2nx\sin nx - (n^2x^2 - 2)\cos nx]$$

$$\int \cos nx\, dx = \frac{1}{n}\sin nx$$

$$\int x\cos nx\, dx = \frac{1}{n^2}[\cos nx + nx\sin nx]$$

$$\int x^2\cos nx\, dx = \frac{1}{n^3}[2nx\cos nx + (n^2x^2 - 2)\sin nx]$$

$$\int e^{ax}\sin nx\, dx = \frac{e^{ax}[a\sin nx - n\cos nx]}{a^2 + n^2}$$

$$\int e^{ax}\cos nx\, dx = \frac{e^{ax}[a\cos nx + n\sin nx]}{a^2 + n^2}$$

$$\int \sin^2 nx\, dx = \frac{1}{n}\left[\frac{nx}{2} - \frac{1}{4}\sin 2nx\right]$$

$$\int \cos^2 nx\, dx = \frac{1}{n}\left[\frac{nx}{2} + \frac{1}{4}\sin 2nx\right]$$

$$\int \sin mx \sin nx\, dx = \frac{\sin(m-n)x}{2(m-n)} - \frac{\sin(m+n)x}{2(m+n)} \qquad m^2 \neq n^2$$

$$\int \sin mx\cos nx\, dx = \frac{\cos(m-n)x}{2(m-n)} - \frac{\cos(m+n)x}{2(m+n)} \qquad m^2 \neq n^2$$

$$\int \cos mx\cos nx\, dx = \frac{\sin(m-n)x}{2(m-n)} + \frac{\sin(m+n)x}{2(m+n)} \qquad m^2 \neq n^2$$

TABLE 2 Identities Concerning Sines and Cosines

$$\sin n\pi = 0$$

$$\cos n\pi = (-1)^n$$

$$\sin\frac{n\pi}{2} = \begin{cases} 0 & n \text{ even} \\ (-1)^{(n-1)/2} & n \text{ odd} \end{cases}$$

$$\cos\frac{n\pi}{2} = \begin{cases} 0 & n \text{ odd} \\ (-1)^{n/2} & n \text{ even} \end{cases}$$

Using Table 2 we can simplify the Fourier series in Equation (3.6) in the following way. Since $\cos n\pi = (-1)^n$, we can write Equation (3.6) in the form

$$f(x) = \frac{\pi}{4} + \sum_{n=1}^{\infty} \left\{ \frac{(-1)^n - 1}{n^2 \pi} \cos nx + \frac{(-1)^{n-1}}{n} \sin nx \right\} \tag{3.7}$$

This series simplifies even further because the Fourier coefficient for $\cos nx$ can be written compactly if we distinguish between n odd and n even. In fact,

$$\frac{(-1)^n - 1}{n^2 \pi} = \begin{cases} 0 & n \text{ even} \\ -\dfrac{2}{n^2 \pi} & n \text{ odd} \end{cases}$$

and Equation (3.7) can be written as

$$f(x) = \frac{\pi}{4} - \frac{2}{\pi} \sum_{n=1,3,5}^{\infty} \frac{1}{n^2} \cos nx + \sum_{n=1}^{\infty} \frac{(-1)^{n-1}}{n} \sin nx \tag{3.8}$$

Even this form can be carried further if you want to avoid summing over odd integers. By letting $n = 2m - 1$, $m = 1, 2, \ldots$ in the first series on the right-hand side, we notice that

$$\frac{2}{\pi} \sum_{n=1,3,5}^{\infty} \frac{1}{n^2} \cos nx = \frac{2}{\pi} \sum_{m=1}^{\infty} \frac{1}{(2m-1)^2} \cos (2m-1)x$$

Finally, remembering that the letter chosen for the index is immaterial, Equation (3.8) becomes

$$f(x) = \frac{\pi}{4} + \sum_{n=1}^{\infty} \left\{ -\frac{2}{(2n-1)^2 \pi} \cos (2n-1)x + \frac{(-1)^{n-1}}{n} \sin nx \right\}$$

EXAMPLE 9. Given the function $f(x) = |x|$ over the interval $[-1, 1]$ and letting F be the periodic extension of f, find the Fourier series expansion of F (see Figure 5).

First we see that $2L = 2$, or $L = 1$. To find a_0, a_n, and b_n, the integration must be done in two parts. Over the interval $[-1, 0]$ we use $-x = |x|$, whereas over $[0, 1]$ we use $x = |x|$. Therefore,

$$\begin{aligned} a_0 &= \frac{1}{1} \int_{-1}^{0} -x \, dx + \frac{1}{1} \int_{0}^{1} x \, dx \\ &= -\frac{x^2}{2}\bigg|_{1}^{0} + \frac{x^2}{2}\bigg|_{0}^{1} = 2 \end{aligned}$$

Next we see that

$$a_n = \int_{-1}^{0} -x \cos n\pi x \, dx + \int_{0}^{1} x \cos n\pi x \, dx$$

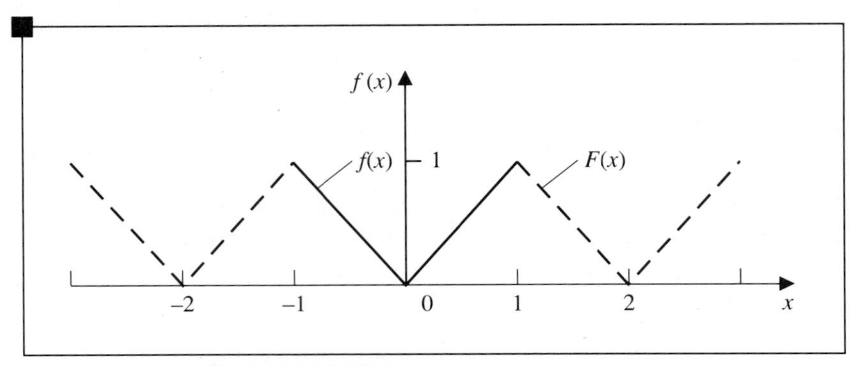

FIGURE 5 **Function in Example 9**

which can be evaluated using Table 1. It follows that

$$a_n = -\frac{1}{n^2\pi^2} \cos n\pi x + n\pi x \sin n\pi x \Big|_{-1}^{0} + \frac{1}{n^2\pi^2} \cos n\pi x + n\pi x \sin n\pi x \Big|_{0}^{1}$$

or

$$a_n = \frac{2}{n^2\pi^2} [(-1)^n - 1]$$

In a similar way b_n can be shown to equal zero. The Fourier series for the periodic extension of $|x|$ over $-1 \leq x \leq 1$ is given by

$$|x| = 1 - \frac{4}{\pi^2} \sum_{n=1,3,5}^{\infty} \frac{\cos n\pi x}{n^2}$$ ■

Another device is sometimes useful when calculating the Fourier coefficients.

THEOREM 5. Suppose $g(x)$ is an integrable periodic function with period $2L$. Then

$$\int_a^{a+2L} g(x)\,dx = \int_{-L}^{L} g(x)\,dx$$

Proof. You are asked to prove Theorem 1 in Exercise 10. ■

This theorem essentially says that when we are integrating a periodic function over an interval whose length is a period, the starting point is immaterial (see Figure 6).

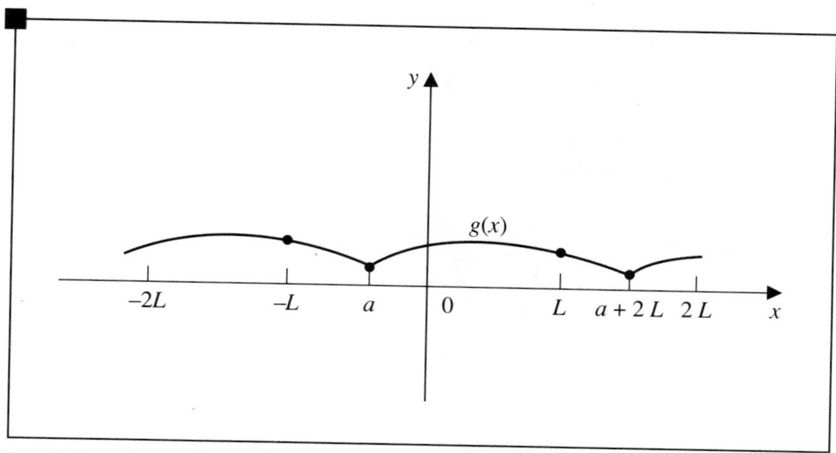

FIGURE 6 **Function as Example for Theorem 5**

EXAMPLE 10. Evaluate the integral of the function f shown below over two different intervals of length 2π and show that the integrals are equal. Let

$$f(x) = \begin{cases} 0 & (2n-1)\pi < x \leq 2n\pi \\ x - 2n\pi & 2n\pi \leq x < (2n+1)\pi \end{cases} \qquad n = 0, \pm 1, \pm 2, \ldots$$

(See Figure 4.) For the interval $-\pi$ to π, we have

$$\int_{-\pi}^{\pi} f(x)\,dx = \int_0^{\pi} x\,dx = \frac{x^2}{2}\bigg|_0^{\pi} = \frac{\pi^2}{2}$$

whereas for $\pi/2$ to $5\pi/2$ we can write

$$\begin{aligned} \int_{\pi/2}^{5\pi/2} f(x)\,dx &= \int_{\pi/2}^{\pi} x\,dx + \int_0^{2\pi} 0\,dx + \int_{2\pi}^{5\pi/2} (x - 2\pi)\,dx \\ &= \frac{x^2}{2}\bigg|_{\pi/2}^{\pi} + \left(\frac{x^2}{2} - 2\pi x\right)\bigg|_{2\pi}^{5\pi/2} \\ &= \frac{\pi^2}{2} \end{aligned}$$

■

We can use Theorem 5 to simplify the work required to evaluate the Fourier coefficients in Example 11.

EXAMPLE 11. Find the Fourier series for the function shown in Figure 7. The period of f is $2L = (3\pi/2) - (-\pi/2) = 2\pi$, or $L = \pi$. If we attempted to find the Fourier coefficients by Definition 4, it would be necessary to find the function representing the straight-line segment to the left of f. (This function is given as $(x/2) + \pi$.) We can simplify the work if we apply Theorem 5. We evaluate a_0,

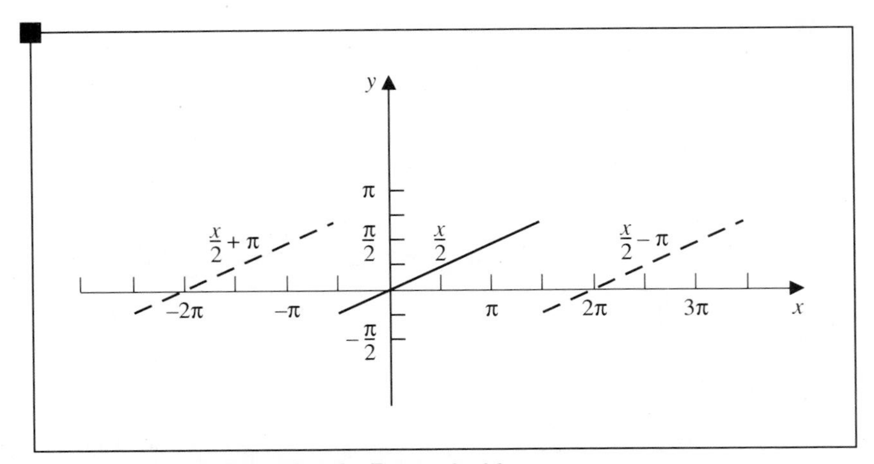

FIGURE 7 **Function in Example 11**

a_n, and b_n as follows:

$$a_0 = \frac{1}{\pi}\int_{-\pi/2}^{3\pi/2} \frac{x}{2}\,dx = \frac{\pi}{2}$$

$$\begin{aligned} a_n &= \frac{1}{\pi}\int_{-\pi/2}^{3\pi/2} \frac{x}{2}\cos nx\,dx = \frac{1}{2\pi n^2}\left[\cos nx + nx\sin nx\right]\Big|_{-\pi/2}^{3\pi/2} \\ &= -\frac{1}{n}\sin\frac{n\pi}{2} \\ &= \begin{cases} 0 & n \text{ even} \\ \dfrac{(-1)^{(n+1)/2}}{n} & n \text{ odd} \end{cases} \end{aligned}$$

whereas

$$\begin{aligned} b_n &= \frac{1}{\pi}\int_{-\pi/2}^{3\pi/2} \frac{x}{2}\sin nx\,dx = -\frac{1}{n}\cos\frac{n\pi}{2} \\ &= \begin{cases} 0 & n \text{ odd} \\ \dfrac{(-1)^{(n+2)/2}}{n} & n \text{ even} \end{cases} \end{aligned}$$

The Fourier series can be written as

$$\begin{aligned} f(x) = \frac{\pi}{4} &+ \sum_{n=1,3,5}^{\infty}\left[\frac{(-1)^{(n+1)/2}}{n}\right]\cos nx \\ &+ \sum_{n=2,4,6}^{\infty}\left[\frac{(-1)^{(n+2)/2}}{n}\right]\sin nx \end{aligned}$$

or

$$f(x) = \frac{\pi}{4} + \sum_{n=1}^{\infty}(-1)^n\left\{\frac{\cos(2n-1)x}{2n-1} - \frac{\sin 2nx}{2n}\right\}$$

■

SECTION 4 EXERCISES

□ ***Note.*** Although at this point we are applying Fourier series to periodic functions, it is less confusing when describing a given function to define it in one period. We assume that the function will be extended as a periodic function along the rest of the real axis. □

9. Prove

(a) $\sin n\pi = 0$
(b) $\cos n\pi = (-1)^n$
for n an integer.

10. Prove $\int_a^{a+2L} g(x)\,dx = \int_{-L}^{L} g(x)\,dx$ if $g(x)$ is an integrable periodic function with period $2L$.

11. Find the Fourier series for the following functions having the indicated period $p = 2L$.

(a) $f(x) = x,\ [-\pi, \pi] \qquad p = 2\pi$
(b) $f(x) = x,\ [-2, 2] \qquad p = 4$

(c) $f(x) = \begin{cases} -x & -1 < x < 0 \\ x & 0 < x < 1 \end{cases} \qquad p = 2$

(d) $f(x) = x^2,\ [-\pi, \pi] \qquad p = 2\pi$

(e) $f(x) = \begin{cases} 0 & -\pi < x < 0 \\ 1 & 0 < x < \pi \end{cases} \qquad p = 2\pi$

12. Find the Fourier series for the following functions. Each periodic function is defined over an interval whose length is one period.

(a) $f(x) = x + 1,\ [0, 3]$

(b) $f(x) = \begin{cases} -1 & -2 < x < 0 \\ 1 & 0 < x < 2 \end{cases}$

(c) $f(x) = \begin{cases} 1 + x & -1 < x < 0 \\ 1 - x & 0 < x < 1 \end{cases}$

(d) $f(t) = \sin 3t,\ [-\pi, \pi]$
(e) $f(x) = e^x,\ [1, 3]$
(f) $f(s) = \sinh s,\ [-1, 1]$

(g) $f(x) = \begin{cases} x & 0 < x < 2 \\ 2 & 2 < x < 4 \end{cases}$

(h) $f(t) = 168 \cos 120\pi t,\ \left[0, \frac{1}{60}\right]$

(i) $f(x) = |x|,\ [-2, 2]$
(j) $f(x) = [x] - x,\ [0, 1] \qquad [\ \ ] =$ greatest integer function
(k) $f(x) = 1,\ [-10, 10]$

(l) $f(r) = \begin{cases} 0 & -\pi < r < 0 \\ 2r & 0 < r < \pi \end{cases}$

13. Find the Fourier series for the following periodic functions. Each function is defined over an interval whose length is one period.

□ ***Hint.*** Use Theorem 5. □

(a) $f(x) = \frac{x}{2}$, $[0, 4]$

(b) $f(y) = \frac{1}{3}(y + 1)$, $\left[-\frac{1}{3}, \frac{2}{3}\right]$

(c) $f(t) = t^2$, $[-1, 2]$

SECTION 5
FOURIER SERIES FOR EVEN AND ODD FUNCTIONS

As we worked through some of the examples of Fourier series in the preceding sections, we observed that the Fourier series of some functions had much simpler representations than others. For example, the Fourier coefficients a_0 and a_n of the square wave described in Example 7 turned out to be 0, leaving only the b_n's as nonzero coefficients. This is not coincidence but results from the fact that the function we were trying to represent was an odd function. In this section we show how to use this property of being even or odd to simplify the computation of a Fourier series.

Definition 5. Let $f(x)$ be defined on the interval $(-b, b)$ where $b > 0$. A function $f(x)$ is said to be **even** on $(-b, b)$ if $f(-x) = f(x)$. Similarly, a function $f(x)$ is said to be **odd** on $(-b, b)$ if $f(-x) = -f(x)$.

EXAMPLE 12

(a) Since $\sin(-x) = -\sin x$, the function $f(x) = \sin x$ is an odd function.

(b) Since $\cos(-x) = \cos x$, the cosine function is even.

(c) The function $g(x) = x^2 - 3x$ is neither even nor odd since

$$g(-x) = (-x)^2 - 3(-x) = x^2 + 3x \neq \pm g(x)$$ ■

Just as we have certain rules concerning even and odd numbers (e.g., the product of two odd numbers is always odd), we can develop similar rules for combining even and odd functions.

□ ***Warning.*** These rules do not always match the corresponding rules for numbers. □

EXAMPLE 13. Show that the product of two odd functions is an even function. Let $f(x)$ and $g(x)$ be two odd functions. Form a new function $h(x) = f(x)g(x)$. Now

$$h(-x) = f(-x)g(-x) = [-f(x)][-g(x)] = f(x)g(x) = h(x)$$

Thus, $h(x)$ is an even function; that is, the product of two odd functions is even. ■

Theorems 6, 7, and 8 show how we can extend these properties to Fourier series.

THEOREM 6. The sum of a finite number of even (odd) functions is even (odd).

Proof. Let $f(x)$ and $g(x)$ be two even functions. Forming a new function $h(x) = f(x) + g(x)$, we can write

$$h(-x) = f(-x) + g(-x) = f(x) + g(x) = h(x)$$

Therefore, the sum of two even functions is even. Repeated use of this result extends the property to any finite number of functions. (The proof for odd functions is similar.) ■

THEOREM 7. If we assume a Fourier series of the form

$$\sum_{n=1}^{\infty} b_n \sin \frac{n\pi x}{L}$$

converges to $f(x)$ for all x, then $f(x)$ is an odd function.

Proof. Since the series converges for all x, we can write

$$f(-x) = \sum_{n=1}^{\infty} b_n \sin \frac{n\pi(-x)}{L} = -\sum_{n=1}^{\infty} b_n \sin \frac{n\pi x}{L} = -f(x)$$

Obviously, $f(x)$ is an odd function. ■

THEOREM 8. If we assume a Fourier series of the form

$$\frac{a_0}{2} + \sum_{n=1}^{\infty} a_n \cos \frac{n\pi x}{L}$$

converges to $f(x)$ for all x, then $f(x)$ is an even function. ■

Extension, Half Fourier Interval

Very often we are in a situation where a function f is defined on the interval $[0, L]$ rather than the interval $[-L, L]$, called the **full Fourier interval**, and we desire a Fourier representation of f over $[0, L]$. The interval $[0, L]$ is called the **half Fourier interval**. The problem is easily solved, for all we have to do is extend or prolong the function f into the interval $[-L, 0]$ in such a way that we can write the Fourier series over $[-L, L]$ and then periodically on the rest of R.

In general, we have the option of extending the function f into $[-L, 0]$ in any reasonable way we like. However, normally it makes sense to extend f

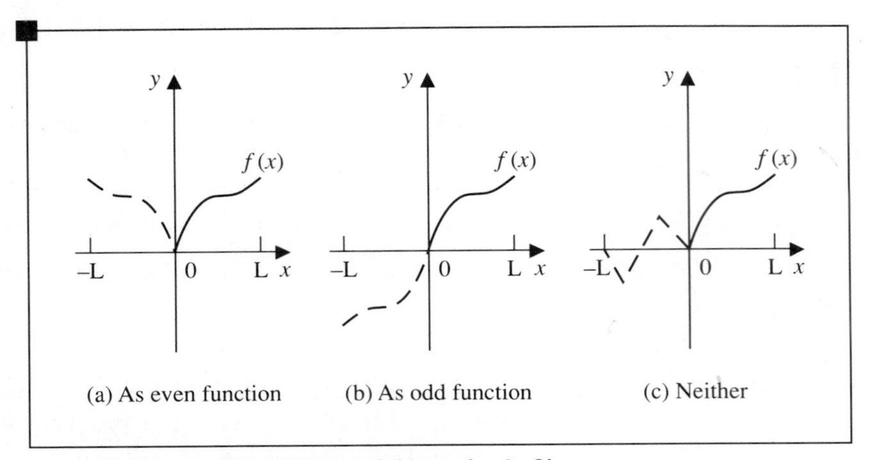

FIGURE 8 **Extensions of f into $(-L, 0)$**

into $[-L, 0]$ so that the new function defined over $[-L, L]$ is either an even or odd function.

On the other hand, when working with boundary value problems we are led to find the Fourier series over the half Fourier interval. But in this case the way in which the function is extended into the other half Fourier interval is determined by the boundary conditions (see Figure 8).

Three different examples of extension are shown in Figure 8. The Fourier series representations for each case will be entirely different, but each series will converge to the same function f in the interval $(0, L)$. However, the difference in the three Fourier series occurs in the half interval $(-L, 0)$. In this interval each Fourier series will represent the function that is shown in Figure 8 by the dashed curve.

Fourier Sine Series

Suppose f is a function that is defined on the half Fourier interval $(0, L)$ and we desire its Fourier series on this interval. If we extend this function into the interval $(-L, 0)$ as an odd function, we can show that its representation is particularly simple.

THEOREM 9. Let f be a function defined on the interval $(0, L)$ that is extended into the interval $(-L, 0)$ as an odd function f_0. If its Fourier series exists, then the series must be of the form

$$f(x) = \sum_{n=1}^{\infty} b_n \sin \frac{n\pi x}{L}$$

where

$$b_n = \frac{2}{L} \int_0^L f(x) \sin \frac{n\pi x}{L}\, dx$$

and

$$a_0 = a_n = 0$$

Proof. From Definition 4 we have

$$\begin{aligned} b_n &= \frac{1}{L}\int_{-L}^{L} f_0(x) \sin \frac{n\pi x}{L}\, dx \\ &= \frac{1}{L}\int_{-L}^{0} f_0(x) \sin \frac{n\pi x}{L}\, dx + \frac{1}{L}\int_{0}^{L} f_0(x) \sin \frac{n\pi x}{L}\, dx \end{aligned} \tag{3.9}$$

In the first integral of Equation (3.9) we let $x = -y$ and see that

$$\frac{1}{L}\int_{-L}^{0} f_0(x) \sin \frac{n\pi x}{L}\, dx = \frac{1}{L}\int_{L}^{0} f_0(-y) \sin \frac{n\pi(-y)}{L}\, d(-y)$$

Since $f_0(x)$ and $\sin(n\pi x/L)$ are odd functions, we continue by writing

$$\begin{aligned} \frac{1}{L}\int_{-L}^{0} f_0(x) \sin \frac{n\pi x}{L}\, dx &= -\frac{1}{L}\int_{0}^{L} -f_0(y)\left(-\sin \frac{n\pi y}{L}\right)(-dy) \\ &= \frac{1}{L}\int_{0}^{L} f_0(y) \sin \frac{n\pi y}{L}\, dy \end{aligned}$$

Since y is a dummy variable, it follows that

$$\int_{-L}^{0} f_0(x) \sin \frac{n\pi x}{L}\, dx = \frac{1}{L}\int_{0}^{L} f_0(x) \sin \frac{n\pi x}{L}\, dx \tag{3.10}$$

Replacing the first integral on the right-hand side of Equation (3.9) by Equation (3.10) we have

$$b_n = \frac{2}{L}\int_{0}^{L} f_0(x) \sin \frac{n\pi x}{L}\, dx = \frac{2}{L}\int_{0}^{L} f(x) \sin \frac{n\pi x}{L}\, dx \qquad \blacksquare$$

Using an argument similar to the one in Theorem 9, we can easily show that $a_0 = a_n = 0$. The series shown in Theorem 9 is called a **Fourier sine series.**

EXAMPLE 14. Redo Example 7 using Theorem 9. The function given in Example 7 was

$$f(x) = \begin{cases} -3 & -5 + 10n < x < 10n \\ 3 & 10n < x < 5 + 10n \end{cases} \qquad n = 0, \pm 1, \pm 2, \ldots$$

First we recall that $2L = 10$, or $L = 5$. Since f is an odd function, we know immediately that $a_0 = a_n = 0$, and therefore we only have to calculate b_n, which is given by

$$b_n = \frac{2}{5}\int_{0}^{5} 3 \sin \frac{n\pi x}{5}\, dx = \frac{6}{5}\frac{5}{n\pi}\left(-\cos \frac{n\pi x}{5}\right)\Bigg|_0^5 = \frac{6}{n\pi}[1 - \cos n\pi]$$

which is the same answer as before. This result can be simplified by noticing that

$$b_n = \frac{6}{n\pi}[1 - \cos n\pi] = \frac{6}{n\pi}[1 - (-1)^n] = \begin{cases} 0 & n \text{ even} \\ \dfrac{12}{n\pi} & n \text{ odd} \end{cases}$$

and our answer can be written compactly in the form

$$f(x) = \frac{12}{\pi} \sum_{n=1,3,5}^{\infty} \frac{1}{n} \sin \frac{n\pi x}{5}$$

■

Fourier Cosine Series

If we proceed similarly as we did for Fourier sine series except that we extend the function f defined on $(0, L)$ into the interval $(-L, 0)$ as an even function, we can state the following theorem.

THEOREM 10. Let f be a function defined on the interval $(0, L)$ that is extended into the interval $(-L, 0)$ as an even function. If its Fourier series exists, then the series must be of the form

$$f(x) = \frac{a_0}{2} + \sum_{n=1}^{\infty} a_n \cos \frac{n\pi x}{L}$$

where

$$a_0 = \frac{2}{L}\int_0^L f(x)\,dx$$

and

$$a_n = \frac{2}{L}\int_0^L f(x) \cos \frac{n\pi x}{L}\,dx$$

and

$$b_n = 0$$

Proof. You are asked to prove Theorem 10 in Exercise 21. A Fourier series of this type is called a **Fourier cosine series.** ■

EXAMPLE 15. Find the Fourier cosine series of the triangular wave formed by extending the function $f(x)$ defined below as a periodic function of period 2 (see Figure 9).

$$f(x) = \begin{cases} 1 + x & -1 \le x \le 0 \\ 1 - x & 0 \le x \le 1 \end{cases}$$

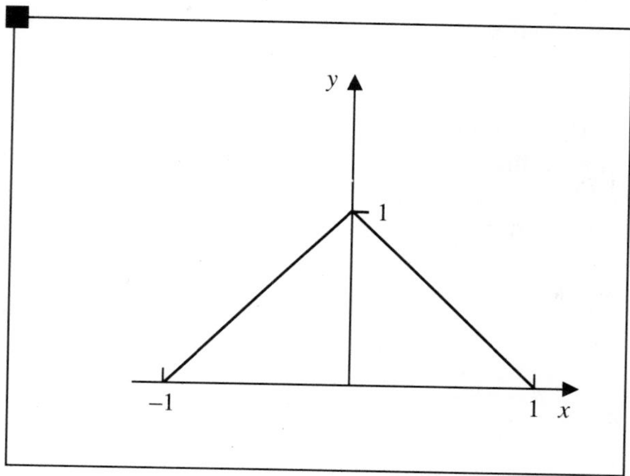

FIGURE 9 **Triangular Wave**

First we see that $2L = 2$, or $L = 1$. Next we calculate

$$a_0 = \frac{2}{1}\int_0^1 (1 - x)\,dx = 1$$

and by using Table 1 of integrals, we find

$$\begin{aligned} a_n &= 2\int_0^1 (1 - x)\cos n\pi x\,dx \\ &= \left\{\frac{2}{n\pi}\sin n\pi x - \frac{2}{n^2\pi^2}[\cos n\pi x + n\pi x \sin n\pi x]\right\}\Bigg|_0^1 \\ &= -\frac{2}{n^2\pi^2}[\cos n\pi - 1] = \frac{2}{n^2\pi^2}[1 - (-1)^n] \\ &= \begin{cases} 0 & n \text{ even} \\ \dfrac{4}{n^2\pi^2} & n \text{ odd} \end{cases} \end{aligned}$$

The Fourier cosine series is given by

$$f(x) = \frac{1}{2} + \frac{4}{\pi^2}\sum_{n=1,3,5}^{\infty} \frac{1}{n^2}\cos n\pi x$$

■

We must observe, however, that when an even function is expressed as a Fourier series, the Fourier cosine series is also the Fourier series. A similar statement can be made for odd functions.

When dealing with functions that are even (odd) over the full Fourier interval, be sure to heed the following warning.

□ ***Warning.*** When a function is given over the half Fourier interval $[0, L]$ and we are to extend it into the other half interval $[-L, 0]$ as an even (odd) function and then write its Fourier series, always use the Fourier cosine series

(Fourier sine series). *Do not return to the Fourier series method* described in Section 4. □

EXAMPLE 16. Extend the function $f(x) = x$ defined on the interval $(0, \pi)$ into the interval $(-\pi, 0)$ as an even function. What is this Fourier series?

Since the function is even, we use the Fourier cosine series representation. First we notice $L = \pi$. Then

$$a_0 = \frac{2}{\pi}\int_0^{\pi} x\,dx = \pi$$

and

$$a_n = \frac{2}{\pi}\int_0^{\pi} x \cos x\,dx = \frac{2(-1)^n}{\pi n^2}$$

The Fourier cosine series is given by

$$f(x) = \frac{\pi}{2} + \frac{2}{\pi}\sum_{n=1}^{\infty}\frac{(-1)^n}{n^2}\cos nx$$ ■

If we attempt to use the Fourier series approach (Theorem 9), we must be careful to recognize that the extended function has the value $-x$ in $(-\pi, 0)$ and *not* x.

Summary of Fourier Series

In this chapter we are concerned with finding the Fourier coefficients a_0, a_n, and b_n, which allow us to write a Fourier series. The Fourier series can be interpreted in two ways. Sometimes we are only interested in expressing a given function f over an interval $[a, b]$ as a Fourier series. The given function does not have to be periodic. If we use this function to find the Fourier coeffi-

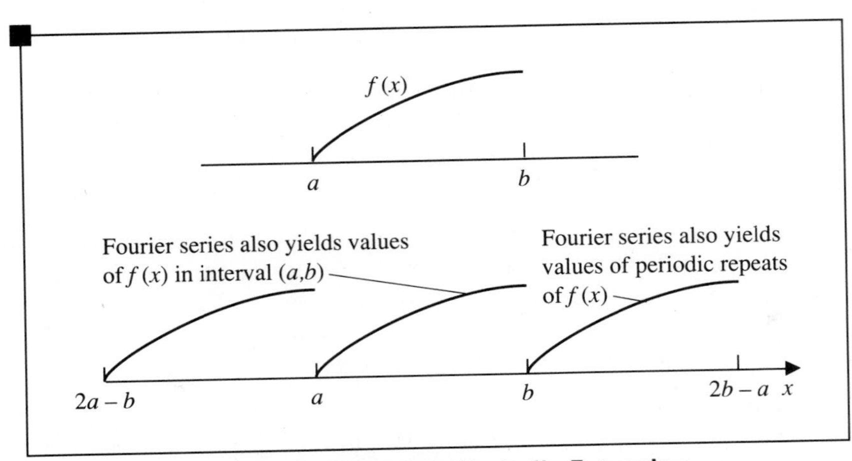

FIGURE 10 **Function f and Its Periodic Extension**

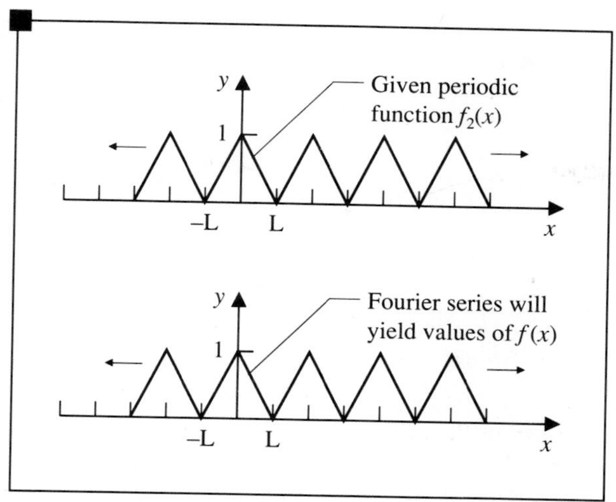

FIGURE 11 **Fourier Series of Periodic Function**

cients and eventually the Fourier series, the series will represent f over $[a, b]$. Actually, the Fourier series gives us more than we ask in this case because it will also represent the periodic function that is a periodic extension of f (see Figure 10).

At other times we are given a periodic function f and wish to find a trigonometric series that represents it for all x. The Fourier series will do this (see Figure 11).

SECTION 5 EXERCISES

14. Show whether the following functions are even, odd, or neither for all real x.

(a) $\sin 2x$

(b) $\cos^3 x$

(c) $\dfrac{\sin^2 x}{\cos x}$

(d) $\cos(x - x^2)$

(e) $\sqrt{x}$

(f) $|x|$

(g) e^{-x^2}

(h) $\sin x \cos x$

15. Suppose for all real x, $f(x)$ is even, $g(x)$ is odd, and $h(x)$ is odd. Show that

(a) $f(x)g(x)$ is odd.

(b) $[g(x)]^2$ is even.

(c) $h(x) - g(x)$ is odd.

(d) $f(x) - g(x)$ is neither even nor odd.

(e) $\dfrac{f(x)}{h(x)}$ is odd.

16. Given the following functions over the half Fourier interval indicated, (1) extend as an even function into the full Fourier interval and write the Fourier cosine series, and (2) extend as an odd function into the full Fourier interval and write the Fourier sine series.

(a) x, $[0, \pi]$
(b) 1, $[0, 2]$
(c) x^2, $[0, 1]$
(d) e^{-x}, $[0, 2]$
(e) $f(x) = \begin{cases} x & 0 < x < 1 \\ 1 & 1 < x < 2 \end{cases}$

17. The voltage $12 \cos 10t$ is passed through a half-wave rectifier. Write the Fourier series of rectified voltage (see Figure 12).

18. The voltage $100 \sin 4\pi t$ as shown in Figure 13 is passed through a full-wave rectifier. Write the Fourier series of rectified voltage.

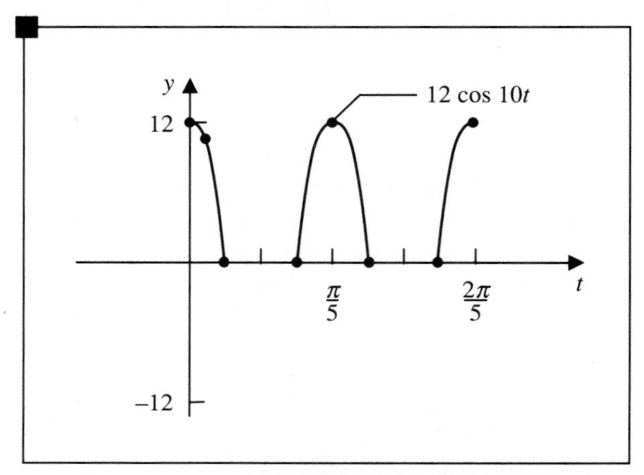

FIGURE 12 **Voltage from Half-Wave Rectifier**

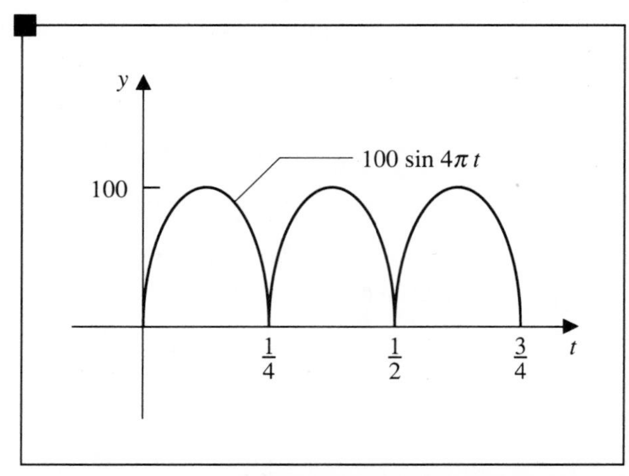

FIGURE 13 **Voltage from Full-Wave Rectifier**

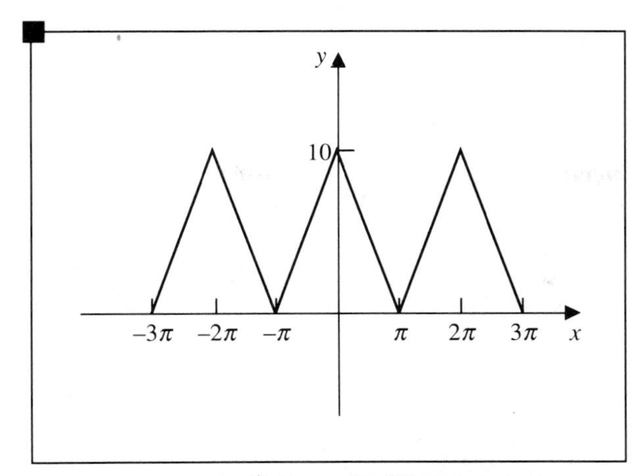

FIGURE 14 **Triangular Wave**

19. Write the Fourier series of a triangular wave as shown in Figure 14.

20. Write the Fourier series of function $f(x)$ defined as

$$f(x) = \begin{cases} 1 & 1 < x < 2 \\ \frac{1}{2} & 0 < x < 1 \\ -\frac{1}{2} & -1 < x < 0 \\ -1 & -2 < x < -1 \end{cases}$$

21. Prove Theorem 10.

SECTION 6*
COMPLEX FORM OF FOURIER SERIES

In general, the Fourier series we have just studied ordinarily consists of a constant term and a series of sine and cosine terms. There are certain times when it would be desirable to write a Fourier series in a more compact form as a single series consisting of one type of function. The complex form of the Fourier series yields just such a type of series.

To find this series, we start with Definition 4 of the Fourier series for a function f of period $2L$, which states that

$$f(x) = \frac{a_0}{2} + \sum_{n=1}^{\infty} \left\{ a_n \cos \frac{n\pi x}{L} + b_n \sin \frac{n\pi x}{L} \right\} \tag{3.11}$$

where

$$a_n = \frac{1}{L} \int_{-L}^{L} f(x) \cos \frac{n\pi x}{L}\, dx \qquad n = 0, 1, 2, \ldots$$

$$b_n = \frac{1}{L} \int_{-L}^{L} f(x) \sin \frac{n\pi x}{L}\, dx \qquad n = 1, 2, \ldots$$

For convenience we will let $\omega_n = (n\pi/L)$ and $\omega_{-n} = (-n\pi/L) = -\omega_n$.

From Euler's formula we know that

$$\cos \omega_n x = \frac{\exp(i\omega_n x) + \exp(-i\omega_n x)}{2}$$

and

$$\sin \omega_n x = \frac{\exp(i\omega_n x) - \exp(-i\omega_n x)}{2i}$$

If we replace $\cos(n\pi x/L)$ and $\sin(n\pi x/L)$ in Equation (3.11) by exponentials, we can write

$$f(x) = \frac{a_0}{2} + \sum_{n=1}^{\infty} \left\{ a_n \frac{\exp(i\omega_n x) + \exp(-i\omega_n x)}{2} + b_n \frac{\exp(i\omega_n x) - \exp(-i\omega_n x)}{2i} \right\} \tag{3.12}$$

Rearranging Equation (3.12) we have

$$f(x) = \frac{a_0}{2} + \frac{1}{2} \sum_{n=1}^{\infty} \{(a_n - ib_n) \exp(i\omega_n x) + (a_n + ib_n) \exp(-i\omega_n x)\} \tag{3.13}$$

Let

$$c_n = a_n - ib_n = \frac{1}{L} \int_{-L}^{L} f(x)(\cos \omega_n x - i \sin \omega_n x)\, dx$$
$$= \frac{1}{L} \int_{-L}^{L} f(x) \exp(-i\omega_n x)\, dx \qquad n = 0, 1, \ldots$$

The conjugate of c_n is

$$\bar{c}_n = a_n + ib_n = \frac{1}{L} \int_{-L}^{L} f(x) \exp(i\omega_n x)\, dx$$

Furthermore, it follows that

$$\bar{c}_{-n} = a_{-n} + ib_{-n} = \frac{1}{L} \int_{-L}^{L} f(x) \exp(i\omega_{-n} x)\, dx$$
$$= \frac{1}{L} \int_{-L}^{L} f(x) \exp(-i\omega_n x)\, dx = c_n$$

Going back to Equation (3.13), we can write this equation in the form

$$f(x) = \frac{a_0}{2} + \frac{1}{2} \sum_{n=1}^{\infty} \{c_n \exp(i\omega_n x) + \bar{c}_n \exp(-i\omega_n x)\}$$
$$= \frac{a_0}{2} + \frac{1}{2} \sum_{n=1}^{\infty} c_n \exp(i\omega_n x) + \frac{1}{2} \sum_{n=1}^{\infty} \bar{c}_n \exp(-i\omega_n x)$$

If we reindex the second summation above by replacing n by $-n$, we have

$$f(x) = \frac{a_0}{2} + \frac{1}{2}\sum_{n=1}^{\infty} c_n \exp(i\omega_n x) + \frac{1}{2}\sum_{-n=1}^{\infty} \bar{c}_{-n} \exp(-i\omega_{-n} x)$$

$$= \frac{a_0}{2} + \frac{1}{2}\sum_{n=1}^{\infty} c_n \exp(i\omega_n x) + \frac{1}{2}\sum_{n=-1}^{-\infty} c_n \exp(i\omega_n x) \tag{3.14}$$

Finally, we see that

$$a_0 = c_0 = c_0 \exp\left(i\,\frac{0\pi x}{L}\right) \tag{3.15}$$

Combining the results of Equations (3.14) and (3.15), we are able to write the complex Fourier series in the following way:

$$f(x) = \frac{1}{2}\sum_{n=-\infty}^{\infty} c_n \exp\left(i\,\frac{n\pi x}{L}\right) \tag{3.16}$$

where

$$c_n = \frac{1}{L}\int_{-L}^{L} f(x) \exp\left(-i\,\frac{n\pi x}{L}\right) dx$$

Some authors let $c_n = 2d_n$. In that case the complex Fourier series takes the form

$$f(x) = \sum_{n=-\infty}^{\infty} d_n \exp\left(i\,\frac{n\pi x}{L}\right)$$

$$d_n = \frac{1}{2L}\int_{-L}^{L} f(x) \exp\left(-i\,\frac{n\pi x}{L}\right) dx$$

EXAMPLE 17. Given the function $f(x) = e^{-x}$ over the interval $(-3, 3)$ and letting f_0 be the periodic extension of f, find the complex Fourier series of f_0.

We find L by noticing that $2L = 6$, therefore $L = 3$. The coefficient

$$c_n = \frac{1}{3}\int_{-3}^{3} e^{-x}e^{-i(n\pi x/3)}\,dx = \frac{1}{3}\int_{-3}^{3} e^{-[1+i(n\pi/3)]x}\,dx$$

$$= \frac{1}{3}\,\frac{1}{-[1+i(n\pi/3)]}\int_{-3}^{3} e^{-[1+i(n\pi/3]x}\left\{-\left[1+i\,\frac{n\pi}{3}\right]\right\}dx$$

$$= \frac{1}{-(3+in\pi)}\left[e^{-(3+in\pi)} - e^{+(3+in\pi)}\right]$$

$$= \frac{1}{+(3+in\pi)}\left[e^{3}(\cos n\pi + i\sin n\pi) - e^{-3}(\cos n\pi - i\sin n\pi)\right]$$

$$= \frac{\cos n\pi}{3+in\pi}(e^3 - e^{-3}) = \frac{2(-1)^n}{3+in\pi}\sinh 3$$

$$= (-1)^n\,\frac{2(3-in\pi)}{9+n^2\pi^2}\sinh 3$$

Replacing c_n in Equation (3.16),

$$e^{-x} = \sum_{n=-\infty}^{\infty} (-1)^n \frac{3 - in\pi}{9 + n^2\pi^2} \sinh 3 \exp\left(i \frac{n\pi x}{3}\right) \qquad \text{on } -3 < x < 3 \quad ■$$

In working with the complex Fourier series, we often deal with complex functions of a real variable. For convenience we summarize some of the definitions and theorems of such functions.

Definition 6. Let $f(x) = u(x) + iv(x)$ where $u(x)$ and $v(x)$ are real functions and $i = \sqrt{-1}$. If $u(x)$ and $v(x)$ are differentiable functions, then

$$\frac{df(x)}{dx} = \frac{du(x)}{dx} + i\frac{dv(x)}{dx}$$

If $u(x)$ and $v(x)$ are integrable, then

$$\int f(x)\,dx = \int u(x)\,dx + i\int v(x)\,dx$$

Using Definition 6 we can prove Theorems 11 and 12.

THEOREM 11. The function e^{ix} is differentiable and

$$\frac{de^{ix}}{dx} = ie^{ix}$$

Proof

$$\frac{d}{dx} e^{ix} = \frac{d}{dx}(\cos x + i \sin x) = -\sin x + i \cos x$$

$$= i(\cos x + i \sin x) = ie^{ix} \quad ■$$

THEOREM 12. The function e^{ix} is integrable and

$$\int e^{ix}\,dx = \frac{e^{ix}}{i} + c$$

Proof. You are asked to prove Theorem 12 in Exercise 24. ■

Notice in particular that the differentiation and integration formulas for e^{ix} obey exactly those rules given in elementary calculus.

As is often the way in mathematics and engineering, if we improve a situation at one point, we sometimes sacrifice some other condition at another point. And so although the complex Fourier series gives us a very compact representation, our work in writing a complex Fourier series is not materially shortened when dealing with even or odd functions.

EXAMPLE 18. Express the odd function f given in Example 7 in the form of a complex Fourier series where

$$f(x) = \begin{cases} -3 & -5 + 10n < x < 10n \\ +3 & 10n < x < 5 + 10n \end{cases} \qquad n = 0, \pm 1, \pm 2, \ldots$$

We already know that $L = 5$, and therefore the coefficient c_n is given by

$$c_n = \frac{1}{5}\int_{-5}^{0} (-3) \exp\left(-i\frac{n\pi x}{5}\right) dx + \frac{1}{5}\int_{0}^{5} 3 \exp\left(-i\frac{n\pi x}{5}\right) dx$$

$$= +\frac{3}{5}\frac{5}{in\pi} \exp\left(-i\frac{n\pi x}{5}\right)\bigg|_{-5}^{0} - \frac{3}{5}\frac{5}{in\pi} \exp\left(-i\frac{n\pi x}{5}\right)\bigg|_{0}^{5}$$

$$= \frac{3}{in\pi}[1 - \exp(in\pi)] - \frac{3}{in\pi}[\exp(-in\pi) - 1]$$

$$= \frac{3}{in\pi}[1 + 1 - \exp(in\pi) - \exp(-in\pi)]$$

$$= \frac{6}{in\pi} - \frac{6}{in\pi}\cos n\pi = \frac{6}{in\pi}[1 - (-1)^n]$$

Substituting this result into Equation (3.16), we find

$$f(x) = \frac{3}{i\pi}\sum_{n=-\infty}^{\infty}[1 - (-1)^n]\frac{1}{n}\exp\left(i\frac{n\pi x}{5}\right)$$

$$= +\frac{6}{i\pi}\sum_{\substack{n=-\infty \\ n \text{ odd}}}^{\infty}\frac{1}{n}\exp\left(i\frac{n\pi x}{5}\right)$$

It is clear from this example that the work required to find the complex Fourier series for this odd function is greatly increased over our earlier effort. ■

SECTION 6 EXERCISES

22. Using Euler's formula $e^{\pm i\theta} = \cos\theta \pm i\sin\theta$, find

(a) $\sin\theta$ in terms of exponentials.
(b) $\cos\theta$ in terms of exponentials.

23. Differentiate

(a) $x^2 + ix^3$
(b) $\cos 5x + i\sin^2 x$
(c) $\ln x + ix$
(d) $\sin nx + i\sinh 4x$
(e) ie^x

24. Prove

$$\int e^{ix}\,dx = \frac{e^{ix}}{i} + c$$

25. Find the indefinite integral

(a) $\int (2x + ix^2)\,dx$

(b) $\int e^x e^{ix}\,dx$

(c) $\int (\sin 3x + i\cos 3x)\,dx$

(d) $\int \left(x - i\frac{1}{x}\right) dx$

(e) $\int e^{ix}\sin x\,dx$

26. Show that if $f(x)$ is a periodic function with period $2L$ and $f(x)$ is defined over the interval $(a, a + 2L)$, then

$$c_n = \frac{1}{L}\int_{-L}^{L} f(x)\exp\left(-i\frac{n\pi}{L}x\right) dx$$
$$= \frac{1}{L}\int_{a}^{a+2L} f(x)\exp\left(-i\frac{n\pi}{L}x\right) dx$$

27. Write the complex form of the Fourier series of the following functions over the interval indicated.

(a) e^x, $(0, 1)$
(b) $e^{x/2}$, $(0, 4)$
(c) $\sin x$, $[-\pi, \pi]$
(d) $\cos x$, $[0, \pi]$
(e) 100, $[-3, 3]$
(f) $5x$, $(0, 8)$
(g) $x + 1$, $(-1, 2)$
(h) x^2, $(-\frac{1}{2}, \frac{1}{2})$

(i) $f(x) = \begin{cases} 0 & -2 < x < 0 \\ 1 & 0 < x < 2 \end{cases}$

(j) $|x|$, $[-1, 1]$

(k) $f(x) = \begin{cases} 0 & -3 < x < 1 \\ x & 1 < x < 3 \end{cases}$

28. In attempting to duplicate a voltage in the shape of a sawtooth wave, a generator yields the graph of the function $e(t) = 100(1 - e^{-t})$ in the interval $(0, 1)$. If E is the periodic extension of e, write the complex Fourier series for E over $(-\infty, \infty)$.

29. (a) Write the complex form of Fourier series for e^x over the interval $(-1, 1)$.

(b) Equating real and imaginary parts of the series found in part (a), show that

$$e^x = (\sinh 1) \sum_{n=-\infty}^{\infty} \frac{\cos n\pi x - n\pi \sin n\pi x}{1 + n^2\pi^2}$$

and

$$\sum_{n=-\infty}^{\infty} \frac{n\pi \cos n\pi x + \sin n\pi x}{1 + n^2\pi^2} = 0$$

SECTION 7

SOME THEORY ABOUT FOURIER SERIES

Introduction

Although we evaluated Fourier coefficients and wrote the corresponding Fourier series in a formal way in Section 6, we still have not justified the derivations of these coefficients or shown if the Fourier series does indeed converge to the given function. A rigorous justification of these operations is beyond the scope of this book. Therefore, we shall be content to state some results that guarantee, for a large class of functions, that the formulas for evaluating coefficients are mathematically correct and that the Fourier series under mild restrictions on the function converge (pointwise) to a given function. At times the terms used in the discussion may be imprecise, but they will give you a feel for what is going on rather than presenting a cold and rigorous proof.

We first define some notation for the discussion that follows. It is convenient to use

$$x \to x_0^+ \qquad \text{to mean} \qquad x \to x_0, \ x > x_0$$

and

$$x \to x_0^- \qquad \text{to mean} \qquad x \to x_0, \ x < x_0$$

Furthermore, the right-hand limit of $f(x)$ as x approaches x_0 is denoted by

$$f(x_0^+) = \lim_{x \to x_0^+} f(x)$$

The left-hand limit of $f(x)$ as x approaches x_0 is given by

$$f(x_0^-) = \lim_{x \to x_0^-} f(x)$$

We are also going to need the limit of the derivative of $f(x)$ as x approaches x_0 from the left or right.

Definition 7. If f' exists in a right neighborhood of x_0, then the limit, if it exists, of the derivative of $f(x)$ as x approaches x_0 from the right is written as

$$f'(x_0^+) = \lim_{x \to x_0^+} f'(x)$$

If f' exists in a left neighborhood of x_0, then the limit, if it exists, of the derivative of $f(x)$ as x approaches x_0 from the left is written as

$$f'(x_0^-) = \lim_{x \to x_0^-} f'(x)$$

EXAMPLE 19. Given the function

$$f(x) = \begin{cases} \dfrac{x}{2} + 1 & -1 < x < 0 \\ x - 1 & 0 < x < 1 \end{cases}$$

what are $f(0^+), f(0^-), f'(0^+)$, and $f'(0^-)$ (see Figure 15).

It follows that

$$f(0^+) = \lim_{x \to 0^+} (x - 1) = -1$$

$$f(0^-) = \lim_{x \to 0^-} \left(\frac{x}{2} + 1\right) = 1$$

$$f'(0^+) = \lim_{x \to 0^+} f'(x) = \lim_{x \to 0^+} 1 = 1$$

$$f'(0^-) = \lim_{x \to 0^-} f'(x) = \lim_{x \to 0^-} \frac{1}{2} = \frac{1}{2}$$

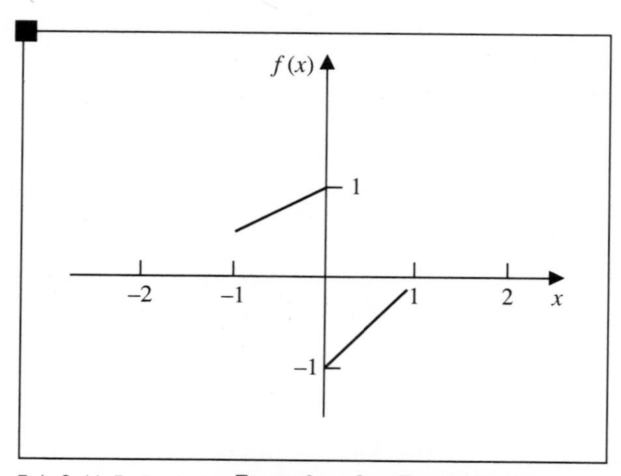

FIGURE 15 **Function for Example 19**

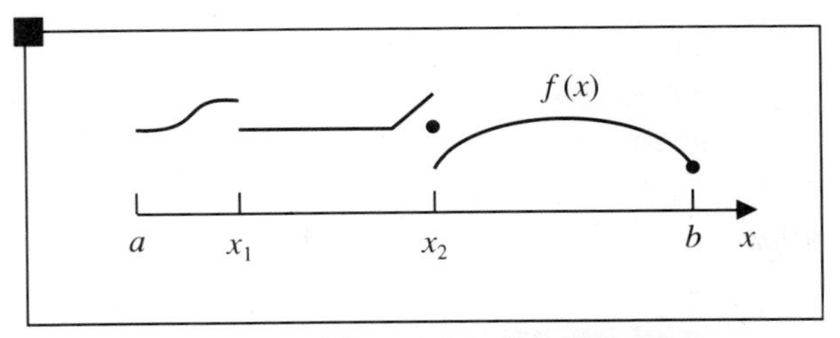

FIGURE 16 **Piecewise Continuous Function**

We are going to need two other properties associated with some functions; one is concerned with continuity and the other pertains to differentiability.

Definition 8. A function f is piecewise continuous over the interval $[a, b]$ if f is continuous on $[a, b]$ except at a finite number of points, and the right- and left-hand limits of f exist at these points. If a and b are also points of discontinuity, only the right-hand (left-hand) limit must exist at $a(b)$ (see Figure 16).

Definition 9. A function f is piecewise differentiable over the interval $[a, b]$ if it is piecewise continuous over the interval $[a, b]$ and differentiable on $[a, b]$ except at a finite number of points. At these points, $f'(x^+)$ and $f'(x^-)$ must exist. If a and b are also points where the derivative does not exist, only $f'(x^+)[f'(x^-)]$ must exist at $a[b]$ (see Figure 17).

Since we are often dealing with periodic functions whose intervals are unbounded, we modify Definitions 8 and 9 to cover such functions.

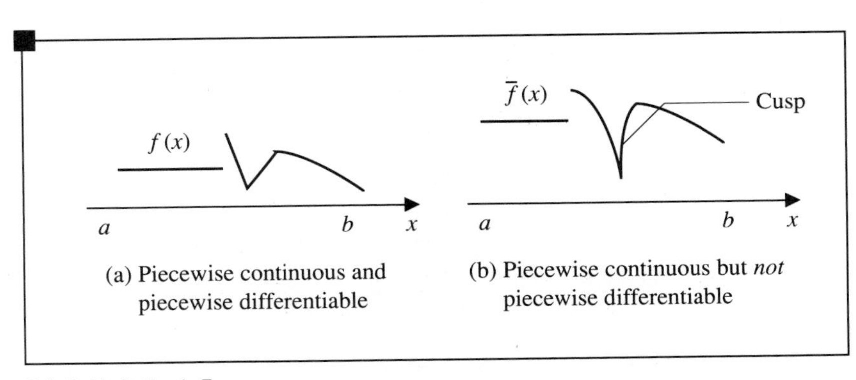

FIGURE 17

Definition 10. If f is a periodic function of period $2L$, then if f is piecewise continuous or piecewise differentiable on $[-L, L]$, we say f is piecewise continuous or piecewise differentiable for all x.

Convergence of Fourier Series

We are now in a position to state a classic theorem, which tells us, for a certain set of functions, that our method of constructing a Fourier series is valid.

THEOREM 13. If f is a periodic function with period $2L$, and f is piecewise differentiable for all x, then the Fourier series

$$\frac{a_0}{2} + \sum_{n=1}^{\infty} \left\{ a_n \cos \frac{n\pi x}{L} + b_n \sin \frac{n\pi x}{L} \right\} = \frac{f(x^+) + f(x^-)}{2}$$

for all x.

If $f(x)$ is continuous at $x = x_0$, then

$$\frac{f(x_0^+) + f(x_0^-)}{2} = f(x_0)$$

On the other hand, if $f(x_0^+) \neq f(x_0^-)$, then the discontinuity is a jump discontinuity and the Fourier series converges to a value halfway between $f(x_0^+)$ and $f(x_0^-)$ (see Figure 18). ■

Although Theorem 13 is a powerful theorem and appears to answer a number of questions, it does have some weaknesses. First, the type of convergence referred to in the theorem is *pointwise* convergence. It is true given a value x that the Fourier series will eventually approach $\frac{1}{2}[f(x^+) + f(x^-)]$ if n is sufficiently large. But the same value of n that moves the partial sums of the Fourier series "closer" to the function at one point may leave the sums at quite a distance from the function at another point.

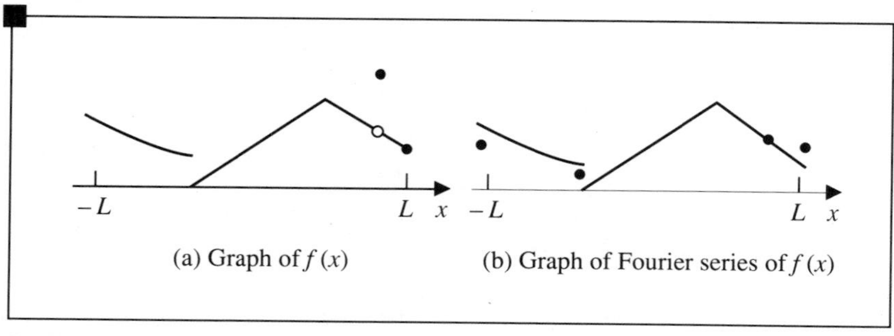

FIGURE 18

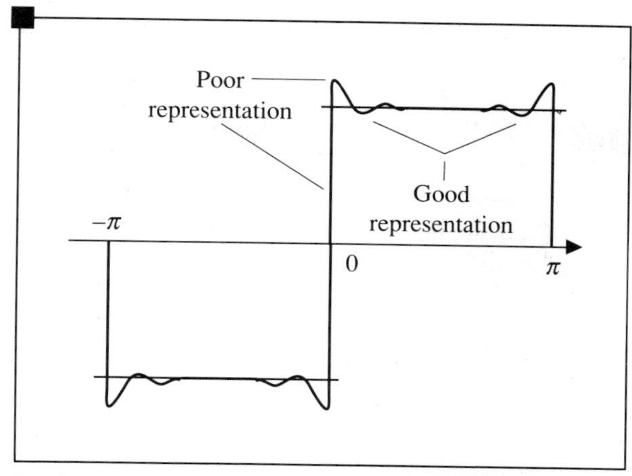

FIGURE 19 **Representation by Partial Sums of a Fourier Series**

EXAMPLE 20. The function

$$\frac{1}{1+x} = \sum_{n=0}^{\infty} (-1)^n x^n$$

although not a Fourier series, demonstrates the difficulty just mentioned. Suppose we use the first 10 terms to approximate the value of $1/(1+x)$ at $x = 0.2$ and $x = 0.9$. From our knowledge of alternating series, the error in using 10 terms at $x = 0.2$ is 1.024×10^{-7}, whereas the error at $x = 0.9$ using the same number of terms is 0.349. In order to get the same degree of accuracy at $x = 0.9$, we would have to take over 150 terms.

Because of this problem of pointwise convergence, if we use the first n terms of a Fourier series to approximate our given function f we may be sadly disappointed with the result. For certain values of x the truncated series may be a very good approximation to the function f. But at other values it will not represent the function well at all (see Figure 19). We also recall that theorems about continuity, integration, and differentiation of series usually require that the series be uniformly convergent. ■

Review of Uniform Convergence

Let $u_i(x)$ be a sequence of functions, defined on some interval $[a, b]$. The sum

$$\sum_{i=1}^{\infty} u_i(x)$$

is called an **infinite series** of the functions $u_i(x)$. What do we mean when we write

$$f(x) = \sum_{i=1}^{\infty} u_i(x) \qquad a \le x \le b$$

In some ways we feel that as $i \to +\infty$, the sum of the functions $u_i(x)$ more closely approximates $f(x)$.

We will examine two ways in which this idea can occur. The first way is called pointwise convergence and is given by Definition 11.

Definition 11. Let $s_n = \sum_{i=1}^{\infty} u_i(x)$ be the nth partial sum of the functions $u_i(x)$ defined on $[a, b]$. For a given value of x, $a \le x \le b$, and a given value of $\varepsilon > 0$, if we can find an $N > 0$ such that $|f(x) - s_n(x)| < \varepsilon$ whenever $n \ge N$ then we say the series $\sum_{i=1}^{\infty} u_i(x)$ **converges pointwise** to $f(x)$ on the interval $[a, b]$. Notice that the value of N depends on both x and ε.

A second and stronger convergence is known as uniform convergence.

Definition 12. Let $s_n = \sum_{i=1}^{\infty} u_i(x)$ be the nth partial sum of the function $u_i(x)$ defined on $[a, b]$. If for all x, $a \le x \le b$, and a given $\varepsilon > 0$, we can find an $N > 0$, independent of x, such that

$$|f(x) - s_n(x)| < \varepsilon \tag{3.17}$$

whenever $n \ge N$ then we say that the series $\sum_{i=1}^{\infty} u_i(x)$ **converges uniformly** to $f(x)$ on the interval $[a, b]$. Here we see that the value of N depends only on ε; that is, that difference Equation (3.17) must hold for all x in $[a, b]$.

EXAMPLE 21. The series $\sum_{n=0}^{\infty} x^n$ can be shown to converge uniformly to the function $f(x) = 1/(1 - x)$ on the interval $0 \le x \le 0.6$. If we approximate f in the given interval by the first four terms of the series, that is, $1 + x + x^2 + x^3 = g(x)$, we see that g is an adequate approximation to f using

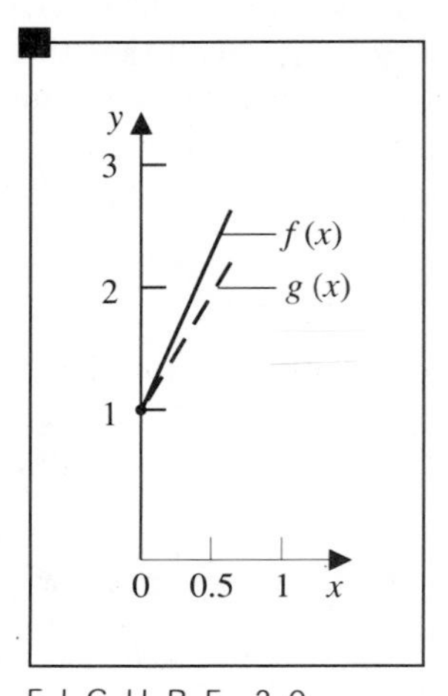

FIGURE 20
Uniform Convergence

the following table:

x	0	0.1	0.2	0.3	0.4	0.5	0.6
$f(x)$	1	1.111	1.25	1.428	1.667	2.000	2.500
$g(x)$	1	1.111	1.248	1.417	1.624	1.875	2.176

(See Figure 20.) ■

Three theorems involving uniform convergence are useful when studying Fourier series.

THEOREM 14. Suppose the functions $u_i(x)(i = 1, 2, \ldots)$ are continuous on $[a, b]$. Then if the series

$$f(x) = \sum_{i=1}^{\infty} u_i(x)$$

is uniformly convergent on $a \leq x \leq b$, f is a continuous function on this interval. ■

THEOREM 15. Suppose the series $\sum_{i=1}^{\infty} u_i(x) = f(x)$ is uniformly convergent on $[a, b]$. If $g(x)$ is a bounded function on $[a, b]$, then $\sum_{i=1}^{\infty} g(x)u_i(x)$ is a uniformly convergent series on $[a, b]$. ■

THEOREM 16. Suppose the functions $u_i(x)(i = 1, 2, \ldots)$ are continuous on $[a, b]$ and the series $\sum_{i=1}^{\infty} u_i(x) = f(x)$ is uniformly convergent on $[a, b]$. Then

$$\int_a^b \left[\sum_{i=1}^{\infty} u_i(x)\right] dx = \sum_{i=1}^{\infty} \left[\int_a^b u_i(x)\, dx\right]$$ ■

Theorems 15 and 16 tell us that if a given Fourier series is uniformly convergent, then the method we used in Section 4 to find the values of the coefficients a_0, a_n, and b_n are all legal.

But even using the formulas for the Fourier coefficients a_0, a_n, and b_n, it is generally difficult to show that a given Fourier series is uniformly convergent. In fact, many of the examples earlier in the chapter are represented by Fourier series that cannot converge uniformly to the given function f. We know from Theorem 14 that if every function $u_i(x)$ is continuous over the interval $[a, b]$ and the series $\sum_{i=1}^{\infty} u_i(x)$ converges uniformly to $f(x)$, then f must be continuous on $[a, b]$. Now every Fourier series

$$\frac{a_0}{2} + \sum_{n=1}^{\infty} \left\{a_n \cos \frac{n\pi x}{L} + b_n \sin \frac{n\pi x}{L}\right\}$$

consists of functions that are continuous over the interval $[-L, L]$. Therefore, if this series converges uniformly on $[-L, L]$, it must converge to a continuous function f. But we have already seen some Fourier series representations that converge (pointwise) to a discontinuous function f. Therefore the convergence *cannot* be uniform.

EXAMPLE 22. The Fourier series for the square wave

$$f(x) = \begin{cases} 1 & 0 < x < \pi \\ -1 & -\pi < x < 0 \\ 0 & x = -\pi, 0, \pi \end{cases}$$

is given by

$$f(x) = \frac{4}{\pi} \sum_{n=1}^{\infty} \frac{\sin (2n-1)x}{2n-1} \tag{3.18}$$

f is a continuous function except at $x = -\pi, 0, \pi$, and, therefore, the Fourier series in Equation (3.18) cannot converge uniformly (see Figure 21). ■

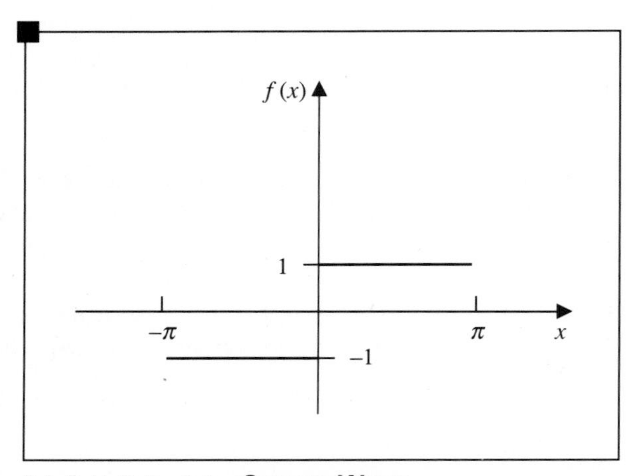

FIGURE 21 **Square Wave**

However, things are not quite as bad as they seem, for there is a theorem that shows that the Fourier series expansion of a continuous function of period $2L$ is uniformly (and absolutely) convergent.

Upon examining these two types of convergence, a series that approaches a function uniformly fulfills our desire that over an interval $[a, b]$, as more terms are taken in the series, the sum of these terms more closely represents the function in the interval. Theorem 17 includes a large class of functions for which the Fourier series is uniformly convergent.

THEOREM 17. Let $f(x)$ be a piecewise differentiable function of period $2L$ that is continuous on $[-L, L]$. Then if

$$\frac{a_0}{2} + \sum_{n=1}^{\infty} \left\{ a_n \cos \frac{n\pi x}{L} + b_n \sin \frac{n\pi x}{L} \right\}$$

is the Fourier expansion of f, the series is uniformly convergent on $[-L, L]$ to $f(x)$. ■

EXAMPLE 23. The triangular wave whose functional representation is given by

$$f(x) = \begin{cases} -\pi - x & -\pi \le x \le -\dfrac{\pi}{2} \\ x & -\dfrac{\pi}{2} \le x \le \dfrac{\pi}{2} \\ \pi - x & \dfrac{\pi}{x} \le x \le \pi \end{cases}$$

is a piecewise smooth function of period 2π that is continuous on $[-\pi, \pi]$. Its Fourier series,

$$f(x) = \frac{4}{\pi} \sum_{n=1}^{\infty} \frac{\sin \dfrac{n\pi}{2}}{n^2} \sin nx = \frac{4}{\pi} \sum_{n=1,3,5}^{\infty} \frac{(-1)^n}{n^2} \sin nx$$

converges uniformly to $f(x)$. (See Figure 22.) ■

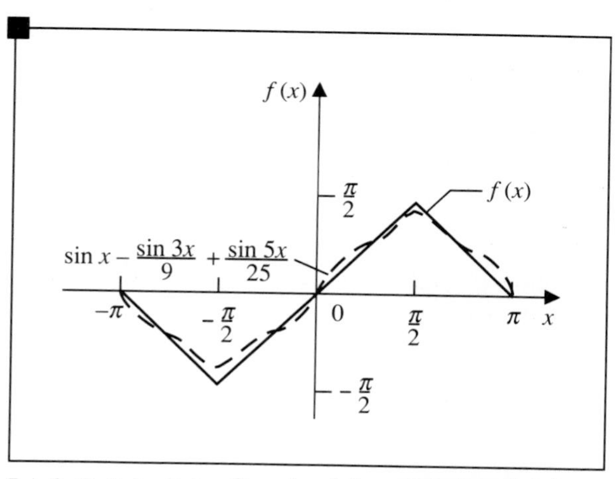

FIGURE 22 **Graph of f and Its Partial Sum**

Unfortunately, when we are trying to represent a periodic function h that is piecewise differentiable but not continuous in $(-L, L)$, the Fourier series will converge to h pointwise. This type of convergence behaves in an unusual way in the neighborhood of a jump discontinuity. At points away from the jump discontinuity, the Fourier series "represents" the function fairly well, and as n gets larger the truncated Fourier series becomes a better approximation to h (see $x = \pi/2$ in Figure 22). But near a point of discontinuity there are always points generated by the Fourier series that for any n are at some distance from the function h. This phenomenon was first noticed and discussed by Gibbs in a letter to a magazine, although his statement was not accompanied by any proof. Today we call this property associated with Fourier series of discontinuous functions the **Gibbs phenomenon**. Bocher investigated this problem in more detail and eventually came to the conclusion that the amount of maximum overshoot of a Fourier series at a jump discontinuity depends only on the amount of the jump.

Furthermore, we cannot reduce this error by taking more and more terms of the Fourier series. However, when we let n increase, the point at which the maximum overshoot occurs moves closer to the point of discontinuity and, therefore, except in an arbitrarily small neighborhood of the point of discontinuity the Fourier series represents the given function reasonably well.

EXAMPLE 24. Let h be the square function given by the equation

$$h(x) = \begin{cases} -1 & -\pi < x < 0 \\ 1 & 0 < x < \pi \end{cases}$$

(see Figure 23).

The Fourier series for h is given by

$$h(x) = \frac{4}{\pi} \sum_{n=1,3,5}^{\infty} \frac{\sin nx}{n}$$

Recall that the convergence of this series is pointwise but not uniform. ∎

Now from the study of Gibbs phenomenon, the amount of overshoot is given by

$$\frac{D(0.2811)}{\pi} \approx 0.09D$$

where D is the amount of jump (i.e., $D = h(0^+) - h(0^-) = 2$). As $n \to +\infty$, the Fourier series representing h near $x = 0$, $x > 0$ overshoots the value 1 by $0.09D = 0.09(2) = 0.18$.

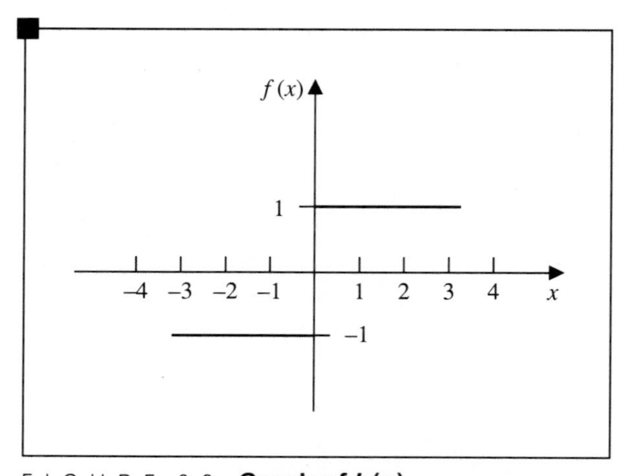

FIGURE 23 **Graph of $h(x)$**

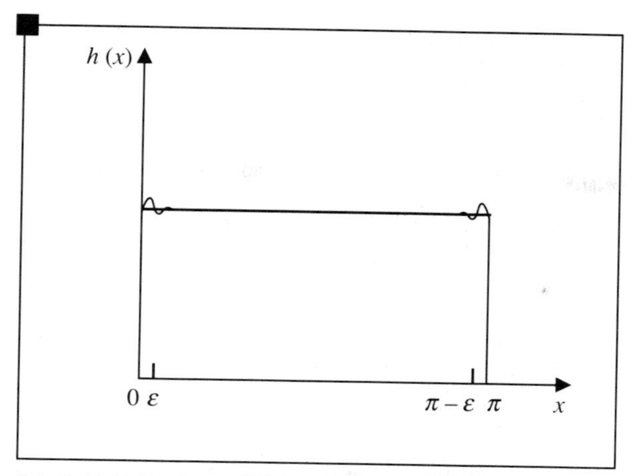

FIGURE 24 **Gibbs Phenomenon**

As we take more terms in the Fourier series to represent h, we notice that the maximum value of the Fourier series in $(0, \pi/2)$ moves from right to left toward the point $x = 0$, but its height does not diminish (see Figure 24). However, we notice as n increases, the Fourier series begins to represent h fairly well in the interval $(\varepsilon, \pi - \varepsilon)$, $\varepsilon > 0$, where ε can be taken smaller as n approaches $+\infty$.

With modern computer graphics, the ideas concerning convergence and Gibbs phenomenon can be clearly seen.

SECTION 7 EXERCISES

30. Are the following functions piecewise continuous over the interval indicated? Give reasons for your answers.

(a) $f(x) = x^3$, $[-1, 1]$

(b) $g(x) = \begin{cases} -5 & -2 < x < -1 \\ 2 & -1 \le x < 3 \end{cases}$

(c) $h(x) = \tan x$, $\quad 0 \le x \le \pi$

(d) $f(x) = \begin{cases} x^2 & -5 < x < -2 \\ -3 & -2 < x \le 2 \\ 2x - 1 & 2 < x < 5 \end{cases}$

(e) $g(x) = \dfrac{1}{x + 1}$, $[-0.9, 0]$

(f) $h(x) = \dfrac{1}{x}$, $(0, 1)$

31. Is Exercise 30, part (f), continuous in the interval defined?

32. Are the following functions piecewise differentiable over the interval indicated? Give reasons for your answers.

(a) $f(x) = \begin{cases} 1 + x & -1 < x \le 0 \\ 1 - x & 0 < x < 1 \end{cases}$

(b) $g(x) = \ln x, \quad 0 < x \le 4$

(c) $h(x) = \begin{cases} \cos x & -\dfrac{\pi}{2} \le x \le \dfrac{\pi}{2} \\ \cos(x - \pi) & \dfrac{\pi}{2} < x \le \dfrac{3\pi}{2} \end{cases}$

(d) $f(x) = \begin{cases} 1 & -1 \le x < 0 \\ \sqrt{1 - x^2} & 0 \le x < 1 \end{cases}$

(e) $G(x) = \begin{cases} 1 & -1 \le x < 0 \\ 1 - x^2 & 0 \le x < 1 \end{cases}$

(f) $h(x) = \begin{cases} 0 & \text{when } x \text{ is a rational number} \\ 1 & \text{when } x \text{ is an irrational number} \end{cases}$ over $[0, 1]$

33. If we apply the voltage coming from a wall receptacle to a full-wave rectifier, the voltage will look like Figure 25. Is this function piecewise differentiable for all t? State your reasons.

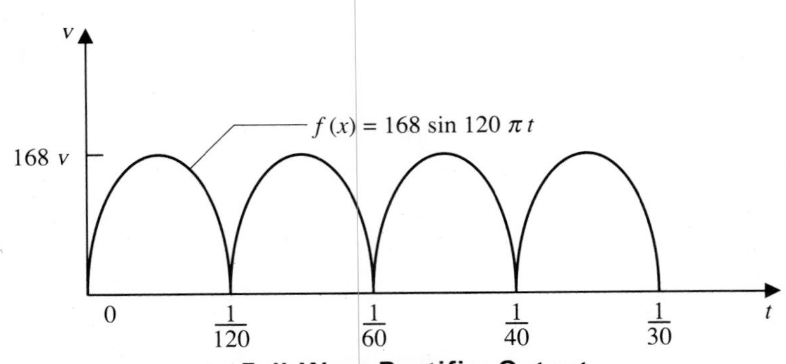

FIGURE 25 **Full-Wave Rectifier Output**

34. If we write the Fourier series for the functions illustrated in Figure 26, sketch the graph of the values to which the Fourier series converges.

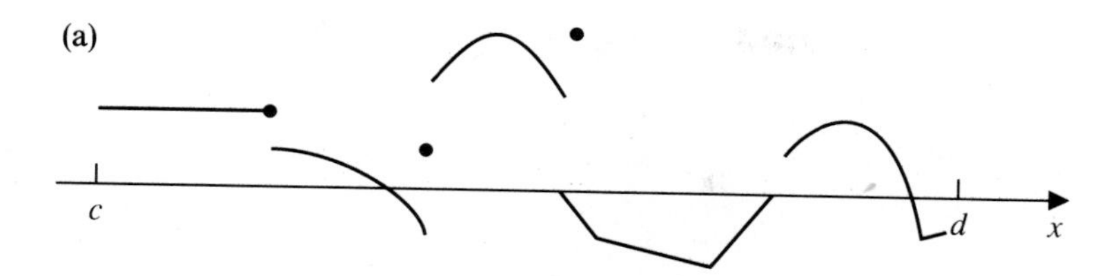

Piecewise differentiable function

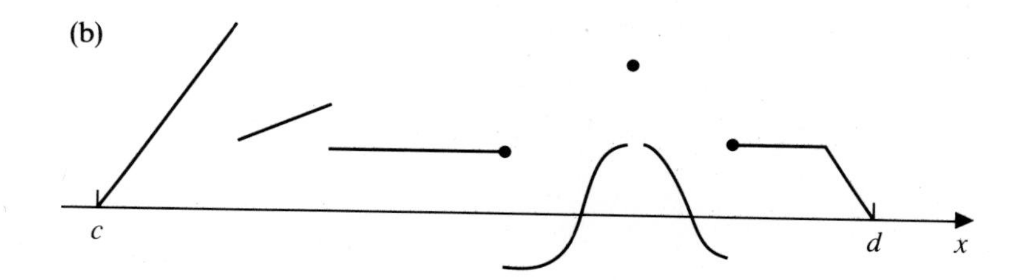

Piecewise differentiable function

FIGURE 26 **Piecewise Differential Functions**

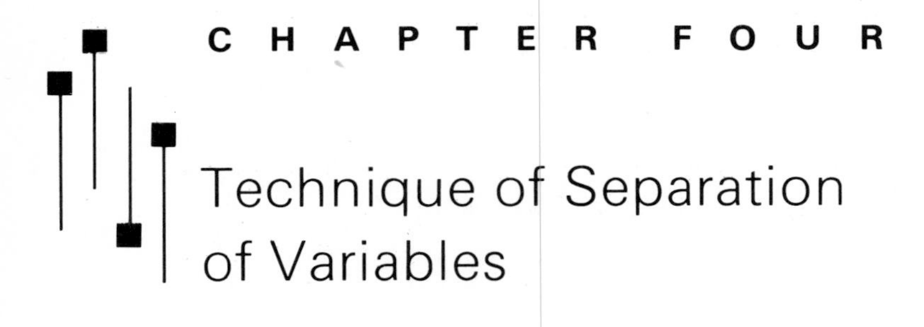

Technique of Separation of Variables

SECTION 1
INTRODUCTION

When it comes to solving boundary value problems involving partial differential equations, a number of approaches are available. The method of separation of variables is very convenient because it draws on many well-known mathematical concepts and it frequently works well.

The general idea of "separating variables" is one of the earliest techniques learned when studying first-order ordinary differential equations. When applied to partial differential equations, the method of separation of variables is similar in some respects to the ordinary differential equation case but eventually takes on a whole new form. Our purpose is to reduce the given partial differential equation into a number of ordinary differential equations. Very often these ordinary differential equations are well known and their solutions are easily found. The whole process follows a logical step-by-step development that terminates in the evaluation of the Fourier coefficients of a Fourier-type series. We suggest that you carefully catalog each step and its position in the stairway to a successful conclusion.

SECTION 2
THE METHOD OF SEPARATION OF VARIABLES BY EXAMPLE

Although the method of separation of variables contains many steps, it does follow a set pattern as we move toward the solution. Probably the best way to understand the method is through observing a number of examples. We start with a straightforward heat flow problem.

EXAMPLE 1. For an introduction to this process, we will seek the solution of the problem

$$\frac{\partial u}{\partial t} = k\frac{\partial^2 u}{\partial x^2} \qquad 0 < x < L,\ 0 < t$$

$$u(0, t) = 0 \qquad 0 < t$$

$$u(L, t) = 0 \qquad 0 < t$$

$$u(x, 0) = x \qquad 0 < x < L$$

This is a one-dimensional heat flow problem in a rod of length L. We are to find the temperature $u(x, t)$ if the right- and left-hand ends are always zero and the initial temperature is x.

Solution

Step 1. We want to seek solutions of the partial differential equation in the special form $u(x, t) = X(x)T(t)$. What we are suggesting is that the variable x occurs only in the function X, whereas T is a function of t only. This device does not always work, but it does solve many engineering problems and is usually a good method to use as a first approach to the problem.

Substitute $u(x, t) = X(x)T(t)$ into the differential equation as follows:

$$X(x)T'(t) = kX''(x)T(t) \tag{4.1}$$

Notice we are able to use the "prime" notation of derivatives because each factor depends only on one variable; that is,

$$T'(t) = \frac{dT(t)}{dt}$$

$$X''(x) = \frac{d^2X}{dx^2}$$

Step 2. We next see if it is possible to separate the variables. Are we able to get all the x's on one side of the equation and all the t's on the other side? Since we are not looking for trivial solutions where either X or T is identically zero, we can divide both sides of Equation (4.1) by $X(x)T(t)$ provided $X(x)T(t) \neq 0$. Thus,

$$\frac{1}{k}\frac{XT'}{XT} = \frac{X''T}{XT}$$

or

$$\frac{1}{k}\frac{T'(t)}{T(t)} = \frac{X''(x)}{X(x)} \tag{4.2}$$

Now pick a fixed value $t = t_0$. Then

$$\frac{1}{k}\frac{T'(t_0)}{T(t_0)} = \text{constant} = \frac{X''(x)}{X(x)}$$

But this is true for any x on the right-hand side of the equation and, therefore,

$$\frac{X''(x)}{X(x)} = \text{constant}$$

In the same way we can show that

$$\frac{T'(t)}{T(t)} = \text{constant}$$

by fixing $x = x_0$. In this case (for convenience) let the constant be equal to $-\lambda$. Equation (4.2) becomes

$$\frac{1}{k}\frac{T'(t)}{T(t)} = \frac{X''(x)}{X(x)} = -\lambda \tag{4.3}$$

Step 3. From Equation (4.3) we see that

$$X''(x) + \lambda X(x) = 0$$
$$T'(t) + \lambda k T(t) = 0$$

which are two ordinary differential equations that are easily solved. These equations are sometimes called the **separate equations**.

□ ***Comment.*** If $X(x)$ or $T(t)$ is zero at $x = x_0$ and $t = t_0$, respectively, the result in Step 3 is still valid. Suppose $X(x_0) = 0$. Returning to Equation (4.1) (before we divide by XT) and substituting $X(x_0) = 0$ into Equation (4.1), we have

$$0 = kX''(x_0)T(t)$$

Since this equation must hold for all $t > 0$, and $T(t)$ is not identically zero, there must be some $t = t_0$ so that $T(t_0) \neq 0$. Since $k > 0$, it follows that $X''(x_0) = 0$, and we can state that

$$X''(x_0) + \lambda X(x_0) = 0$$

is satisfied.

The argument for showing that

$$T'(t_0) + \lambda k T(t_0) = 0$$

for $T(t_0) = 0$ is similar. □

Step 4. Before solving these equations, let us find the boundary conditions that go with the ordinary differential equations. Now

$$u(0, t) = X(0)T(t) = 0$$

If $T(t) = 0$, then $u(x, t) = X(x)T(t) = X(x) \cdot 0 = 0$. But if $u(x, t) \equiv 0$, it is known as the **trivial solution**. Since we are looking for nontrivial solutions, we must avoid setting $T(t) = 0$. Therefore,

$$X(0) = 0$$

is one boundary condition.

In a similar fashion we can show that

$$X(L) = 0$$

is another boundary condition. In this problem there are no initial conditions for the first-order equation in t.

□ ***Warning.*** Do not attempt this type of argument on a nonhomogeneous condition. For example, the initial condition $u(x, 0) = X(x)T(0) = x$ cannot be solved uniquely for $T(0)$. Only when the product of two quantities equals zero can we conclude that one or both of the factors is zero. □

Step 5. Now we have the following ordinary differential equations plus boundary conditions

$$X'' + \lambda X = 0 \qquad X(0) = 0 \qquad X(L) = 0$$
$$T' + \lambda k T = 0$$

The equation with the two boundary conditions is known as an **eigenvalue** problem and is solved first. Unfortunately, λ is unknown to us at the moment, and therefore we must look at three problems because the character of the solution changes with each case. Theorem 1 (see Section 6) tells us that λ must be real because our example is a Sturm–Liouville problem. We shall look at the eigenvalue problem for $\lambda > 0$, $\lambda = 0$, and $\lambda < 0$.

Case 1. $\lambda > 0$. Let $\alpha^2 = \lambda > 0$ (again for convenience). Then the general solution to $X''(x) + \alpha^2 X(x) = 0$ is

$$X(x) = A \cos \alpha x + B \sin \alpha x$$

Now since $X(0) = 0$ we have

$$X(0) = 0 = A \cdot 1 + B \cdot 0$$

which tells us that

$$A = 0$$

and

$$X(x) = B \sin \alpha x$$

Using the second boundary condition $X(L) = 0$, we have

$$X(L) = 0 = B \sin \alpha L$$

Now if $B = 0$, then since $A = 0$, $X(x) \equiv 0$, which implies $u(x, t) = X(x)T(t) \equiv 0$, which is the trivial solution we are trying to avoid. Therefore, $\sin \alpha L = 0$. But recall from trigonometry that $n\pi$ ($n = 0, \pm 1, \pm 2, \ldots$), are the angles that make the sine be zero, and therefore we write

$$\alpha L = n\pi$$

so that

$$\alpha = \frac{n\pi}{L}$$

and

$$\lambda = \lambda_n = \alpha_n^2 = \frac{n^2\pi^2}{L^2} \qquad \text{for } n = \pm 1, \pm 2, \ldots$$

□ ***Note.*** The value $n = 0$ is dropped since this implies $\alpha = 0$ and we are assuming $\alpha > 0$. This case is covered next. □

The values of λ in Case 1 are called the **eigenvalues** of the boundary value problem.

Now the solutions of the ordinary differential equation and its boundary conditions corresponding to these eigenvalues λ_n are

$$X_n = \sin \frac{n\pi x}{L} \qquad n = \pm 1, \pm 2, \ldots$$

These solutions are called **eigenfunctions**.

Case 2. $\lambda = 0$. The general solution of $X'' = 0$ is

$$X(x) = A + Bx$$

Since

$$X(0) = 0 = A$$

and

$$X(L) = BL = 0$$

implies $B = 0$, we see that the only solution to this problem is $X(x) \equiv 0$, which is the trivial solution. $\lambda = 0$ is *not* an eigenvalue.

□ ***Note.*** Eigenfunctions corresponding to an eigenvalue λ must be **nontrivial**. □

Case 3. $\lambda < 0$. Let $\lambda = -\alpha^2 < 0$. The differential equation $X'' + \alpha^2 X = 0$ has the general solution

$$X(x) = C_1 e^{\alpha x} + C_2 e^{-\alpha x}$$

Although this solution may be used, it is more convenient to write the solution in terms of hyperbolic functions instead. Since

$$\sinh \alpha x = \frac{e^{\alpha x} - e^{-\alpha x}}{2} \qquad \text{and} \qquad \cosh \alpha x = \frac{e^{\alpha x} + e^{-\alpha x}}{2}$$

which are linear independent functions, we can write the general solution as

$$X(x) = A \cosh \alpha x + B \sinh \alpha x$$

Now since $X(0) = 0$ we have

$$X(0) = 0 = A \cdot 1 + B \cdot 0$$

which shows that

$$A = 0$$

and $X(x) = B \sinh \alpha x$.

Using $X(L) = B \sinh \alpha L = 0$, we find either $B = 0$, which leads to the trivial solution or $\sinh \alpha L = 0$. But $\sinh 0 = 0$ is the only zero of the hyperbolic sine function. Therefore, $\alpha L = 0$. And since $L \neq 0$, it follows that $\alpha = 0$. But this is a contradiction since we are assuming $-\alpha^2 < 0$.

There are no nontrivial solutions under this case, which tells us there are no negative eigenvalues.

This long discourse completes the solution of the eigenvalue problem posed in Step 5.

Step 6. We now shift to the differential equation

$$T' + \lambda k T = T' + \frac{n^2\pi^2}{L^2} kT = 0 \tag{4.4}$$

Notice that the eigenvalues are now known to us from Step 5 and it is possible to solve Equation (4.4) up to an arbitrary constant. The general solution to the first-order equation is

$$T_n(t) = C_n \exp(-kn^2\pi^2/L^2)t \qquad n = \pm 1, \pm 2, \ldots$$

In solving boundary value problems by this approach, we find it convenient to set all C_n's $= 1$ because we will be multiplying each solution by an arbitrary constant later on.

Step 7. To continue with the solution, we multiply X_n by T_n because in Step 1 we assumed a solution of the partial differential equation of the form $u(x, t) = X(x)T(t)$. Therefore, we can write

$$u_n(x, t) = \exp(-kn^2\pi^2/L^2)t \sin \frac{n\pi x}{L} \qquad n = \pm 1, \pm 2, \ldots$$

For $n = \pm 1, \pm 2, \ldots$, $u_n(x, t)$ satisfies the partial differential equation and the two boundary conditions.

We still have to match the initial condition $u(x, 0) = x$. To do this, we form an infinite linear combination of all solutions $u_n(x, t)$. Basing our ideas on the superposition principle stated in Chapter 2, we can write the general solution to the partial differential equation and boundary conditions as

$$u(x, t) = \sum_{n=1}^{\infty} b_n u_n(x, t) = \sum_{n=1}^{\infty} b_n \exp(-kn^2\pi^2/L^2)t \sin \frac{n\pi x}{L} \tag{4.5}$$

□ ***Note.*** When dealing with sines and cosines, we question whether it is necessary to consider n for negative integers. Since $\sin(-n\pi x/L) =$

$-\sin(n\pi x/L)$ and $\cos(-n\pi x/L) = \cos(n\pi x/L)$, each solution derived from a negative n can be combined with the corresponding solution derived from a positive n. From here on we will assume this is done and let $n = 1, 2, \ldots$. □

Step 8. Finally, using the condition $u(x, 0) = x$, $0 < x < a$, we must satisfy

$$u(x, 0) = x = \sum_{n=1}^{\infty} b_n \sin \frac{n\pi x}{L}$$

since the time exponential is 1. But this is just a Fourier sine series when x on $(0, L)$ is extended into $(-L, 0)$ as an odd function, and it follows that

$$\begin{aligned} b_n &= \frac{2}{L}\int_0^L x \sin \frac{n\pi x}{L}\, dx \\ &= \frac{2L^2}{Ln^2\pi^2}\left\{\sin \frac{n\pi x}{L} - \frac{n\pi x}{L}\cos \frac{n\pi x}{L}\right\}\Bigg|_0^L \\ &= \frac{2L}{n\pi}\{-\cos n\pi\} = -\frac{2L}{n\pi}(-1)^n \end{aligned}$$

or

$$b_n = (-1)^{n+1}\frac{2L}{n\pi} \tag{4.6}$$

We have now determined all the constants b_n. Substituting their value from Equation (4.6) in Equation (4.5), we see that the final solution is

$$u(x, t) = \frac{2L}{\pi}\sum_{n=1}^{\infty}\frac{(-1)^{n+1}}{n}\exp\left(-k\frac{n^2\pi^2}{L^2}\right)t \sin \frac{n\pi x}{L}$$ ■

It is possible to garner some information about the solution $u(x, t)$ without too much effort. True, if we have access to a computer, we could numerically evaluate a large number of terms of the series to construct a table of values or even plot a graph of an approximation to the solution. Of course, how well this can be done depends on the convergence properties of the series, as we discussed at the end of Chapter 3.

The first thing we observe is that $u(x, t)$ is a **transient** solution; that is, $u(x, t) \to 0$ as $t \to +\infty$. This idea follows easily from the fact that each term in the series is of the form $e^{-\alpha t}$ where $\alpha = kn^2\pi^2/L^2$. Carrying this idea a bit further, we notice that n enters into the exponent as n^2, which indicates that $e^{-\alpha t}$ decreases very rapidly as n increases. Therefore, most of the information about the solution is carried in the first few terms of the series.

EXAMPLE 2. Let $k = 1$, $L = \pi$, and $t = 1$, and compare the $e^{-\alpha t}$ terms for $n = 1, 2, 3$.

n	αt	$e^{-\alpha t}$
1	1	0.3679
2	4	0.0183
3	9	0.0001

Notice that the series is an alternating series, which should make it relatively easy to estimate the error induced by using a finite number of terms. ■

We now look at an example in which $\lambda = 0$ is an eigenvalue.

EXAMPLE 3. Consider a high-frequency transmission line whose length is 40 meters. Suppose the rate of change of voltage with respect to x at both ends is zero. If the initial voltage is $f(x)$ and the rate of change of voltage with respect to t is zero, find the solution as a function of x and t.

This problem can be written formally as

$$\frac{\partial e}{\partial x}(0, t) = 0 \qquad 0 < t$$

$$\frac{\partial e}{\partial x}(40, t) = 0 \qquad 0 < t$$

$$e(x, 0) = f(x) \qquad 0 < x < 40$$

$$\frac{\partial e}{\partial t}(x, 0) = 0 \qquad 0 < x < 40$$

The differential equation is

$$\frac{\partial^2 e}{\partial x^2} = LC\,\frac{\partial^2 e}{\partial t^2} \qquad 0 < x < 40,\ 0 < t$$

Steps 1 and 2. Let $e(x, t) = X(x)T(t)$. Substituting into the differential equation, we find

$$\frac{X''}{X} = LC\,\frac{T''}{T} = -\lambda$$

Step 3. This result leads to the two ordinary differential equations

$$X'' + \lambda X = 0$$

$$T'' + \frac{\lambda}{LC}\,T = 0$$

Step 4. The three homogeneous boundary and initial conditions yield the following conditions for the ordinary differential equations:

$$X'(0) = 0, \qquad X'(40) = 0, \qquad T'(0) = 0$$

Steps 5 and 6. When we solve the differential equation in x, we again consider three cases: $\lambda > 0$, $\lambda = 0$, and $\lambda < 0$.

Case 1. $\lambda > 0$. Let $\alpha^2 = \lambda > 0$. The general solution of $x'' + \alpha^2 X = 0$ is then

$$X(x) = A \cos \alpha x + B \sin \alpha x$$

and

$$X'(x) = -\alpha A \sin \alpha x + \alpha B \cos \alpha x$$

Applying the boundary conditions, we find

$$X'(0) = -\alpha A \cdot 0 + \alpha B \cdot 1 = 0$$

for which

$$\alpha B = 0$$

But since $\alpha > 0$, $B = 0$ and $X'(x) = -\alpha A \sin \alpha x$. Then $X'(40) = -\alpha A \sin 40\alpha = 0$. Neither α nor A can equal zero, so we can write

$$\sin 40\alpha = 0$$

from which it follows that

$$40\alpha = n\pi \qquad n = 1, 2, \ldots$$

$$\alpha = \frac{n\pi}{40}$$

The positive eigenvalues are

$$\lambda_n = \alpha^2 = \left[\frac{n\pi}{40}\right]^2$$

and the eigenfunctions are

$$X_n(x) = \cos \frac{n\pi x}{40}$$

Solving the other differential equation, we have

$$T_n(t) = D_n \cos \frac{n\pi t}{40\sqrt{LC}} + E_n \sin \frac{n\pi t}{40\sqrt{LC}}$$

and

$$T_n'(t) = -D_n \frac{n\pi}{40\sqrt{LC}} \sin \frac{n\pi t}{40\sqrt{LC}} + E_n \frac{n\pi}{40\sqrt{LC}} \cos \frac{n\pi t}{40\sqrt{LC}}$$

Using the condition

$$T_n'(0) = E_n \frac{\pi n}{40\sqrt{LC}} = 0$$

we observe that

$$E_n = 0$$

Therefore,

$$T_n(t) = \cos \frac{n\pi t}{40\sqrt{LC}} \qquad n = 1, 2, \ldots$$

Combining X_n and T_n we find

$$e_n(x, t) = \cos \frac{n\pi x}{40} \cos \frac{n\pi t}{40\sqrt{LC}} \qquad n = 1, 2, \ldots$$

Case 2. $\lambda = 0$. In this case, $X'' = 0$ and $X(x) = A + Bx$. It follows that

$$X'(x) = B$$

and

$$X'(0) = B = 0$$

But now $X'(x) = 0$ for all x and in particular for $x = 40$; that is, $X'(40) = 0$. Therefore, both boundary conditions are satisfied: $\lambda = 0$ is an eigenvalue and $X(x) = A$ (a constant) is an eigenfunction. We find it convenient to let $A = 1$ and use $X(x) = 1$ as our eigenfunction. To identify the eigenvalue and eigenfunction associated with $\lambda = 0$, we write $\lambda_0 = 0$ and $X_0 = 1$. Now solving $T'' = 0$ we have $T(t) = Dt + E$ and $T'(t) = D$. Using the initial condition $T'(0) = 0$, we find $D = 0$ and $T_0 = 1$. Therefore, $e_0(x, t) = X_0 T_0 = 1$.

Case 3. $\lambda < 0$. In a way similar to Case 3 of Example 1 we can show that there are no negative eigenvalues.

Step 7. To finish the problem we form an infinite linear combination of all the solutions e_0 and e_n, which yields

$$e(x, t) = \frac{a_0}{2} e_0(x, t) + \sum_{n=1}^{\infty} a_n e_n(x, t) \tag{4.7}$$

$$= \frac{a_0}{2} + \sum_{n=1}^{\infty} a_n \cos \frac{n\pi x}{40} \cos \frac{n\pi t}{40\sqrt{LC}}$$

Step 8. We use our final condition, which tells us that

$$e(x, 0) = f(x) = \frac{a_0}{2} + \sum_{n=1}^{\infty} a_n \cos \frac{n\pi x}{40} \qquad \text{on } (0, 40)$$

Since this is just a Fourier cosine series, we know that

$$a_0 = \frac{1}{20} \int_0^{40} f(x)\, dx$$

and

$$a_n = \frac{1}{20} \int_0^{40} f(x) \cos \frac{n\pi x}{40} \, dx$$

Substituting these values back into Equation (4.7), we have the solution to our problem:

$$e(x, t) = \frac{1}{40} \int_0^{40} f(x) \, dx + \frac{1}{20} \sum_{n=1}^{\infty} \left[\int_0^{40} f(s) \cos \frac{n\pi s}{40} \, ds \right] \cos \frac{n\pi x}{40} \cos \frac{n\pi t}{40\sqrt{LC}}$$

We recognize immediately that the solution $e(x, t)$ is an even function in both x and t. The integral used for evaluating a_0 is the average value for $f(x)$ over the half Fourier interval whose length is 40. Half the value of this integral, that is, $a_0/2$, is equal to the DC component of the series answer.

The frequency f_n of the components of the voltage $e(x, t)$ with respect to time is

$$f_n = \frac{n\pi}{2\pi(40\sqrt{LC})} = \frac{n}{80\sqrt{LC}}$$

Therefore, the fundamental frequency is $1/80\sqrt{LC}$. ■

Examples 1 and 3 cover two of the three major classes of second-order partial differential equations: the parabolic type and the hyperbolic type. We now investigate the solution of the elliptic type, which introduces us to some new techniques used in solving boundary value problems.

EXAMPLE 4. Recall that the temperature in a plate insulated above and below must satisfy the differential equation,

$$\frac{\partial u}{\partial t} = k\left[\frac{\partial^2 u}{\partial x^2} + \frac{\partial^2 u}{\partial y^2}\right]$$

If the temperature is independent of t (i.e., steady state), this differential equation becomes

$$\frac{\partial^2 u}{\partial x^2} + \frac{\partial^2 u}{\partial y^2} = 0$$

which is Laplace's equation in two variables.

Now suppose we have a rectangular plate whose dimensions are a by b and whose boundary conditions are given by

$$\begin{aligned} &u(x, 0) = f_1(x) && 0 < x < a \\ &u(a, y) = 0 && 0 < y < b \\ &u(x, b) = f_2(x) && 0 < x < a \\ &u(0, y) = 0 && 0 < y < b \end{aligned}$$

Find the temperature u as a function of x and y (see Figure 1).

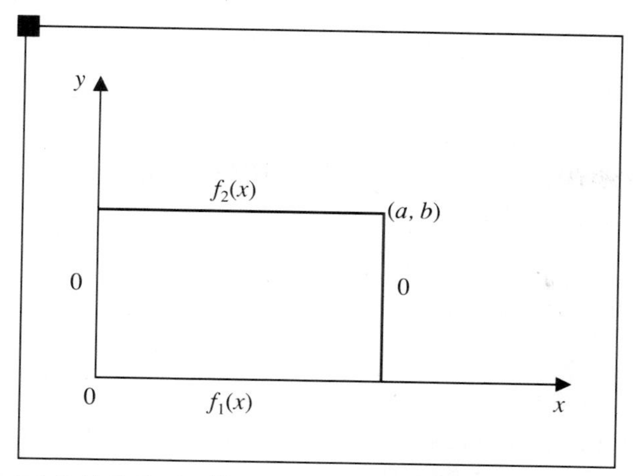

FIGURE 1 **Temperature in Plate**

In order to carry out specific operations, let us examine the special case where $f_1(x) = 0$ and $f_2(x) = x$.

Steps 1, 2, and 3. Assume $u(x, y) = X(x)Y(y)$. Computing the necessary derivatives and substituting into the differential equation, we have

$$X''Y + XY'' = 0$$

or

$$\frac{X''}{X} = -\frac{Y''}{Y} = -\lambda$$

which yields the two ordinary differential equations

$$X''(x) + \lambda X(x) = 0$$
$$Y''(y) - \lambda Y(y) = 0$$

Steps 4 and 5. From our assumption and the given boundary conditions, we arrive at the following boundary conditions associated with the ordinary differential equations:

$$X(0) = 0, \qquad X(a) = 0$$

and

$$Y(0) = 0$$

Case 1. $\lambda > 0$. Let $\lambda = \alpha^2 > 0$. Then we can show as before that the eigenvalues are

$$\lambda_n = \alpha_n^2 = \frac{n^2\pi^2}{a^2} \qquad n = 1, 2, \ldots$$

and the eigenfunctions are

$$X_n(x) = \sin \frac{n\pi x}{a}$$

Cases 2 and 3. $\lambda = 0$ and $\lambda < 0$. It is easily shown that there are no nonpositive eigenvalues.

Step 6. One new idea introduced in this problem occurs when we wish to solve the second differential equation,

$$Y'' - \frac{n^2\pi^2}{a^2} Y = 0$$

Normally a beginning student would solve this equation using exponentials as follows:

$$Y_n(y) = C_n \exp \frac{n\pi y}{a} + D_n \exp\left(-\frac{n\pi y}{a}\right) \tag{4.8}$$

where $\exp y = e^y$. Because neither exponential in Equation (4.8) vanishes, it is necessary to solve a system of two equations to find D in terms of C. Therefore, it is much more convenient to use hyperbolic functions, and we can write

$$Y_n(y) = C_n \cosh \frac{n\pi y}{a} + D_n \sinh \frac{n\pi y}{a}$$

Using the boundary condition $Y(0) = 0$, we see that

$$Y_n(0) = 0 = C_n \cdot 1 + D_n \cdot 0$$

or

$$C_n = 0$$

and that

$$Y_n(y) = \sinh \frac{n\pi y}{a}$$

is a solution to our ordinary differential equation in y and its one boundary condition. For convenience we have again set the constant $D = 1$.

Step 7. Combining the two families of solutions $X_n(x)$ and $Y_n(y)$, we construct solutions to the partial differential equation and the three homogeneous boundary conditions that take the form

$$u_n(x, y) = X_n(x)Y_n(y) = \sin \frac{n\pi x}{a} \sinh \frac{n\pi y}{a}$$

Step 8. We still have to meet the final nonhomogeneous boundary condition $u(x, b) = x$. In order to do this we use the superposition principle and form the infinite linear combination

$$u(x, y) = \sum_{n=1}^{\infty} b_n \sin \frac{n\pi x}{a} \sinh \frac{n\pi y}{a} \tag{4.9}$$

expecting that we can find values for b_n such that the nonhomogeneous condition is met. Letting $x = b$ in Equation (4.9), we see that

$$u(x, b) = x = \sum_{n=1}^{\infty} b_n \sinh \frac{n\pi b}{a} \sin \frac{n\pi x}{a} \tag{4.10}$$

If we let

$$B_n = b_n \sinh \frac{n\pi b}{a}$$

Equation (4.10) becomes

$$x = \sum_{n=1}^{\infty} B_n \sin \frac{n\pi x}{a}$$

which is a Fourier sine series. Therefore,

$$\begin{aligned} B_n = b_n \sinh \frac{n\pi b}{a} &= \frac{2}{a} \int_0^a x \sin \frac{n\pi x}{a}\, dx \\ &= \frac{2}{a} \frac{a^2}{n^2\pi^2} \left[\sin \frac{n\pi x}{a} - \frac{n\pi x}{a} \cos \frac{n\pi x}{a} \right]\Bigg|_0^a \\ &= \frac{2a}{n^2\pi^2} [-n\pi \cos n\pi] = (-1)^{n+1} \frac{2a}{n\pi} \end{aligned}$$

The constants b_n become

$$b_n = (-1)^{n+1} \frac{2a}{n\pi \sinh \dfrac{n\pi b}{a}}$$

and the final answer can be written as

$$u(x, y) = \frac{2a}{\pi} \sum_{n=1}^{\infty} (-1)^{n+1} \frac{\sin (n\pi x/a) \sinh (n\pi y/a)}{n \sinh (n\pi b/a)}$$ ■

□ ***Remark.*** When solving the ordinary differential equation

$$y''(x) - \alpha^2 y(x) = 0$$

it is useful to remember that there are three practical ways in which to express the solution.

If one of the boundary conditions is of the form

$$y(0) = 0 \qquad \text{or} \qquad y'(0) = 0$$

then the general solution

$$y(x) = A \cosh \alpha x + B \sinh \alpha x \tag{4.11}$$

is most convenient to use because the value of A or B is quickly determined. On the other hand, if

$$y(L) = 0 \qquad \text{or} \qquad y'(L) = 0$$

is given as a boundary condition, one of the following forms of the solution is useful:

$$y(x) = E \sinh \alpha(x - d) \qquad \text{for } y(L) = 0$$
$$y(x) = E \cosh \alpha(x - d) \qquad \text{for } y'(L) = 0$$

where E and d are arbitrary constants. These forms of the solutions are not as well known as the form in Equation (4.11), but they can easily be shown to be equivalent for the particular boundary condition given using the well-known hyperbolic identities

$$\sinh (\alpha - \beta) = \sinh \alpha \cosh \beta - \cosh \alpha \sinh \beta$$

and

$$\cosh (\alpha - \beta) = \cosh \alpha \cosh \beta - \sinh \alpha \sinh \beta$$

Finally, the classic solution

$$y(x) = Ae^{\alpha x} + Be^{-\alpha x}$$

finds its greatest use in solving boundary value problems over semi-infinite intervals and the study of Fourier integrals. □

SECTION 2 EXERCISES

1. The temperature $u(x, t)$ in a laterally insulated rod of length L satisfies the following boundary value problem:

$$\frac{\partial^2 u}{\partial x^2} = \frac{1}{k}\frac{\partial u}{\partial t} \qquad 0 < x < L,\ 0 < t$$
$$u(0, t) = 0 \qquad 0 < t$$
$$u(L, t) = 0 \qquad 0 < t$$
$$u(x, 0) = 100 \qquad 0 < x < L$$

Use the technique of separation of variables to find $u(x, t)$.

2. Use the technique of separation of variables to solve for the temperature $u(x, t)$ if

$$\frac{\partial^2 u}{\partial u^2} = \frac{1}{k}\frac{\partial u}{\partial t} \qquad 0 < x < 10,\ 0 < t$$
$$\frac{\partial u}{\partial x}(0, t) = 0 \qquad 0 < t$$
$$\frac{\partial u}{\partial x}(10, t) = 0 \qquad 0 < t$$
$$u(x, 0) = 1 - x \qquad 0 < x < 10$$

3. The voltage $e(x, t)$ along a submarine cable 3000 kilometers long satisfies the boundary value problem

$$
\begin{aligned}
&e_{xx} = RCe_t && 0 < x < 3000,\ 0 < t \\
&e(0, t) = 0 && 0 < t \\
&e(3000, t) = 0 && 0 < t \\
&e(x, 0) = \sin \frac{x}{100} && 0 < x < 3000
\end{aligned}
$$

Use the technique of separation of variables to find $e(x, t)$.

4. The current $i(x, t)$ along a submarine cable of length L satisfies

$$
\begin{aligned}
&i_{xx} = RCi_t && 0 < x < L,\ 0 < t \\
&\frac{\partial i}{\partial x}(0, t) = 0 && 0 < t \\
&\frac{\partial i}{\partial x}(L, t) = 0 && 0 < t \\
&i(x, 0) = 2 && 0 < x < L
\end{aligned}
$$

Use the technique of separation of variables to find the current $i(x, t)$ in the cable.

5. Find the temperature $u(x, t)$ by the technique of separation of variables in a laterally insulated rod of length 20 meters that has a heat source given by $\delta u(x, t)$ joules/m^3. The other conditions are

$$
\begin{aligned}
&u(0, t) = 0 \\
&u(20, t) = 0 \\
&u(x, 0) = x(20 - x)
\end{aligned}
$$

6. The voltage $e(x, t)$ satisfies the differential equation $e_{xx} = RCe_t + Ae_x$. Using separation of variables find $e(x, t)$ if

$$
\begin{aligned}
&\frac{\partial e}{\partial x}(0, t) = 0 && 0 < t \\
&\frac{\partial e}{\partial x}(L, t) = 0 && 0 < t \\
&e(x, 0) = 50 && 0 < x < L
\end{aligned}
$$

7. (a) A vibrating string is fastened to air bearings situated on vertical rods at $x = 0$ and $x = 2$. Find the displacement $y(x, t)$ if the conditions are

$$
\begin{aligned}
&y_x(0, t) = 0 && 0 < t \\
&y_x(2, t) = 0 && 0 < t \\
&y(x, 0) = x && 0 < x < 2 \\
&\frac{\partial y}{\partial t}(x, 0) = 0 && 0 < x < 2
\end{aligned}
$$

(b) Sketch $y(x, t)$ over $0 < x < 2$ for $t = 0$, $t = 1$, and $t = 2$.

□ ***Hint.*** Use only the first two terms of the series solution. □

8. Two parallel wires with $L = 0.01h/m$, $C = 1\mu fd/m$, 2 meters long are connected to a high-frequency generator. Find the current $i(x, t)$ if the line is open at $x = 0$ and $x = 2$. The initial current is zero and the rate of change of current is 1 A/sec.

9. The pressure $p(x, t)$ in an organ pipe satisfies the differential equation

$$p_{xx} = \frac{1}{c^2} p_{tt}$$

If the pipe is L meters long and open at both ends, find the pressure $p(x, t)$ if $p(x, 0) = 0$ and $(\partial p/\partial t)(x, 0) = 40$.

10. (a) Given the telegraph equation for finding voltage $e(x, t)$ on a transmission line of length a along with the boundary and initial conditions, we can write

$$e_{xx} = LCe_{tt} + (RC + GL)e_t + RGe \qquad 0 < x < a,\ 0 < t$$

$$\frac{\partial e}{\partial x}(0, t) = 0 \qquad 0 < t$$

$$\frac{\partial e}{\partial x}(a, t) = 0 \qquad 0 < t$$

$$e(x, 0) = 100 \qquad 0 < x < a$$

$$\frac{\partial e}{\partial t}(x, 0) = 0 \qquad 0 < x < a$$

Solve for $e(x, t)$ using separation of variables.

(b) What is the frequency of the third harmonic?

11. The length of a guitar string is 65 centimeters. If the string is plucked 15 centimeters from the bridge (i.e., at the end of the wire) by raising it 3 millimeters, find the displacement $u(x, t)$ using separation of variables.

12. The length of a piano string is 1 meter. When a pupil strikes a key, the following velocity is imparted to the string:

$$\frac{\partial u}{\partial t} = \text{velocity} = \begin{cases} 0 & 0 < x < 49 \text{ cm} \\ 1 & 49 < x < 51 \text{ cm} \\ 0 & 51 < x < 100 \text{ cm} \end{cases}$$

Find the displacement $u(x, t)$ using separation of variables. Do not evaluate the Fourier coefficients of the final solution.

SECTION 3
FURTHER EXAMPLES OF METHOD OF SEPARATION OF VARIABLES

Examples 1, 3, and 4 were chosen especially so that you would see the method of separation of variables applied to the three classes of second-order differential equations: parabolic, hyperbolic, and elliptic. All were set in terms of rectangular coordinates. In Example 5 we will examine a boundary value problem that is more easily solved in polar coordinates.

EXAMPLE 5. Let us consider a problem similar to Example 4 except that the shape of the plate is circular rather than rectangular. We wish to find the steady-state temperature $u(r, \theta)$ throughout a circular plate of radius c that is insulated laterally. The temperature on the circumference is $100°$ C over one semicircle and $0°$ over the other (see Figure 2).

Expressed formally, $u(r, \theta)$ must satisfy

$$r^2 \frac{\partial^2 u}{\partial r^2} + r \frac{\partial u}{\partial r} + \frac{\partial^2 u}{\partial \theta^2} = 0$$

$$u(c, \theta) = \begin{cases} 100 & 0 < \theta < \pi \\ 0 & \pi < \theta < 2\pi \end{cases}$$

We assume the solution to this problem can be written in the form

$$u(r, \theta) = R(r)\Theta(\theta)$$

Substituting in the differential equation and using the method of separation of variables, we are led to the two ordinary differential equations

$$\Theta'' + \lambda\Theta = 0 \tag{4.12}$$

$$r^2 R'' + rR' - \lambda R = 0 \tag{4.13}$$

When we attempt to solve Equation (4.12) (which we have done previously in Examples 1, 3, and 4), we notice we have no boundary conditions since there are no exposed radial edges. This situation was covered in Example

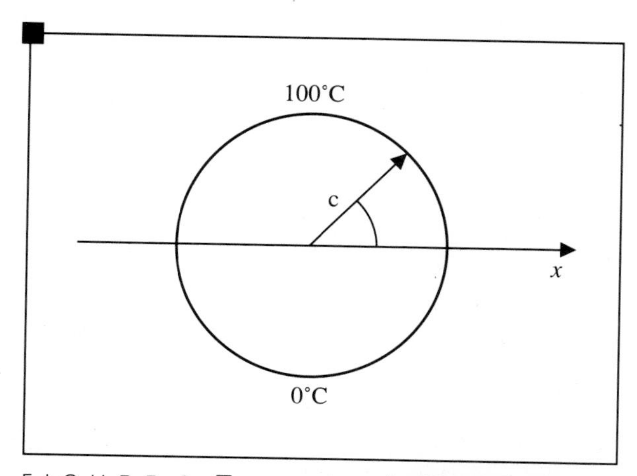

FIGURE 2 **Temperature in Circular Plate**

15 of Section 3 of Chapter 2, where we discovered that the boundary conditions can be written as

$$u(r, -\pi) = u(r, \pi)$$
$$\frac{\partial u}{\partial \theta}(r, -\pi) = \frac{\partial u}{\partial \theta}(r, \pi) \qquad 0 < r < c \tag{4.14}$$

Such boundary conditions are called **periodic boundary conditions.**

As before, we must consider the three cases $\lambda > 0$, $\lambda = 0$, and $\lambda < 0$. If we let $\lambda = \alpha^2 > 0$, the differential equation in Equation (4.12) becomes

$$\Theta'' + \alpha^2\Theta = 0$$

where the solution is

$$\Theta(\theta) = A \cos \alpha\theta + B \sin \alpha\theta$$

and **(4.15)**

$$\Theta'(\theta) = -\alpha A \sin \alpha\theta + \alpha B \cos \alpha\theta$$

From the periodic boundary conditions in Equation (4.14) we see that

$$\Theta(-\pi) = \Theta(\pi)$$
$$\Theta'(-\pi) = \Theta'(\pi) \tag{4.16}$$

Substituting the solutions in the equations in (4.15) into the equations in (4.16), we have

$$A \cos \alpha(-\pi) + B \sin \alpha(-\pi) = A \cos \alpha\pi + B \sin \alpha\pi$$
$$-\alpha A \sin \alpha(-\pi) + \alpha B \cos \alpha(-\pi) = -\alpha A \sin \alpha\pi + \alpha B \cos \alpha\pi$$

which yields

$$B \sin \alpha\pi = 0 \qquad \text{and} \qquad A \sin \alpha\pi = 0$$

Since A and B cannot both be zero, it follows that

$$\sin \alpha\pi = 0$$

or

$$\alpha = n \qquad n = 1, 2, \ldots$$

or

$$\lambda_n = n^2$$

and

$$\Theta_n = A_n \cos n\theta + B_n \sin n\theta$$

If $\lambda = 0$, then $\Theta'' = 0$ has the solution

$$\Theta = A\theta + B$$

Substituting in the boundary conditions in Equation (4.16), we find

$$A(-\pi) + B = A\pi + B$$

or $A = -A$, which implies $A = 0$. Therefore, $\lambda = 0$ is an eigenvalue and the corresponding eigenfunction is an arbitrary constant that we will choose as 1.

It can be shown that there are no negative eigenvalues.

We now move to the solution of the other ordinary differential equation

$$r^2R'' + rR' - n^2R = 0 \tag{4.17}$$

We recognize that this is a Cauchy–Euler differential. To obtain a solution to this differential equation, we attempt a solution of the form $R(r) = r^\alpha$, from which it follows that $R'(r) = \alpha r^{\alpha-1}$ and $R''(r) = \alpha(\alpha - 1)r^{\alpha-2}$. Substituting these values in Equation (4.17), we find that

$$r^\alpha[\alpha(\alpha - 1) + \alpha - n^2] = r^\alpha[\alpha^2 - n^2] = 0$$

This equation must hold for all r, $0 < r < c$; therefore,

$$\alpha^2 - n^2 = 0 \qquad \text{or} \qquad \alpha = \pm n$$

Since r^n and r^{-n} are linearly independent solutions to Equation (4.17), the general solution of the Cauchy–Euler equation is

$$R(r) = Cr^n + Dr^{-n} \qquad n = 1, 2, 3, \ldots$$

Once again there is no explicit boundary condition, but we observe that since $r = 0$ is the center of our circular plate, the

$$\lim_{r\to 0^+} r^{-n} = +\infty$$

Since we are only considering bounded solutions, we must set $D = 0$ and

$$R_n(r) = r^n \qquad n = 1, 2, \ldots$$

For $\lambda = 0$, the Cauchy–Euler equation becomes

$$r^2R'' + rR' = 0 \tag{4.18}$$

whose solution is

$$R(r) = C + D \ln r$$

Since we are considering only bounded solutions, D must equal zero since

$$\lim_{r\to 0} \ln r = -\infty$$

and therefore we take $R_0 = 1$ as the solution of Equation (4.18).

The solution to the partial differential equation is

$$u(r, \theta) = \frac{A_0}{2} + \sum_{n=1}^{\infty} r^n(A_n \cos n\theta + B_n \sin n\theta)$$

Using the nonhomogeneous condition we can write

$$u(c, \theta) = \frac{A_0}{2} + \sum_{n=1}^{\infty} c^n(A_n \cos n\theta + B_n \sin n\theta)$$

where

$$A_0 = \frac{1}{\pi} \int_0^\pi 100 \, d\theta = 100$$

$$c^n A_n = \frac{1}{\pi} \int_0^\pi 100 \cos n\theta \, d\theta = \frac{100}{n\pi} \sin n\theta \Big|_0^\pi = 0$$

and

$$\begin{aligned} c^n B_n &= \frac{1}{\pi} \int_0^\pi 100 \sin n\theta \, d\theta = -\frac{100}{n\pi} \cos n\theta \Big|_0^\pi \\ &= \frac{100}{n\pi} [1 - (-1)^n] \\ &= \begin{cases} 200/n\pi & n \text{ odd} \\ 0 & n \text{ even} \end{cases} \end{aligned}$$

The solution to our problem is

$$u(r, \theta) = 50 + \frac{200}{\pi} \sum_{n=1, 3, 5}^{\infty} \left(\frac{r}{c}\right)^n \frac{\sin n\theta}{n}$$

If we set $\theta = 0$ or π, we see immediately that the temperature $u(r, \theta)$ along the diameter $\theta = 0$ or π is 50 degrees. At points of symmetry with respect to this diameter, the temperature equals 50 plus or minus the value of the series. Notice that only odd harmonics of the fundamental frequency $\sin \theta$ exist. ■

Flow Around a Cylinder

The steady-state flow of a fluid in a direction transverse to a long metal cylinder can be approximated at low velocities by Laplace's equation in two dimensions. We will see that the boundary conditions are of the Neumann type.

If φ is the **velocity potential**, then

$$\nabla^2 \varphi = r^2 \frac{\partial^2 \varphi}{\partial r^2} + r \frac{\partial \varphi}{\partial r} + \frac{\partial^2 \varphi}{\partial \theta^2} = 0$$

where the velocity $\mathbf{v} = \text{grad } \varphi = \nabla \varphi$. For our purposes it is natural to locate the cylinder at the origin. Its equation is then $x^2 + y^2 = c^2$. Far from the cylinder (i.e., as $r^2 = x^2 + y^2 \to +\infty$), the velocity $\mathbf{v}$ of the fluid is constant and parallel to the x-axis and can be written as $\mathbf{v}(x, y) = a\mathbf{i}$ as $r^2 \to +\infty$ (see Figure 3). To solve this problem we introduce polar coordinates and note that in these coordinates the velocity is given by

$$\mathbf{v} = \nabla \varphi = \varphi_r \mathbf{e}_r + \frac{1}{r} \varphi_\theta \mathbf{e}_\theta \tag{4.19}$$

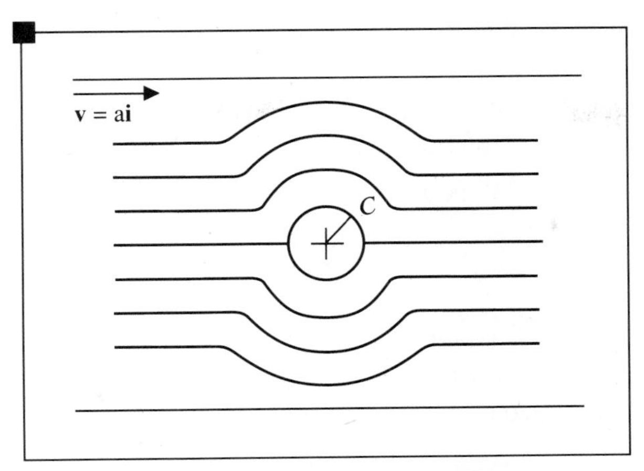

FIGURE 3 **Flow Around Cylinder**

where $\mathbf{e}_r$ is the unit vector outward from the origin and $\mathbf{e}_\theta$, a unit vector, is perpendicular to $\mathbf{e}_r$ in a counterclockwise direction. Looking at Figure 4 we observe that the boundary conditions at $r = +\infty$ are

$$\varphi_r = a \cos \theta, \qquad \frac{1}{r} \varphi_\theta = -a \sin \theta \tag{4.20}$$

Since the fluid cannot penetrate the cylinder, we must have $\varphi_r(c, \theta) = 0$. Finally, since the velocity at θ and $\theta + 2\pi$ is the same, we must have from Equation (4.19),

$$\varphi_r(r, \theta + 2\pi) = \varphi_r(r, \theta)$$
$$\varphi_\theta(r, \theta + 2\pi) = \varphi_\theta(r, \theta)$$

The method used in this problem is similar to that used in Example 5. We assume $\varphi(r, \theta) = R(r)\Theta(\theta)$, which after substituting into Laplace's equation

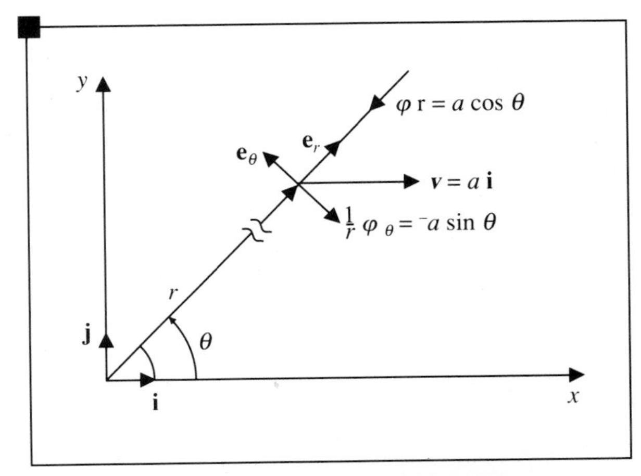

FIGURE 4 **Flow Around Cylinder**

in polar form yields the two ordinary differential equations and boundary conditions

$$\Theta'' + \lambda\Theta = 0, \qquad \Theta(\theta + 2\pi) = \Theta(\theta), \qquad \Theta'(\theta + 2\pi) = \Theta'(\theta) \tag{4.21}$$

$$r^2R'' + R' - \lambda R = 0, \qquad R'(c) = 0 \tag{4.22}$$

If $\lambda = \alpha^2 > 0$, then the general solution of $\Theta'' + \alpha^2\Theta = 0$ is

$$\Theta(\theta) = A \cos \alpha\theta + \beta \sin \alpha\theta$$

and

$$\Theta'(\theta) = -\alpha A \sin \alpha\theta + \alpha B \cos \alpha\theta$$

Substituting these two equations into the periodic conditions in Equation (4.20), we find (after some effort) that

$$\alpha = n \qquad n = 1, 2, 3, \ldots$$

or

$$\lambda_n = n^2$$

and

$$\Theta_n(\theta) = A_n \cos n\theta + B_n \sin n\theta$$

With these eigenvalues the differential equation in Equation (4.22) becomes

$$r^2R'' + rR' - n^2R = 0 \qquad n = 1, 2, \ldots$$

whose solution is

$$R_n(r) = C_n r^n + D_n r^{-n}$$

Using the boundary conditions in Equation (4.22), we have

$$R'(c) = 0 = nC_n c^{n-1} - nD_n c^{-n-1}$$

from which it follows that

$$D_n = c^{2n}C_n$$

For $\lambda = 0$, $\Theta'' = 0$, and its solution is

$$\Theta_0 = A_0 + B_0\theta$$

The solution to the differential equation in Equation (4.22) is

$$R_0 = C_0 \ln r + D_0$$

Now since $\ln r \to +\infty$ as $r \to +\infty$, we must set $C_0 = 0$ and we can let

$$R_0 = D_0 = 1$$

Since there are no negative eigenvalues, we expect the solution to take the form

$$\varphi(r, \theta) = A_0 + B_0\theta + \sum_{n=1}^{\infty} [r^n + c^{2n}r^{-n}][A_n \cos n\theta + B_n \sin n\theta] \tag{4.23}$$

We still have two conditions [Equation (4.20)] that must be met. Differentiating Equation (4.23) with respect to r and θ, we have

$$\varphi_r = \sum_{n=1}^{\infty} [nr^{n-1} - nc^{2n}r^{-n-1}][A_n \cos n\theta + B_n \sin n\theta] \tag{4.24}$$

and

$$\varphi_\theta = B_0 + \sum_{n=1}^{\infty} [r^n + c^{2n}r^{-n}]n[- A_n \sin n\theta + B_n \cos n\theta] \tag{4.25}$$

Now the only term in Equation (4.24) that is bounded occurs when $n = 1$; therefore, A_n, $B_n = 0$ for $n = 2, 3, \ldots$ and

$$\varphi_r \underset{r \to +\infty}{(r, \theta)} = a \cos \theta = A_1 \cos \theta + B_1 \sin \theta$$

or

$$A_1 = a$$
$$B_1 = 0$$

Under these conditions Equation (4.25) becomes

$$\frac{1}{r}\varphi_\theta = \frac{B_0}{r} + \frac{1}{r}[r + c^2 r^{-1}][- A_1 \sin \theta]$$
$$= \frac{B_0}{r} + \left[1 + \frac{c^2}{r^2}\right][- a \sin \theta]$$

As $r \to +\infty$,

$$\frac{1}{r}\varphi_\theta = - a \sin \theta$$

and the second condition in Equation (4.20) is satisfied.

Combining all these conditions in Equation (4.23), we write the solution as

$$\varphi(r, \theta) = A_0 + B_0 \theta + a\left[r + \frac{c^2}{r}\right]\cos \theta$$

where A_0 and B_0 are arbitrary constants.

As we investigate this solution we must recall that $\varphi(r, \theta)$ is the velocity *potential*, not the velocity. The velocity field is given by

$$\mathbf{v} = \varphi_r \mathbf{e}_r + \frac{1}{r}\varphi_\theta \mathbf{e}_\theta = a\left[1 - \frac{c^2}{r^2}\right]\cos \theta \mathbf{e}_r - a\left[1 + \frac{c^2}{r^2}\right]\sin \theta \mathbf{e}_\theta$$

When we measure the velocity far from the axis of the cylinder,

$$\mathbf{v} \approx a \cos \theta \mathbf{e}_r - a \sin \theta \mathbf{e}_\theta = a\mathbf{i}$$

In other words, the effect on the velocity due to the cylinder becomes less noticeable as we move away from the axis. On the other hand, when we are near the cylindrical obstruction, that is, $r \approx c$,

$$\mathbf{v} \approx -2a \sin \theta \mathbf{e}_\theta$$

which shows that the flow follows the shape of the cylinder when $r \approx c$ and approaches zero as we get nearer the x-axis.

Vibrations in a Sphere

In all our examples up to this point it has been relatively easy to evaluate the eigenvalues and eigenfunctions. We will now examine a problem in which this is not the case. We wish to study the vibrations or pressure of air within a sphere of radius c such as might be caused by an exploding firecracker.

The pressure equation in space is given by

$$(\rho^2 p_\rho)_\rho + \frac{1}{\sin^2 \theta} p_{\varphi\varphi} + \frac{1}{\sin \theta} (p_\theta \sin \theta)_\theta = \frac{\rho^2}{a^2} p_{tt} \tag{4.26}$$

If we assume the pressure is independent of θ and φ (i.e., in the radial direction only), Equation (4.26) becomes

$$\frac{1}{\rho^2} (\rho^2 p_\rho)_\rho = \frac{1}{a^2} p_{tt} \tag{4.27}$$

Since the pressure under consideration is within the sphere, the directional derivative of p in the direction ρ must be zero on the boundary or

$$p_\rho(c, t) = 0 \qquad t > 0$$

Choosing a trial solution of the form $p(\rho, t) = R(\rho)T(t)$, we arrive at the two ordinary differential equations using the method of separation of variables on Equation (4.27):

$$\rho^2 R'' + 2\rho R' + \lambda \rho^2 R = 0 \qquad R'(c) = 0 \tag{4.28}$$

$$T'' + a^2 \lambda T = 0 \tag{4.29}$$

The general solution of Equation (4.28) for $\lambda = \alpha^2 > 0$ is

$$R(\rho) = A \frac{\cos \alpha\rho}{\rho} + B \frac{\sin \alpha\rho}{\rho}$$

This solution can be found by replacing $R(\rho)$ by $S(\rho)/\rho$, which transforms the differential equation into one with constant coefficients (see Exercise 16). Since $p(\rho, t)$ must be bounded in the sphere, it is necessary that $A = 0$.

In order to satisfy the boundary condition at $\rho = c$ it follows that

$$R'(c) = \alpha c \cos \alpha c - \sin \alpha c = 0$$

or

$$\tan \mu = \mu \tag{4.30}$$

where $\mu = \alpha c$.

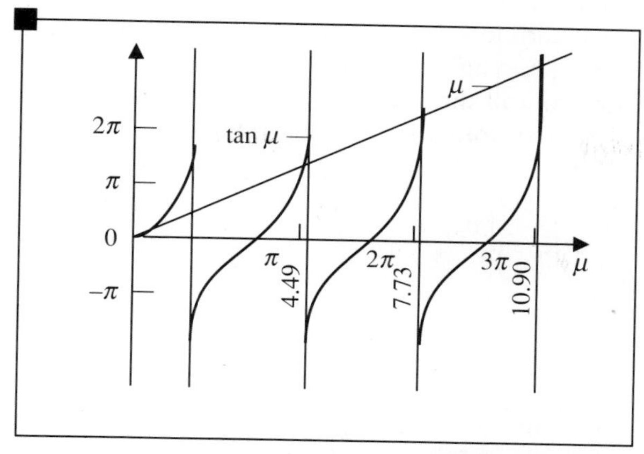

FIGURE 5 **Eigenvalues Satisfying tan μ = μ**

Unlike the previous examples this trigonometric equation is difficult to solve in order to find the eigenvalues. We can easily see that there are an infinite number of eigenvalues by solving the equation graphically, as shown in Figure 5. The first nonzero root is 4.49, the second 7.73, and so on. Notice in particular that as $\mu = \alpha c \to \infty$, the intersection points approach $(2n + 1)\pi/2$, $n = 1, 2, \ldots$. Table 1 lists the first eight nonzero solutions done numerically. Using these values for μ we see that the eigenvalues λ satisfy

$$\lambda_n = \alpha_n^2 = \frac{\mu_n^2}{c^2} \qquad n = 1, 2, \ldots$$

and the eigenfunctions are

$$R_n(\rho) = \frac{\sin(\mu_n \rho/c)}{\rho}$$

TABLE 1 **Solutions of tan μ = μ**

1	2	3	4	5	6	7	8
4.49	7.73	10.90	14.07	17.22	20.37	23.52	26.66

To see if $\lambda = 0$ is an eigenvalue, we solve the differential equation

$$R'' + \frac{2}{\rho} R' = 0$$

whose general solution is

$$R(\rho) = \frac{A}{\rho} + B$$

Once again since $R(\rho)$ must be bounded in the sphere, $A = 0$ and $R(\rho) = B$. Since $R'(\rho) = 0$ for any ρ, it certainly follows that $R'(c) = 0$. Therefore, $\lambda = 0$ is an eigenvalue and its corresponding eigenfunction is $R_0(\rho) = 1$. This ends our search for eigenvalues, for you can show there are no negative eigenvalues.

We now proceed to the solution of Equation (4.29). For $\lambda = \mu_n^2/c^2$, $n = 1, 2, \ldots$, its solution is

$$T_n(t) = C_n \cos \frac{a\mu_n t}{c} + D_n \sin \frac{a\mu_n t}{c}$$

and for $\lambda = 0$,

$$T_0(t) = C_0 + D_0 t$$

Combining the solutions R and T, we see that the answer to our boundary value problem takes the form

$$p(\rho, t) = C_0 + D_0 t + \sum_{n=1}^{\infty} \frac{\sin(\mu_n \rho/c)}{\rho} \times \left(C_n \cos \frac{a\mu_n t}{c} + D_n \sin \frac{a\mu_n t}{c}\right) \tag{4.31}$$

In order to solve for constants C_n and D_n, $n = 0, 1, 2, \ldots$, we need two initial conditions:

$$p(\rho, 0) = f(\rho) \qquad \text{and} \qquad \frac{\partial p(\rho, 0)}{\partial t} = g(\rho) \qquad 0 < \rho < c$$

When we attempt to solve for the coefficients, the series is not the Fourier series. But since the problem falls into Sturm–Liouville theory, the family of functions does possess an orthogonal property and we can solve for the coefficients as follows:

$$\begin{aligned} C_0 &= \frac{3}{c^3} \int_0^c \rho^2 f(\rho)\, d\rho, \qquad D_0 = \frac{3}{c^3} \int_0^c \rho^2 g(\rho)\, d\rho \\ C_n &= \frac{2(1 + \mu_n^2)}{c\mu_n^2} \int_0^c \rho f(\rho) \sin \frac{\mu_n \rho}{c}\, d\rho \\ D_n &= \frac{2(1 + \mu_n^2)}{a\mu_n^3} \int_0^c \rho g(\rho) \sin \frac{\mu_n \rho}{c}\, d\rho \end{aligned} \tag{4.32}$$

This problem can also be solved in terms of Bessel functions, which we discuss in Chapter 5.

SECTION 3 EXERCISES

13. You wish to find the temperature $u(\rho, \theta)$ in a laterally insulated pie-shaped region of radius c. The temperature satisfies the differential equation $\rho^2 u_{\rho\rho} + \rho u_\rho + u_{\theta\theta} = 0$

and the boundary conditions

$$u(\rho, 0) = 0 \qquad 0 < \rho < c$$

$$u\left(\rho, \frac{\pi}{6}\right) = 0 \qquad 0 < \rho < c$$

$$u(c, \theta) = \theta \qquad 0 < \theta < \frac{\pi}{6}$$

Use the technique of separation of variables to find the temperature $u(\rho, \theta)$.

14. Find the steady-state temperature $u(\rho, \theta)$ in a circular plate insulated laterally of radius 10 if the temperature on the circumference is $3 - \theta$.

15. Find the electrostatic potential $\Phi(\rho, \theta)$ on the plate shown in Figure 6. The numbers along the outside edges indicate potential.

16. Show that Equation (4.28) can be transformed into one with constant coefficients by using the substitution $R(\rho) = S(\rho)/\rho$.

17. Evaluate and graph the velocity $\mathbf{v}$ of the fluid around the cylinder discussed in the previous section for $a = 1$, $r = 2c$, $\theta = 0, \pi/4, \pi/2$ radians.

18. Apply the method of separation of variables to Laplace's equation in Example 5 to find the two ordinary differential equations in Equations (4.12) and (4.13).

19. Show that there are no negative eigenvalues to be found in the boundary value problem in Equation (4.21).

20. Prove that

$$\int_0^c \rho \sin \frac{\mu_n \rho}{c}\, d\rho = 0$$

where $\tan \mu_n = \mu_n$.

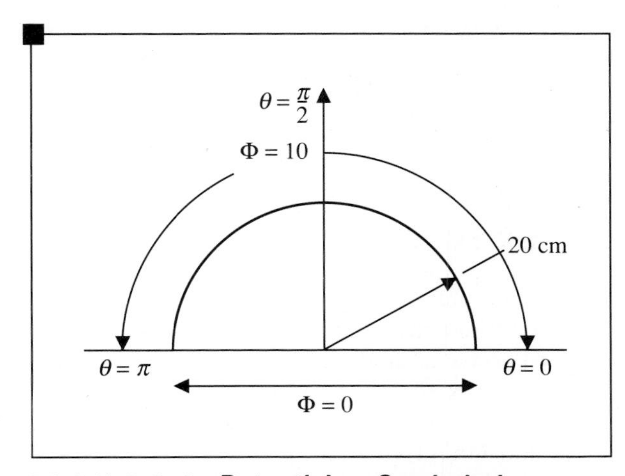

FIGURE 6 **Potential on Semi-circle**

21. Show that

(a) $\displaystyle\int_0^c \sin\frac{\mu_n\rho}{c}\sin\frac{\mu_m\rho}{c}\,d\rho = 0 \qquad m \neq n$

(b) $\displaystyle\int_0^c \sin^2\frac{\mu_n\rho}{c}\,d\rho = \frac{c}{2}\left(\frac{\mu m^2}{1+\mu m^2}\right) \qquad n = 1, 2, \ldots$

where $\tan\mu_n = \mu_n$.

22. Using the series in Equation (4.31) and the fact that $p(\rho, 0) = f(\rho)$, justify C_0 in Equation (4.32).

□ ***Hint.*** Multiply both sides of Equation (4.29) by $\rho^2 d\rho$ and integrate from 0 to c. □

23. Using the conditions in Exercise 22, justify C_n in Equation (4.32).

□ ***Hint.*** Multiply both sides of Equation (4.31) by $\rho\sin(\mu_m\rho/c)\,d\rho$ and integrate from 0 to c. □

24. The coefficient D_n can be found easily from the equation for C_n in Equation (4.32). Using the condition $\partial p(\rho, 0)/\partial t = g(\rho)$, prove D_n in Equation (4.32).

25. Find $p(\rho, t)$ if $c = 1$ and

$$p_\rho(1, t) = 0 \qquad t > 0$$

$$p(\rho, 0) = 0 \qquad 0 < \rho < 1$$

$$p_t(\rho, 0) = \begin{cases} \rho & 0 < \rho < \frac{1}{2} \\ 0 & \frac{1}{2} < \rho < 1 \end{cases}$$

26. Write out the boundary value problem for finding the pressure $p(\rho, t)$ in a spherical region if the pressure is zero at 100 meters from the center and the initial pressure is zero, and the rate of change of pressure with respect to t is $h(\rho)$.

27. Consider two concentric spheres of radius a and b, $a < b$, respectively. If the pressure on the inner sphere is 1 and the rate of change of pressure with respect to ρ on the outer sphere is 0, and the initial pressure is zero and the rate of change of pressure with respect to t is ρ, write the partial differential equation and boundary and initial conditions satisfied.

SECTION 4

MORE GENERAL INITIAL CONDITIONS

We will now examine the method of solution of boundary value problems where there are several initial conditions and two or more are nonhomogeneous. There are two well-known methods for solving such a problem.

The first approach consists of solving two boundary value problems, as we have done previously. Suppose we wish to solve a vibrating string problem where both ends are fixed and initially the string is not only displaced but a velocity is imparted to the string (see Figure 7). Formally, this problem is

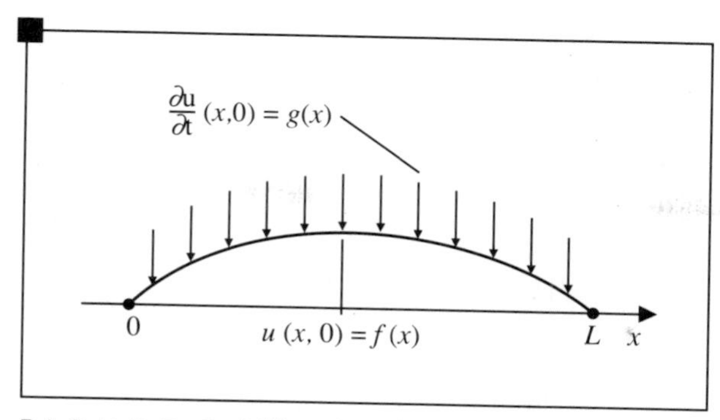

FIGURE 7 **Vibrating String**

written as

$$\left.\begin{array}{ll}\dfrac{\partial^2 u}{\partial t^2} = c^2 \dfrac{\partial^2 u}{\partial x^2} & 0 < x < L,\ 0 < t \\ u(0, t) = 0 & 0 < t \\ u(L, t) = 0 & 0 < t \\ u(x, 0) = f(x) & 0 < x < L \\ \dfrac{\partial u}{\partial t}(x, 0) = g(x) & 0 < x < L \end{array}\right\} \tag{4.33}$$

We now assume that the solution to this problem can be written in the form $u(x, t) = v(x, t) + w(x, t)$ where v and w satisfy the following boundary value problems, respectively:

$$\begin{array}{lll}\dfrac{\partial^2 v}{\partial t^2} = c^2 \dfrac{\partial^2 v}{\partial x^2} & \dfrac{\partial^2 w}{\partial t^2} = c^2 \dfrac{\partial^2 w}{\partial x^2} & \\ v(0, t) = 0 & w(0, t) = 0 & 0 < t \\ v(L, t) = 0 & w(L, t) = 0 & 0 < t \\ v(x, 0) = f(x) & w(x, 0) = 0 & 0 < x < L \\ \dfrac{\partial v}{\partial t}(x, 0) = 0 & \dfrac{\partial w}{\partial t}(x, 0) = g(x) & 0 < x < L \end{array}$$

It follows from the method of superposition that if v and w satisfy their boundary value problem, u satisfies the problem in Equation (4.33).

The second way of solving might be called a straightforward attack. We solve Equation (4.33) directly as we have done in earlier examples (see Example 3). The eigenvalue problem will be the same as before, but when we come to solve the other ordinary differential equation we will have no initial conditions. When we multiply X_n by T_n and then form the infinite linear combination to find $u(x, t)$, we will find it necessary to solve for two sets of arbitrary constants. Using one initial condition at a time, we are led to two Fourier series problems from which we evaluate our two sets of constants.

EXAMPLE 6. There is a vibrating string fastened to air bearings that move along two parallel rods 4 meters apart. Find the displacement $u(x, t)$ if the initial displacement is 1 meter and the initial velocity is x meter per second.

Solution. The formal statement of this problem is given as

$$\frac{\partial^2 u}{\partial x^2} = c^2 \frac{\partial^2 u}{\partial t^2} \qquad 0 < x < 4, 0 < t$$

$$\frac{\partial u}{\partial x}(0, t) = 0 \qquad 0 < t$$

$$\frac{\partial u}{\partial x}(4, t) = 0 \qquad 0 < t$$

$$u(x, 0) = 1 \qquad 0 < x < 4$$

$$\frac{\partial u}{\partial t}(x, 0) = x \qquad 0 < x < 4$$

Using the method of separation of variables with $u(x, t) = X(x)T(t)$, we see easily that

$$X'' + \lambda X = 0, \qquad X'(0) = 0, \qquad X'(4) = 0 \tag{4.34}$$

$$T'' + \frac{\lambda}{c^2} T = 0 \tag{4.35}$$

Solving the eigenvalue problem for Equation (4.34), we find the eigenvalues are

$$\lambda_n = \frac{n^2\pi^2}{16}$$

and the eigenfunctions are

$$X_n = \cos\frac{n\pi x}{4} \qquad n = 0, 1, 2, \ldots$$

Our next step is to solve the differential equation in Equation (4.35). Since λ is known, the solution of this equation is

$$T_0(t) = A_0 + B_0 t$$

$$T_n(t) = A_n \cos\frac{n\pi t}{4c} + B_n \sin\frac{n\pi t}{4c} \qquad n = 1, 2, \ldots$$

Unlike previous examples, no initial conditions are attached to the differential equation in Equation (4.35) and therefore we cannot evaluate the A_n's or B_n's at this stage. We continue by assuming a solution of the form

$$u(x, t) = \frac{1}{2}(A_0 + B_0 t) + \sum_{n=1}^{\infty}\left[A_n \cos\frac{n\pi t}{4c} + B_n \sin\frac{n\pi t}{4c}\right]\cos\frac{n\pi x}{4} \tag{4.36}$$

from which it follows that

$$\frac{\partial u}{\partial t}(x, t) = \frac{B_0}{2} + \sum_{n=1}^{\infty} \left(\frac{n\pi}{4c}\right)\left(-A_n \sin \frac{n\pi t}{4c} + B_n \cos \frac{n\pi t}{4c}\right) \cos \frac{n\pi x}{4}$$

Using the initial condition $u(x, 0) = 1$, we have

$$1 = \frac{A_0}{2} + \sum_{n=1}^{\infty} A_n \cos \frac{n\pi x}{4}$$

which is a Fourier cosine series whose coefficients are $A_0 = 2$ and $A_n = 0$. In the same way, knowing that $(\partial u/\partial t)(x, 0) = x$, we can write

$$x = \frac{B_0}{2} + \sum_{n=1}^{\infty} \frac{n\pi}{4c} B_n \cos \frac{n\pi x}{4}$$

on the interval (0, 4). This is a Fourier cosine series representing an even function by extending x into $(-4, 0)$ as $-x$. Our calculations yield

$$B_0 = 4 \qquad \text{and} \qquad B_n = \begin{cases} -\dfrac{64c}{n^3\pi^3} & n \text{ odd} \\ 0 & n \text{ even} \end{cases}$$

Substituting these coefficients into Equation (4.36), the solution to our problem is

$$u(x, t) = 1 + 2t - \frac{64c}{\pi^3} \sum_{1, 3, 5}^{\infty} \frac{1}{n^3} \sin \frac{n\pi t}{4c} \cos \frac{n\pi x}{4} \tag{4.37}$$

■

We see from the solution to Equation (4.37) that the center of the string, that is, $x = 2$, has the displacement $u(2, t) = 1 + 2t$. This fact tells us that once the string is put in motion the center moves away from $u(2, 0) = 1$ at a constant velocity of 2 meters per second. When $t = 4cq$ where $q = 0, 1, 2, \ldots$, the string becomes parallel to the x-axis.

The factor $\sin(n\pi t/4c)$ in Equation (4.37) allows us to determine the frequency of a vibrating string. The period λ_n of the individual terms in the series (4.37) is given by

$$\lambda_n = \text{period} = \frac{2\pi}{\dfrac{n\pi}{4a}} = \frac{8c}{n}$$

from which we can find the frequency f_n which is related to the period by

$$f_n = \frac{1}{\lambda_n} = \frac{n}{8c}$$

Now the smallest n allowed in the series (4.37), in this case $n = 1$, determines the **fundamental frequency**, that is

$$f_1 = \frac{1}{8c}$$

Notice that all higher frequencies are integer multiples of the fundamental frequency and are called **harmonics**. The value $n = 2$ is the second harmonic, $n = 3$ the third harmonic, and so on.

SECTION 5

NONHOMOGENEOUS DIFFERENTIAL EQUATIONS AND BOUNDARY CONDITIONS

Recall from Chapters 1 and 2 that in some boundary value problems the differential equation and the boundary conditions may be nonhomogeneous. In general, there is no straightforward way to solve these problems. However, if the nonhomogeneous part of the differential equation is a function of x and the boundary conditions are constants, there is a step-by-step way to solve such a problem. The method is best shown by example.

EXAMPLE 7. Suppose we wish to find the temperature $u(x, t)$ in a laterally insulated rod of length π whose initial temperature is $f(x)$. For $t > 0$, the left end of the rod is fixed at 500° and the right end is fixed at a temperature of 100°. Furthermore, for $t > 0$, an electric current is made to pass through the rod heating it, which introduces the $\sin x$ term in the partial differential equation.

Solution. Formally stated, this boundary value problem looks like

$$\frac{\partial u}{\partial t} = k\frac{\partial^2 u}{\partial x^2} + \sin x$$

$$u(0, t) = 500 \qquad 0 < t$$

$$u(\pi, t) = 100 \qquad 0 < t$$

$$u(x, 0) = f(x) \qquad 0 < x < \pi$$

We assume the solution can be broken into two parts; that is,

$$u(x, t) = v(x, t) + h(x) \tag{4.38}$$

Our plan of attack is to choose constants of integration so that eventually the $v(x, t)$ term will be a solution of a **homogeneous boundary value problem**.

Substituting the right-hand side of Equation (4.38) into the differential equation, we have

$$\frac{\partial v}{\partial t} = k\frac{\partial^2 v}{\partial x^2} + kh''(x) + \sin x$$

Therefore to follow our plan, $kh''(x) = -\sin x$. Solving this simple differential equation, we find

$$h(x) = C_1 x + C_2 + \frac{\sin x}{k} \tag{4.39}$$

Next we look at the boundary conditions, which can be written as

$$v(0, t) + h(0) = 500$$
$$v(\pi, t) + h(\pi) = 100$$

In order that $v(0, t)$ and $v(\pi, t)$ equal zero,

$$h(0) = 500 \qquad \text{and} \qquad h(\pi) = 100$$

We use these conditions to find the specific values of C_1 and C_2 in Equation (4.39). Therefore,

$$h(0) = 500 = C_1(0) + C_2 + \frac{\sin 0}{k}$$

which implies that $C_2 = 500$. Then,

$$h(\pi) = 100 = C_1\pi + 500 + \frac{\sin \pi}{k}$$

or

$$C_1 = -\frac{400}{\pi}$$

Substituting these constants in the solution [Equation (4.39)], it follows that

$$h(x) = -\frac{400}{\pi}x + 500 + \frac{\sin x}{k}$$

We have completely determined $h(x)$. Our next task is to find $v(x, t)$, which now satisfies the boundary value problem

$$\frac{\partial v}{\partial t} = k\frac{\partial^2 v}{\partial x^2}$$
$$v(0, t) = 0 \qquad 0 < t$$
$$v(\pi, t) = 0 \qquad 0 < t$$
$$v(x, 0) = u(x, 0) - h(x) = f(x) - \frac{\sin x}{k} + \frac{400}{\pi}x - 500 \qquad 0 < x < \pi$$

If $f(x) = (\sin x/k) + 500$, then $v(x, 0) = (400/\pi)x$. Using the method of separation of variables and setting $v(x, t) = X(x)T(t)$, we are led to the two differential equations and boundary conditions

$$X''(x) + \lambda X(x) = 0 \qquad X(0) = X(\pi) = 0$$
$$T'(t) + \lambda k T(t) = 0$$

The eigenvalues are $\lambda = n^2$, $n = 1, 2, \ldots$, and the eigenfunctions are $X_n(x) = \sin nx$. Solving the other equation, find $T_n(t) = e^{-kn^2t}$.

Combining this information, we expect our solution to be of the form

$$v(x, t) = \sum_{n=1}^{\infty} b_n e^{-kn^2t} \sin nx$$

And since

$$v(x, 0) = \frac{400}{\pi} x = \sum_{n=1}^{\infty} b_n \sin nx$$

we see that

$$b_n = \frac{2}{\pi} \int_0^{\pi} \frac{400_x}{\pi} \sin nx$$

$$= (-1)^{n+1} \frac{800}{n\pi}$$

The solution to the homogeneous boundary value problem is

$$v(x, t) = \frac{800}{\pi} \sum_{n=1}^{\infty} \frac{(-1)^n}{n} e^{-kn^2 t} \sin nx$$

The solution to the nonhomogeneous problem using the principle of superposition is

$$u(x, t) = \frac{800}{\pi} \sum_{n=1}^{\infty} \frac{(-1)^n}{n} e^{-kn^2 t} \sin nx - \frac{400}{\pi} x + 500 + \frac{\sin x}{k}$$ ■

This problem is one that commonly occurs in heat flow problems. We observe that the answer for $u(x, t)$ consists of two parts: the infinite series, which depends on x and t, and the remaining part $h(x)$, which depends only on x. Because of the exponential term in $v(x, t)$ we see that as $t \to +\infty$ the value of the series approaches zero. This part of the solution is called the **transient** solution because it passes away quickly. The other part of the solution $h(x)$ does not vary with time and is called the **steady-state** solution.

Once again we notice that because the factor n^2 appears in the exponential term in the series, the terms decrease quite rapidly in size for even small values of t. Therefore, we can often get a good approximation of the transient term by taking only one or two terms.

SECTION 5 EXERCISES

28. Given the wave equation $y_{tt} = c^2 y_{xx}$ and the conditions

$$\begin{aligned} &y(0, t) = 0 && 0 < t \\ &y(L, t) = 0 && 0 < t \\ &y(x, 0) = 10 && 0 < x < L \\ &y_t(x, 0) = -5 && 0 < x < L \end{aligned}$$

use the technique of separation of variables to find $y(x, t)$.

29. Find the voltage $e(x, t)$ on a high-frequency line of length 20 centimeters if both ends of the line are shorted. The initial voltage $e(x, 0)$ is $20(1 - x)$; $e_t(x, 0)$ is $20x$.

30. Find the longitudinal displacement of a rod of length L whose ends are free if the initial displacement is $x(L - x)$ and $u_t(x, 0)$ is 2.

31. Given the wave equation and the conditions

$$u_{tt} = c^2 u_{xx} + e^{-x}$$

$$u(0, t) = 100 \qquad 0 < t$$

$$u(L, t) = 50 \qquad 0 < t$$

$$u(x, 0) = 0 \qquad 0 < x < L$$

$$u_t(x, 0) = x \qquad 0 < x < L$$

and letting $u(x, t) = v(x, t) + \Phi(x)$, solve for Φ completely and write (but do not solve) the boundary value problem for $v(x, t)$.

32. The voltage $e(x, t)$ satisfies the differential equation $e_{xx} = RCe_t + Ax$. Using separation of variables, find e if

$$\frac{\partial e}{\partial x}(0, t) = 0 \qquad 0 < t$$

$$\frac{\partial e}{\partial x}(L, t) = 0 \qquad 0 < t$$

$$e(x, 0) = 50 \qquad 0 < x < L$$

33. The current in a submarine cable of length 1000 meters is given by

$$i_{xx} = RCi_t$$

If the conditions are

$$i(0, t) = 2a \qquad 0 < t$$

$$i(1000, t) = 0.1a \qquad 0 < t$$

$$i(x, 0) = 0a \qquad 0 < x < 1000$$

find the current $i(x, t)$.

34. Solve the heat equation with decomposition

$$u_{xx} - ku_t + A = 0$$

with conditions

$$u(0, t) = 0 \qquad 0 < t$$

$$u(L, t) = 0 \qquad 0 < t$$

$$u(x, 0) = 0 \qquad 0 < x < L$$

35. The equation for a vibrating string with external force is given by $y_{tt} = c^2 y_{xx} + F/\delta$, where $F = \delta x$ and where the left-hand end $x = 0$ is fixed and the slope of the right-hand end $x = L$ is zero. The initial displacement is $f(x)$ while initial velocity is zero. Let $y(x, t) = z(x, t) + \Phi(x)$. Solve for $\Phi(x)$ completely and write (but do not solve) the boundary value problem for $z(x, t)$.

36. Poisson's equation is given by $u_{xx} + u_{yy} = y$. If the conditions are

$$u(0, y) = \frac{5}{6} + \frac{y^3}{6} \qquad 0 < y < 1$$

$$u(1, y) = \frac{y^3}{6} - \frac{1}{6} \qquad 0 < y < 1$$

$$u(x, 0) = -\frac{1}{6} \qquad 0 < x < 1$$

$$u(x, 1) = 0 \qquad 0 < x < 1$$

solve for $u(x, y)$.

□ ***Hint.*** Let $u(x, y) = w(x, y) + g(y)$. □

37. The equation of a vibrating string with an external force is given by $u_{tt} = c^2 u_{xx} + \sin x$. The boundary and initial conditions are $u(0, t) = 0$, $u_x(L, t) = -1$, $u(x, 0) = 0$, and $u_t(x, 0) = f(x)$. Let $u(x, t) = w(x, t) + \Phi(x)$. Find $\Phi(x)$ so that $w(x, t)$ satisfies a homogeneous boundary value problem. Write out the differential equation and conditions satisfied by $w(x, t)$ but do not solve the equation for w.

SECTION 6*
STURM–LIOUVILLE THEORY

When we first discussed the method of separation of variables earlier in this chapter, we said that we need to consider only real eigenvalues. We now justify the conditions under which this statement is true.

Consider the general second-order boundary value problem, which can be written as follows:

$$[r(x)y'(x)]' + [p(x) + \lambda s(x)]y(x) = 0 \qquad \textbf{(4.40)}$$
$$a_1 y(a) + a_2 y'(a) = 0$$
$$b_1 y(b) + b_2 y'(b) = 0$$

Such a problem is called a **Sturm–Liouville problem**. We assume that the following conditions hold in this problem:

(a) r, p, and s are continuously differentiable on $[a, b]$.

(b) $r(x) > 0$ and $s(x) > 0$ on $[a, b]$.

(c) a_1, a_2, b_1, and b_2 are real constants. **(4.41)**

(d) At least a_1 or a_2 is not zero.

(e) At least b_1 or b_2 is not zero.

THEOREM 1. If Φ is a solution of the Sturm–Liouville problem corresponding to the eigenvalue λ, then λ is real.

Proof. Suppose λ is a complex number (i.e., $\lambda = \alpha + i\beta$). Under these circumstances Φ could be a complex function of a real variable. From the theory of such functions it is known that $\overline{\Phi'(x)} = \bar{\Phi}'(x)$. If we take the conjugate of the differential equation and boundary conditions in the Sturm–Liouville problem, we have

$$
\begin{aligned}
&[r(x)\bar{\Phi}'(x)]' + [p(x) + \bar{\lambda}s(x)]\bar{\Phi}(x) = 0 \\
&a_1\bar{\Phi}(a) + a_2\,\bar{\Phi}'(a) = 0 \\
&b_1\bar{\Phi}(b) + b_2\,\bar{\Phi}'(b) = 0
\end{aligned}
\tag{4.42}
$$

Thus, if Φ is a solution of the Sturm–Liouville problem corresponding to the eigenvalue λ, $\bar{\Phi}$ is also a solution corresponding to $\bar{\lambda}$.

Now if we multiply Equations (4.40) and (4.42) by $\bar{\Phi}(x)$ and $\Phi(x)$, respectively, and subtract one from the other, we find

$$(\lambda - \bar{\lambda})s(x)\Phi(x)\bar{\Phi}(x) = \Phi(x)[r(x)\bar{\Phi}'(x)]' - \bar{\Phi}(x)[r(x)\Phi'(x)]'$$

Integrating both sides of this equation from a to b, we have

$$(\lambda - \bar{\lambda}) \int_a^b s(x)\Phi(x)\bar{\Phi}(x)\,dx = \int_a^b \{\Phi(x)[r(x)\bar{\Phi}'(x)]' - \bar{\Phi}(x)[r(x)\Phi'(x)]'\}\,dx$$

The integral on the right-hand side can be evaluated by using integration by parts, from which it follows that

$$
\begin{aligned}
(\lambda - \bar{\lambda}) \int_a^b s(x)\Phi(x)\bar{\Phi}(x)\,dx &= r(x)[\Phi(x)\bar{\Phi}'(x) - \bar{\Phi}(x)\Phi'(x)]\Big|_a^b \\
&= r(b)[\Phi(b)\bar{\Phi}'(b) - \bar{\Phi}(b)\Phi'(b)] \\
&\quad - r(a)[\Phi(a)\bar{\Phi}'(a) - \bar{\Phi}(a)\Phi'(a)]
\end{aligned}
\tag{4.43}
$$

Now if we look at the boundary conditions that Φ and $\bar{\Phi}$ satisfy, we notice that since a_1 and a_2 are not both zero, nor are b_1 and b_2, it is necessary that

$$\Phi(b)\bar{\Phi}'(b) - \bar{\Phi}(b)\Phi'(b) = 0$$

and

$$\Phi(a)\bar{\Phi}'(a) - \bar{\Phi}(a)\Phi'(a) = 0$$

Using these results in Equation (4.43), we conclude that

$$(\lambda - \bar{\lambda}) \int_a^b s(x)\Phi(x)\bar{\Phi}(x)\,dx = 0 \tag{4.44}$$

Since $\Phi\bar{\Phi} = |\Phi|^2 > 0$ (Φ, $\bar{\Phi}$ are nontrivial solutions) and $s(x) > 0$ on $[a, b]$, the integral

$$\int_a^b s(x)\Phi(x)\bar{\Phi}(x)\,dx = 0$$

Therefore, it follows that from Equation (4.44),

$$\lambda - \bar{\lambda} = 0$$

or

$$\lambda = \bar{\lambda}$$

But if a number and its conjugate are equal, the number must be real. ■

A second important consequence of the Sturm–Liouville problem is contained in Theorem 2.

THEOREM 2. If Φ_m, Φ_n are solutions of the Sturm–Liouville problem whose respective eigenvalues λ_m, λ_n are not equal, then Φ_m, Φ_n satisfy the orthogonal condition

$$\int_a^b s(x)\Phi_m(x)\Phi_n(x)\,dx = 0 \qquad m \neq n$$

Proof. The proof is similar to that of Theorem 1. You are asked to prove Theorem 2 in Exercise 43. ■

Finally, Theorems 1 and 2 hold for less stringent boundary conditions of the Sturm–Liouville problem (see Table 2).

TABLE 2 **Alternative Boundary Conditions for Sturm–Liouville Problem**

Conditions	Type	Example
$r(a) = 0$ $b_1 y(b) + b_2 y'(b) = 0$	Singularity at $x = a$	Bessel's equation
$r(b) = 0$ $a_1 y(a) + a_2 y'(a) = 0$	Singularity at $x = b$	
$r(a) = 0$, $r(b) = 0$	Singularity at $x = a$ and $x = b$	Legendre equation
$r(a) = r(b)$ $y(a) = y(b)$ $y'(a) = y'(b)$	Periodic on (a, b)	Circular plate

We have shown that if the hypotheses of the Sturm–Liouville problem are satisfied then the eigenvalues are real and possess an orthogonal condition. Another property concerning the eigenvalues and eigenfunctions comes from the study of Sturmian theory. We will state a typical result.

THEOREM 3. Given the Sturm–Liouville problem in Equation (4.40), if the conditions in Equation (4.41) are met, then the following statements are true:

(a) There exists a real infinite set of eigenvalues $\lambda_0, \lambda_1, \lambda_2, \ldots, \lambda_n, \ldots$ such that $\lim_{n\to\infty} \lambda_n = \infty$.

(b) If Φ_n is the eigenfunction associated with the eigenvalue λ_n, $n = 0, 1, 2, \ldots$, then Φ_n has exactly n zeros on the interval $a < x < b$. ■

EXAMPLE 8. Let us examine the Sturm–Liouville problem

$$y''(x) + \lambda y(x) = 0$$
$$y'(0) = 0$$
$$y(\pi) = 0$$

For convenience, let $\lambda = \alpha^2 > 0$. The general solution of the differential equation is

$$y(x) = A \cos \alpha x + B \sin \alpha x \tag{4.45}$$

Differentiating the solution we find

$$y'(x) = -\alpha A \sin \alpha x + \alpha B \cos \alpha x \tag{4.46}$$

Applying the boundary condition to Equation (4.46), we can write

$$y'(0) = 0 = -\alpha A(0) + \alpha B$$

or

$$B = 0$$

Similarly, using the other boundary condition with Equation (4.45), we see that

$$y(\pi) = 0 = A \cos \alpha\pi$$

A cannot be zero since we are looking for nontrivial solutions. Therefore,

$$\alpha\pi = \frac{(2n+1)\pi}{2} \qquad n = 0, 1, 2, \ldots$$

and

$$\alpha_n = \frac{2n+1}{2}$$

or

$$\lambda_n = \alpha_n^2 = \left(\frac{2n+1}{2}\right)^2$$

These are the only eigenvalues, for it can be shown that the problem has no solution for $\lambda \leqq 0$.

Obviously, the eigenvalues are real and the $\lim_{n\to\infty} \lambda_n = \infty$, which corroborates part (a) of Theorem 3. The eigenfunctions are

$$\Phi_n(x) = \cos\left(\frac{2n+1}{2}\right)x \qquad n = 0, 1, 2, \ldots$$

Let us apply part (b) of Theorem 3 to $\Phi_3(x) = \cos(7/2)x$. We see from Figure 8 that there are exactly three zeros in $(0, \pi)$. ■

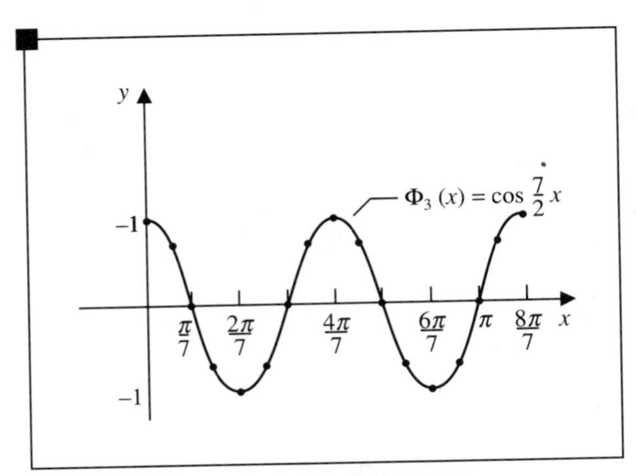

FIGURE 8 **Zeros of cos (7/2)x**

SECTION 6 EXERCISES

38. In order to use the theorems concerning Sturm–Liouville theory, the second-order linear differential equation must be in self-adjoint form, which is given by

$$[r(x)y'(x)]' + p(x)y(x) = 0$$

Show that if

$$a_0\, y''(x) + a_1 y'(x) + a_2\, y(x) = 0 \tag{4.47}$$

where $a_0(x) \neq 0$ and is differentiable over some interval (a, b), then the differential equation can be changed into self-adjoint form by multiplying both sides by

$$\frac{1}{a_0(x)} \exp \int \frac{a_1(x)}{a_0(x)}\, dx$$

□ ***Hint.*** Multiply Equation (4.47) by $u(x)$ and then use the fact that the coefficient of the first-order term is the derivative of the second-order term. □

39. Write the following differential equations in self-adjoint form.

(a) $y'' + 2y' + 3y = 0$
(b) $xy'' + y' + (x - \lambda)y = 0$
(c) $x^2y'' + xy' + 10y = 0$
(d) $x^2y'' + xy' + (x^2 - v^2)y = 0$ Bessel's equation
(e) $(1 - x^2)y'' - 2xy' + n(n + 1)y = 0$ Legendre's equation
(f) $y'' + \lambda y = 0$
(g) $y'' - xy = 0$ Airy's equation
(h) $(1 - x^2)y'' - xy' + n^2y$ Chebyshev's equation
(i) $(1 - x^2)y'' - 3xy' + n(n + 2)y$ Chebyshev's equation
(j) $xy'' + (1 - x)y' + ny = 0$ Laguerre's equation
(k) $y'' - 2xy' + 2ny = 0$ Hermite's equation

40. Show that the eigenvalues of the following boundary value problem are real.

$$x^2y'' + xy' + \lambda x^2 y = 0 \qquad \text{Bessel's equation}$$
$$y(c) = 0$$
$$|y(x)| < M \qquad \text{on } [0, c]$$

41. Show that the eigenvalues of the following boundary value problem are real.

$$(1 - x^2)y'' - 2xy' + \lambda y = 0 \qquad \text{Legendre's equation}$$
$$|y(x)| < M \qquad \text{on } [-1, 1]$$

42. (a) Show that the eigenvalues of the following boundary value problem are real and $\lim_{n\to\infty} \lambda_n = \infty$.

$$y'' + \lambda y = 0$$
$$y(0) = 0$$
$$y(3) = 0$$

(b) Make a sketch of y_0, y_2 and show that they have the requisite number of zeros in (0, 3).

43. Prove Theorem 2.

44. Given the following eigenvalue problems, find all eigenvalues and eigenfunctions. In certain cases find the equation to be solved to find eigenvalues but do not solve it.

(a) $y'' + 2y' - 3\lambda y = 0 \qquad y(0) = 0,\ y(L) = 0$
(b) $y'' + 2y' - 3\lambda y = 0 \qquad y'(0) = 0,\ y(L) = 0$
(c) $y'' - (1 + \lambda)y = 0 \qquad y(0) = 0,\ y(L) = 0$
(d) $y'' + \lambda y = 0 \qquad y(0) + y'(0) = 0,\ y'(L) = 0$
(e) $y'' + \lambda y = 0 \qquad y(0) + y'(0) = 0,\ y(1) - y'(1) = 0$
(f) $x^2y'' + xy' + \lambda y = 0 \qquad y(1) = 0,\ y(2) = 0$

SECTION 7
D'ALEMBERT'S SOLUTION

Using the method of separation of variables, we can obtain series solutions to many boundary value problems. This representation does not usually give us a direct insight into the properties of the solution, although we can use it to obtain an approximation of the exact solution by summing a finite number of terms in the series.

In some instance, however, we can obtain a closed-form solution to a boundary value problem. It is well-known that the wave equation in one dimension falls into this category. The method used to solve this equation is known as d'Alembert's solution. We illustrate this method in Example 9.

EXAMPLE 9. Solve in closed form the following boundary value problem:

$$\frac{1}{c^2}\frac{\partial^2 u}{\partial t^2} = \frac{\partial^2 u}{\partial x^2} \qquad 0 < x < a,\ 0 < t \tag{4.48}$$

$$u(0, t) = u(a, t) = 0 \tag{4.49}$$

$$u(x, 0) = f(x), \qquad \frac{\partial u}{\partial t}(x, 0) = g(x) \tag{4.50}$$

Solution. First apply a change of variables,

$$w = x + ct, \qquad z = x - ct \tag{4.51}$$

to the wave equation. Letting $\bar{u}(w, z) = u(x, t)$ and using the chain rule for multivariate functions, we obtain

$$\frac{\partial u}{\partial x} = \frac{\partial \bar{u}}{\partial w}\frac{\partial w}{\partial x} + \frac{\partial \bar{u}}{\partial z}\frac{\partial z}{\partial x} = \frac{\partial \bar{u}}{\partial w} + \frac{\partial \bar{u}}{\partial z}$$

Letting $h_1 = \partial\bar{u}/\partial w$ and $h_2 = \partial\bar{u}/\partial z$, we compute

$$\frac{\partial^2 u}{\partial x^2} = \frac{\partial h_1}{\partial x} + \frac{\partial h_2}{\partial x} = \frac{\partial h_1}{\partial w} + \frac{\partial h_1}{\partial z} + \frac{\partial h_2}{\partial w} + \frac{\partial h_2}{\partial z} = \frac{\partial^2 \bar{u}}{\partial w^2} + 2\,\frac{\partial^2 \bar{u}}{\partial w \partial z} + \frac{\partial^2 \bar{u}}{\partial z^2}$$

Similarly, we can show that

$$\frac{1}{c^2}\frac{\partial^2 u}{\partial t^2} = \frac{\partial^2 \bar{u}}{\partial w^2} - 2\,\frac{\partial^2 \bar{u}}{\partial w \partial z} + \frac{\partial^2 \bar{u}}{\partial z^2}$$

It follows then that, in these new coordinates, Equation (4.48) is equivalent to

$$\frac{\partial^2 \bar{u}}{\partial z \partial w} = 0 \tag{4.52}$$

In order to solve this differential equation, we use partial integration twice. Integrating once with respect to z, we have

$$\frac{\partial \bar{u}}{\partial w} = \int \frac{\partial^2 \bar{u}}{\partial z \partial w}\, dz = \int 0 \, dz = \Phi(w)$$

Integrating again with espect to w, we find

$$\bar{u} = \int \Phi(w)\, dw = F(w) + H(z)$$

where F and H are any twice differentiable functions, or in terms of x and t,

$$u(x, t) = F(x + ct) + H(x - ct)$$

Making use of the two initial conditions in Equation (4.50), we have

$$u(x, 0) = f(x) = F(x) + H(x) \tag{4.53}$$

$$\frac{\partial u}{\partial t}(x, 0) = g(x) = c[F'(x) - H'(x)] \tag{4.54}$$

Integrating Equation (4.54) leads to

$$F(x) - H(x) = \frac{1}{c}\int_0^x g(s)\, ds + C_1 \tag{4.55}$$

Equations (4.53) and (4.55) can be solved for $F(x)$ and $H(x)$, which gives

$$F(x) = \frac{f(x)}{2} + \frac{1}{2c}\int_0^x g(s)\, ds + \frac{C_1}{2} \qquad 0 < x < a \tag{4.56}$$

and

$$H(x) = \frac{f(x)}{2} - \frac{1}{2c}\int_0^x g(s)\,ds - \frac{C_1}{2}$$

These formulas determine F and H only on $[0, a]$. However, to apply Equation (4.56) we need these functions defined on $-\infty < x < \infty$. To determine the appropriate continuation of these functions outside the interval $[0, a]$, let

$$G(x) = \frac{1}{2c}\int_0^x g(s)\,ds$$

and apply the two remaining boundary conditions,

$$\begin{aligned} u(0, t) = 0 &= F(ct) + H(-ct) \\ &= \frac{1}{2}[f(ct) + f(-ct)] + \frac{1}{2}[H(ct) - H(-ct)] \end{aligned} \tag{4.57}$$

and

$$u(a, t) = 0 = F(a + ct) + H(a - ct) \tag{4.58}$$

Since the two initial conditions are independent of each other (i.e., the choice of f does not constrain the choice of g), we infer from Equation (4.57) that

$$f(ct) = -f(-ct), \qquad H(ct) = H(-ct) \tag{4.59}$$

Thus, the continuation of f and H outside $[0, a]$ (which we continue to denote as f and H) must be an odd and even function, respectively. Since f and g are known on $[0, a]$, we can use the relations in Equation (4.59) to determine f and H explicitly on $[-a, a]$.

Finally, the second boundary condition [Equation (4.58)] implies that

$$f(a + ct) + f(a - ct) + H(a + ct) - H(a - ct) = 0$$

or

$$f(a + ct) = -f(a - ct), \qquad H(a + ct) = H(a - ct)$$

because f and g are independent of each other. Since this must hold for all t, we infer that f and H must be continued as periodic functions of $2a$.

We have determined F, G, and H for all real x and have obtained a complete solution to the boundary value problem which can be written as

$$u(x, t) = \frac{f(x + ct) + f(x - ct)}{2} + \frac{1}{2c}\int_{x-ct}^{x+ct} g(s)\,ds \tag{4.60}$$

■

Observe that in this method we first apply the initial conditions and then the boundary conditions. This approach is in contrast to the method of separation of variables, in which we apply the boundary conditions first and then the initial conditions.

We now look at two interesting examples.

EXAMPLE 10. Find a solution graphically of the boundary value problem

$$\frac{\partial^2 u}{\partial t^2} = \frac{\partial^2 u}{\partial x^2} \qquad c = 1$$

$$u(0, t) = u(12, t) = 0$$

$$u(x, 0) = \begin{cases} 2x & 0 \le x \le 2 \\ \dfrac{24}{5} - \dfrac{2}{5}x & 2 \le x \le 12 \end{cases}$$

$$\frac{\partial u}{\partial t}(x, 0) = 0$$

(see Figure 9).

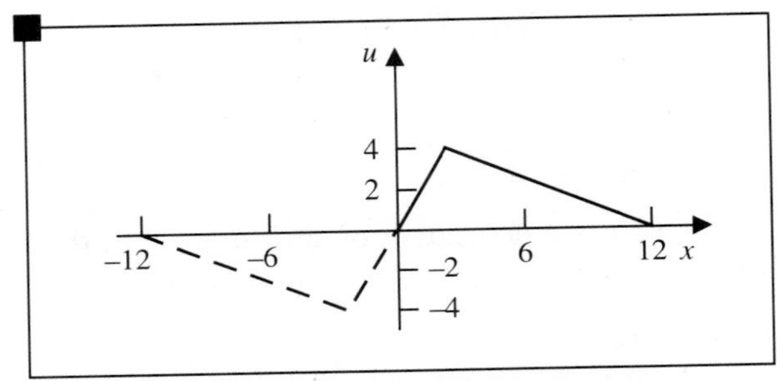

FIGURE 9 **Initial Displacement of String in Example 10**

Solution. In order to find the solution (i.e., the shape of the vibrating string) when $t = 4$, we proceed as follows. Using Equation (4.60) we write

$$u(x, 4) = \frac{f(x + 4) + f(x - 4)}{2}$$

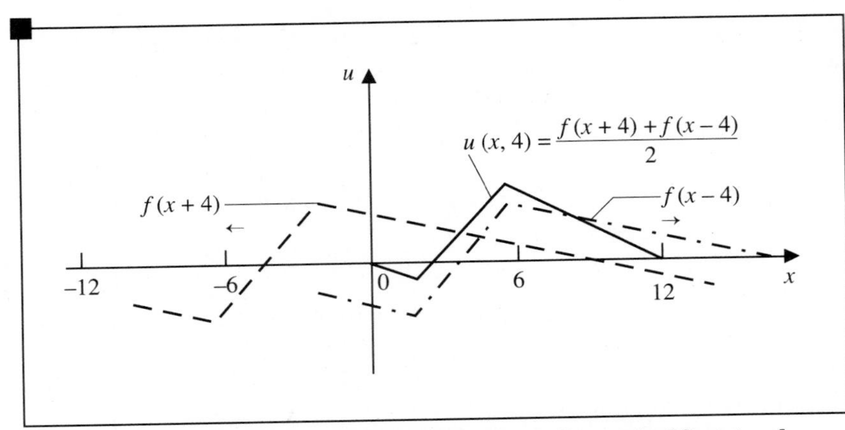

FIGURE 10 **Displacement of String in Example 10 at $t = 4$**

We can easily draw $\frac{1}{2}f(x+4)$ and $\frac{1}{2}f(x-4)$ by reducing the height of $f(x)$ by 2 and then translating this curve 4 units to the left and 4 units to the right. Our final answer is found by adding $\frac{1}{2}f(x+4)$ and $\frac{1}{2}f(x-4)$ graphically (see Figure 10).

EXAMPLE 11. Suppose we are looking for the solution to the problem

$$\frac{\partial^2 u}{\partial t^2} = \frac{\partial^2 u}{\partial x^2} \qquad c = 1$$

$$u(0, t) = u(12, t) = 0$$

$$u(x, 0) = 0$$

$$\frac{\partial u}{\partial t}(x, 0) = \begin{cases} 0 & 0 \leq x < 5 \\ 1 & 5 < x < 7 \\ 0 & 7 < x \leq 12 \end{cases}$$

(see Figure 11).

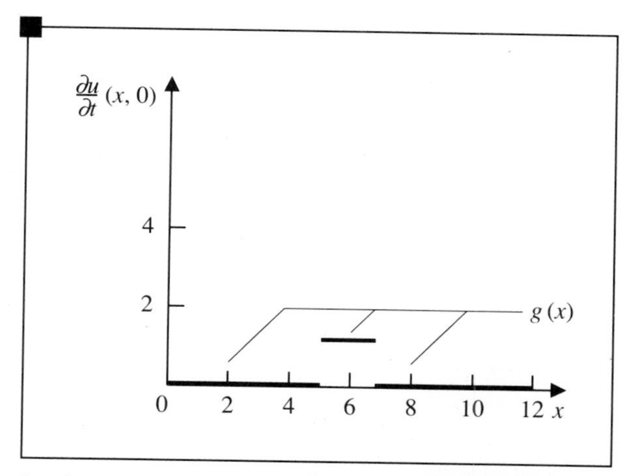

FIGURE 11 **Initial Velocity of String in Example 11**

Solution. The displacement of the vibrating string when $t = 2$ is established by the following method and is shown in Figure 11. We see from Equation (4.60) that

$$u(x, 2) = \frac{1}{2}\int_{x-2}^{x+2} g(s)\, ds \tag{4.61}$$

In what follows we must remember that if $f(x) \equiv 0$, then $g(x)$ must be extended as an odd function since $H(x)$ is even. To find the displacement at specific values of x, say $x = 0, 1, 2$, and so on, we evaluate

$$u(0, 2) = \frac{1}{2}\int_{-2}^{2} g(s)\, ds = \frac{1}{2}\int_{-2}^{2} 0\, ds = 0$$

$$u(1, 2) = \frac{1}{2}\int_{-1}^{3} g(s)\, ds = \frac{1}{2}\int_{-1}^{3} 0\, ds = 0$$

$$u(2, 2) = u(3, 2) = 0$$

$$u(4, 2) = \frac{1}{2}\int_{2}^{6} g(s)\, ds = \frac{1}{2}\int_{5}^{6} ds = \frac{1}{2}$$

$$u(5, 2) = \frac{1}{2}\int_{3}^{7} g(s)\, ds = \frac{1}{2}\int_{3}^{7} ds = 1$$

In a similar fashion we can show that

$$u(6, 2) = u(7, 2) = 1$$

$$u(8, 2) = \frac{1}{2}$$

$$u(9, 2) = u(10, 2) = u(11, 2) = u(12, 2) = 0$$

Plotting these points as shown in Figure 12, we can sketch the form of the string at $t = 2$. ■

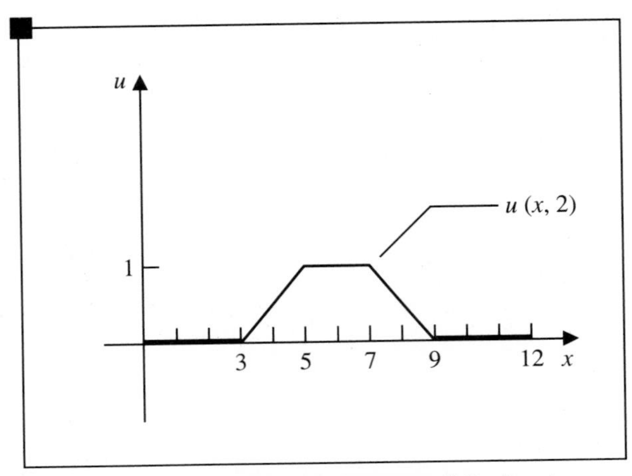

FIGURE 12 **Displacement of String in Example 11 at $t = 2$**

SECTION 7 EXERCISES

45. Carry out the transformation from Equation (4.48) to Equation (4.49).

46. Sketch the wave form using d'Alembert's solution for $t = 1, 2, 4$ and $c = 1$ if the length of wire is 16 and

$$u(0, t) = 0$$

$$u(16, t) = 0$$

$$u(x, 0) = f(x) = \begin{cases} x & 0 \le x \le 4 \\ -\dfrac{x}{3} + \dfrac{16}{3} & 4 \le x \le 16 \end{cases}$$

$$\frac{\partial u(x, 0)}{\partial t} = 0$$

47. Sketch the wave form on an infinite string for $t = 0, 2, 8, 20$ and $c = 0.1\pi$ if

$$u(x, 0) = \begin{cases} \cos x & -\dfrac{\pi}{2} \le x \le \dfrac{\pi}{2} \\ 0 & \text{elsewhere} \end{cases}$$

$$\frac{\partial u}{\partial t}(x, 0) = 0$$

48. Sketch the wave form on a string for $t = 0, 2, 12$ and $c = 1$ if the length of the string is 12 and

$$u(-6, t) = 0$$

$$u(6, t) = 0$$

$$u(x, 0) = \begin{cases} \dfrac{x}{2} + 3 & -6 < x \le -4 \\ 1 & -4 \le x \le 4 \\ -\dfrac{x}{2} + 3 & 4 \le x \le 6 \end{cases}$$

$$\frac{\partial u}{\partial t}(x, 0) = 0$$

49. Using d'Alembert's solution, sketch the wave form on a string for $t = 1, 2, 4$ and $c = 1$ if the length of the string is 8 and

$$u(0, t) = 0$$

$$u(8, t) = 0$$

$$u(x, 0) = 0$$

$$\frac{\partial u}{\partial t}(x, 0) = \begin{cases} x & 0 \le x \le 4 \\ 8 - x & 4 \le x \le 8 \end{cases}$$

50. An infinite string at rest is given an initial velocity $g(x)$ where

$$g(x) = \begin{cases} 1 & -5 \leq x \leq 5 \\ 0 & \text{elsewhere} \end{cases}$$

Sketch the wave form for $t = 0, 0.5, 1, 4$ if $c = 2$.

51. The string on a piano is 1 meter in length and c is 20 meters per second. A pianist strikes a key so as to impart a velocity $g(x)$ where

$$g(x) = \begin{cases} 10 & 0.6 < x < 0.7 \\ 0 & 0 \leq x < 0.6,\ 0.7 < x \leq 1 \end{cases}$$

Sketch the wave form of the piano string for $t = 2, 5, 8$ seconds.

52. If $f(x)$ and $G(x)$ are odd and even functions, respectively, show that if

$$f(a + ct) = -f(a - ct), \qquad G(a + ct) = G(a - ct)$$

then f and G are periodic functions of period $2a$.

53. Derive the solution $u(x, t)$ found in Equation (4.60) using the fact that $u(x, t) = F(x + ct) + G(x - ct)$ and F and G are defined for all real numbers.

54. If $f(x) \equiv 0$ and $G(x)$ is an even function, show that $g(x)$ defined on $[0, a]$ must be extended over $[-a, a]$ as an odd function.

CHAPTER FIVE

Bessel Functions

SECTION 1
INTRODUCTION

In Chapter 4 we looked at boundary value problems that were set in one-dimensional space and one-dimensional time, or steady-state problems set in rectangular coordinates. Using the method of separation of variables, we eventually found it necessary to find the Fourier coefficients associated with the orthogonal set of functions made up of sines and cosines.

When we solve boundary value problems whose shape is circular, we are often led to an ordinary differential equation known as Bessel's equation, whose solutions are called Bessel functions. In this chapter we will show that the Bessel functions form an orthogonal set that is useful in solving this type of problem. In Chapter 6 we will examine problems whose basic shape is spherical; and in solving them we are led to Legendre's equation and Legendre polynomials.

SECTION 2
TEMPERATURE IN A DISK

Suppose there is a homogeneous circular plate of radius c, whose specific heat and thermal conductivity are constant. This plate is heated in such a way that its initial temperature is a function of r alone. At $t = 0$ the plate is placed between two layers of perfect insulation and the temperature on the circumference is set equal to zero. We wish to find the temperature $u(r, t)$, assuming it depends only on r and t (see Figure 1).

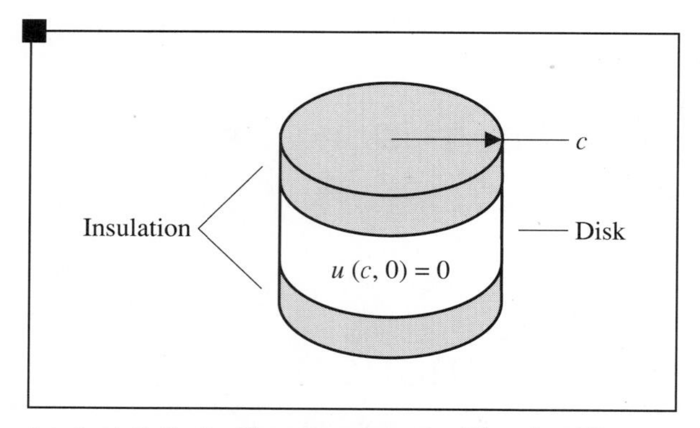

FIGURE 1 **Temperature in Circular Plate**

More formally, the problem can be stated as the solution to the differential equation

$$\frac{\partial u}{\partial t} = k\left[\frac{\partial^2 u}{\partial r^2} + \frac{1}{r}\frac{\partial u}{\partial r}\right] \qquad 0 < r < c,\ t > 0$$

subject to the boundary condition

$$u(c, t) = 0 \qquad t > 0$$

and the initial condition

$$u(r, 0) = f(r) \qquad 0 < r < c$$

We proceed as in Chapter 4 by assuming the solution is of the form $u(r, t) = R(r)T(t)$. When we substitute u into the differential equation, we find

$$RT' = k\left(R''T + \frac{1}{r}R'T\right)$$

or

$$\frac{1}{k}\frac{T'}{T} = \frac{R''}{R} + \frac{1}{1}\frac{R'}{R} = -\lambda$$

from which we ascertain the two ordinary differential equations

$$r^2R'' + rR' + \lambda r^2 R = 0 \qquad \textbf{(5.1)}$$

$$T' + \lambda kT = 0 \qquad \textbf{(5.2)}$$

It is easy to show that the boundary condition attached to Equation (5.1) is $R(c) = 0$.

When we attempt to solve Equation (5.1), we will find it is associated with Bessel's equation. In the next sections we will review the method used in solving this equation and some of the properties associated with these solutions.

SECTION 3
SOLUTION OF BESSEL'S EQUATION

The ordinary differential equation

$$x^2y''(x) + xy'(x) + (x^2 - \nu^2)y(x) = 0 \tag{5.3}$$

is known as **Bessel's equation of order ν** (ν can be any real or complex number). The parameter ν is normally zero, or positive if ν is real.

In order to solve this equation, we recognize that $x = 0$ is a **regular singular point*** and, therefore, we can solve the differential equation using Frobenius's series

$$y(x) = x^r \sum_{n=0}^{\infty} c_n x^n$$

Substituting this series into Equation (5.3)‡, we arrive first at the **indicial equation**, which in this problem is

$$r^2 - \nu^2 = 0$$

or

$$r = \pm\nu$$

From the general theory for solving a differential equation using a Frobenius series, we know that if the difference between the two roots is *not* zero or an integer, then we can find two linearly independent solutions, the first by using $r = \nu$ and the second by using $r = -\nu$.

On the other hand, if the difference between the two roots is zero or an integer, then both roots yield essentially the same solution. In order to find the second linearly independent solution in this case, we can use the method of reduction of order.

In all cases as we continue solving, we are lead to the **recursion formula**

$$c_n = -\frac{1}{(n + r + \nu)(n + r - \nu)} c_{n-2}$$

Now if $\nu = n$ is zero or a positive integer, a solution of Bessel's equation is

$$J_n(x) = \sum_{k=0}^{\infty} \frac{(-1)^k}{k!(k+n)!} \left(\frac{x}{2}\right)^{n+2k} \qquad n = 0, 1, 2, \ldots$$

---

* Given the second-order ordinary differential equation

$$a_0(x)y''(x) + a_1(x)y'(x) + a_2(x)y(x) = 0$$

a real value of $x = c$ is a **singular** point if $a_0(c) = 0$. Furthermore, it is **regular** if

$$\lim_{x \to c} \frac{1}{x} a_1(x) \qquad \text{and} \qquad \lim_{x \to c} \frac{1}{x^2} a_2(x)$$

exist.

‡ This method can be found in any good beginning text on ordinary differential equations.

On the other hand, if $v = -n$ where n is zero or a positive integer,

$$J_{-n}(x) = \sum_{k=0}^{\infty} \frac{(-1)^k}{k!(k-n)!} \left(\frac{x}{2}\right)^{-n+2k}$$

$J_n(x)$ and $J_{-n}(x)$ are known as **Bessel functions of the first kind of order n and $-n$**, respectively.

EXAMPLE 1. What is the series expansion of $J_0(x)$?

Solution

$$J_0(x) = 1 - \left(\frac{x}{2}\right)^2 + \frac{1}{4}\left(\frac{x}{2}\right)^4 - \frac{1}{36}\left(\frac{x}{2}\right)^6 + \cdots \qquad \blacksquare$$

The other linear independent solution of Bessel's equation is denoted by $Y_n(x)$ and is called a **Bessel function of the second kind of order n**. Now all Y_n's have the property that $\lim_{x\to 0} Y_n(0) = \pm\infty$; that is, $Y_n(x)$ is unbounded when $x \to 0$ on the interval $(0, c)$ where c is any positive real number. The graphs of $J_0(x)$, $J_1(x)$, $Y_0(x)$, and $Y_1(x)$ are shown in Figure 2. The general solution of Bessel's equation if $v = n$ (zero or a positive integer) is

$$y(x) = AJ_n(x) + BY_n(x)$$

SECTION 3 EXERCISES

1. Use the method of Frobenius to find the indicial equation for Bessel's equation,

$$x^2y''(x) + xy'(x) + (x^2 - v^2)y(x) = 0$$

2. Continuing from Exercise 1, find the recursion formula for Bessel's equation.
3. Using the recursion formula, write out the first four terms of
   (a) J_1
   (b) J_4
   (c) J_{10}
4. Using the series solution for $J_n(x)$, write the first three terms of
   (a) J_1
   (b) J_5
   (c) J_{20}
5. Prove $J_n(x)$ is an odd function if n is odd.
6. Prove $J_n(x)$ is an even function if n is even.
7. Show $J_0(0) = 1$.
8. Show $J_n(0) = 0, \quad n = 1, 2, 3, \ldots$.

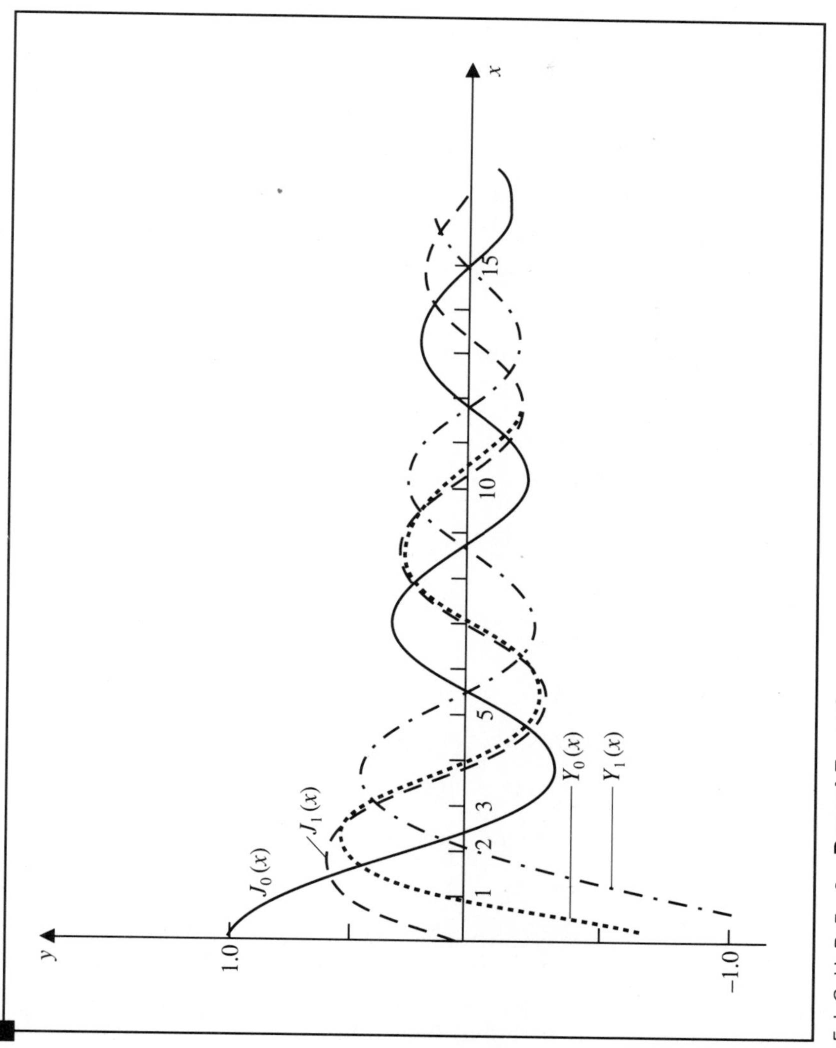

FIGURE 2 **Bessel Functions**

SECTION 4
THE GAMMA FUNCTION

To write the solution of Bessel's equation if v is not equal to zero or an integer, we must use the **gamma function**, which is defined as

$$\Gamma(v) = \int_0^\infty e^{-t}t^{v-1}\,dt \qquad v > 0$$

The parameter v must be positive in order that the improper integral will converge. We can compute some values directly, such as

$$\Gamma(1) = \int_0^\infty e^{-t}t^0\,dt = \lim_{b\to\infty}\int_0^b e^{-t}\,dt$$

$$= \lim_{b\to\infty} -e^{-t}\Big|_0^b = \lim_{b\to+\infty}(1 - e^{-b}) = 1$$

Using integration by parts, we can derive the fundamental identity concerning gamma functions. If we let

$$dv = e^{-t}\,dt \qquad u = t^{v-1}$$

$$v = -e^{-t} \qquad du = (v-1)t^{v-2}\,dt$$

then

$$\Gamma(v) = \int_0^\infty e^{-t}t^{v-1}\,dt = -t^{v-1}e^{-t}\Big|_0^\infty + (v-1)\int_0^\infty e^{-t}t^{v-2}\,dt \qquad v > 1$$

$$\Gamma(v) = (v-1)\Gamma(v-1) \qquad v > 1 \tag{5.4}$$

If we apply this identity over and over again, we arrive at the result

$$\Gamma(v) = (v-1)(v-2)\ldots(v-r)\Gamma(v-r) \qquad v > r$$

Now if $v = n$ is a positive integer,

$$\Gamma(n) = (n-1)(n-2)\ldots 1 = (n-1)! \tag{5.5}$$

Because of this result, $\Gamma(v)$ is sometimes known as the **generalized factorial.**

Using the formula in Equation (5.5), it is possible to find the value of $\Gamma(n)$ where n is any positive integer. But how do we find $\Gamma(v)$ when v is a positive nonintegral value? In order to find such values it is necessary to evaluate $\Gamma(v)$ by numerical or series methods for a sufficient quantity of real numbers between two integers. The integers chosen are usually 1 and 2. In Figure 3 notice that the gamma function is continuous between 1 and 2 and the range is between 0 and 1. Such an interval makes a convenient table. A table of gamma function values is given in Table 1. It is possible to find the value of the gamma function for any real number greater than zero by repeated use of Equation (5.4) along with the appropriate choice from Table 1.

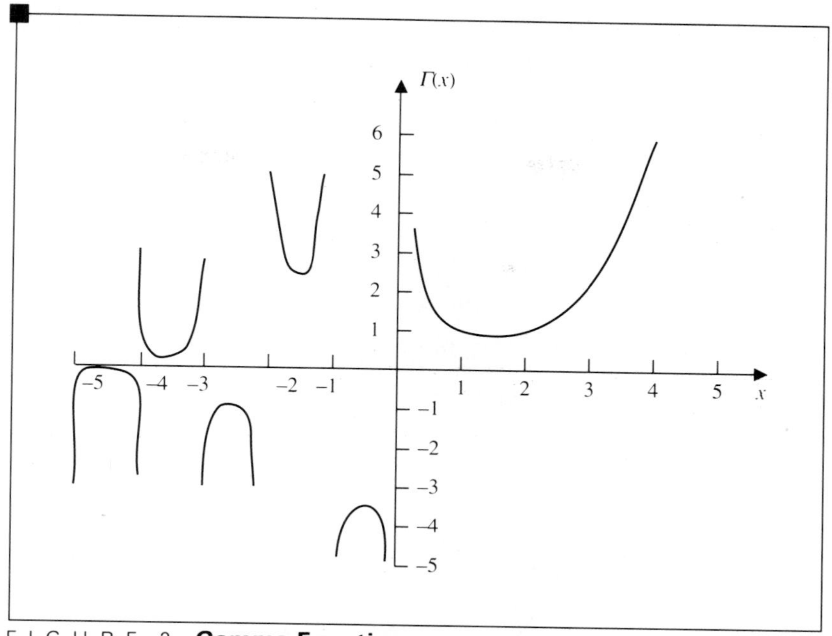

FIGURE 3 **Gamma Function**

TABLE 1 **Values of Gamma Function $\Gamma(v)$**

$$\Gamma(v) = \int_0^\infty e^{-t} t^{v-1}\, dt$$

v	$\Gamma(v)$
1.00	1.000
1.05	0.974
1.10	0.951
1.15	0.933
1.20	0.918
1.25	0.906
1.30	0.897
1.35	0.891
1.40	0.887
1.45	0.886
1.50	0.886
1.55	0.889
1.60	0.894
1.65	0.900
1.70	0.909
1.75	0.919
1.80	0.931
1.85	0.946
1.90	0.962
1.95	0.980
2.00	1.000

gamma function for any real number greater than zero by repeated use of Equation (5.4) along with the appropriate choice from Table 1.

EXAMPLE 2. Evaluate $\Gamma(3)$, $\Gamma(10)$, $\Gamma(3.7)$, and $\Gamma(0.7)$.

Solution.

$$\Gamma(3) = 2.1 = 2$$

$$\Gamma(10) = 9!$$

To find $\Gamma(3.7)$ it is necessary to use Equation (5.4). Therefore,

$$\Gamma(3.7) = 2.7\Gamma(2.7) = (2.7)(1.7)\Gamma(1.7) = (2.7)(1.7)(0.908) = 4.168$$

To find $\Gamma(0.7)$ we write Equation (5.4) in the form

$$\Gamma(v - 1) = \frac{\Gamma(v)}{v - 1}$$

Therefore,

$$\Gamma(0.7) = \frac{\Gamma(1.7)}{0.7} = \frac{0.008}{0.7} = 1.30$$ ■

What is $\Gamma(v)$ when v is a negative number (not an integer)? Earlier we stated that the integral definition of $\Gamma(v)$ applies when $v > 0$. In order to extend the definition of $\Gamma(v)$ to negative numbers, we require that the identity

$$\Gamma(v) = (v - 1)\Gamma(v - 1)$$

hold for v negative and not an integer. It is easier to use this identity if it is written in the form

$$\Gamma(v) = \frac{\Gamma(v + 1)}{v} \qquad v \neq 0 \tag{5.6}$$

EXAMPLE 3. What is the value of $\Gamma(-1.3)$?

Solution. Using the formula in Equation (5.6), we can write

$$\Gamma(-1.3) = \frac{\Gamma(-0.3)}{-1.3} = \frac{\Gamma(0.7)}{(-1.3)(-0.3)} = \frac{\Gamma(1.7)}{(-1.3)(-0.3)(0.7)}$$

$$= \frac{0.908}{(-1.3)(-0.3)(0.7)} = 3.33$$ ■

Negative integers and zero for v take special handling. The value of $\Gamma(0)$ is defined as

$$\Gamma(0) = \lim_{\varepsilon \to 0} \Gamma(\varepsilon) = \lim_{\varepsilon \to 0} \frac{\Gamma(1 + \varepsilon)}{\varepsilon} = \pm\infty$$

We relate any negative integer to $\Gamma(0)$ in Example 4 using the identity in Equation (5.6).

EXAMPLE 4. Evaluate $\Gamma(-2)$.

Solution

$$\Gamma(-2) = \frac{\Gamma(1-2)}{-2} = \frac{\Gamma(0)}{1 \cdot 2} = \pm\infty$$

Therefore, we define $\Gamma(v) = \pm\infty$ when v equals a negative integer. ■

We have now assigned a value to $\Gamma(v)$ for every real value of v. The graph of $\Gamma(v)$ is shown in Figure 3.

SECTION 4 EXERCISES

9. Evaluate the following gamma functions without using tables.

(a) $\Gamma(13)$
(b) $\Gamma(6)$
(c) $\Gamma(0)$
(d) $\Gamma(-5)$
(e) $\dfrac{\Gamma(100)}{\Gamma(99)}$

10. Evaluate the following gamma functions using Table 1.

(a) $\Gamma(3.4)$
(b) $\Gamma(0.2)$
(c) $\Gamma(-1.3)$
(d) $\Gamma(-4.5)$
(e) $\Gamma(\pi)$
(f) $\Gamma(e)$

11. Show that $\Gamma(k+1)\Gamma(k+\frac{3}{2}) = \Gamma(2k+2)$, k = integer.

12. (a) Show that $\Gamma(\frac{1}{2}) = \sqrt{\pi}$.

□ ***Hint.*** Evaluate $[\Gamma(\frac{1}{2})]^2$. □

(b) Use this result to compute $\Gamma(n+\frac{1}{2})$ for $n = 1, 2, 3, \ldots$.

13. Prove the following identity:

$$\Gamma(x)\Gamma(1-x) = \frac{\pi}{\sin \pi x}$$

□ ***Hint.*** $\Gamma(x)\Gamma(1-x) = \int_0^\infty \int_0^\infty e^{-(r+s)} r^{-x} s^{x-1}\, dr\, ds$. Now introduce $w = r + s$ and $v = r/s$ to compute this integral □

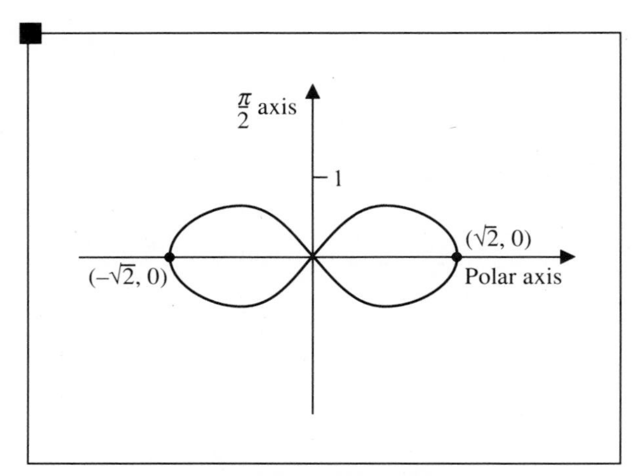

FIGURE 4 **Leminscate $r^2 = 2 \cos 2\theta$**

14. Show that the arc length of the leminscate $r^2 = 2 \cos 2\theta$ is $[\Gamma(\frac{1}{4})]^2/\sqrt{\pi}$ (see Figure 4)

□ ***Hint.*** Arc length in polar coordinates is given by

$$L = \int_\alpha^\beta \sqrt{r^2 + \left(\frac{dr}{d\theta}\right)^2}\, d\theta \quad \square$$

15. (a) Show that the logarithmic derivative of $\Gamma(x)$ is

$$\varphi(x) = \frac{\Gamma'(x)}{\Gamma(x)}$$

(b) Show that $\varphi(x)$ satisfies $\varphi(x + 1) = \varphi(x) + (1/x)$.

□ ***Hint.*** Use the identity $\Gamma(z + 1) = z\Gamma(z)$. □

□ ***Remark.*** Note that $-\varphi(1) = -\Gamma'(1) = 0.5722 \ldots$ and is called the **Euler constant**. □

16. The beta function is defined as

$$\beta(r, s) = \int_0^1 t^{r-1}(1 - t)^{s-1}\, dt$$

(a) By substituting $t = \sin^2 \theta$, show that

$$\beta(r, s) = 2 \int_0^{\pi/2} \sin^{2r-1} \theta \cos^{2s-1} \theta\, d\theta$$

(b) Use part (a) to show that

$$\beta(r, s) = \frac{\Gamma(r) \cdot \Gamma(s)}{\Gamma(r + s)}$$

□ ***Hint.*** Evaluate $\beta(r, s)\Gamma(r + s)$. □

17. Show that $\Gamma(p) = 2 \int_0^\infty e^{-t^2} t^{2p-1}\, dt$.

SECTION 5
BESSEL FUNCTIONS CONTINUED

We now return to the Bessel function. If v is any number, then

$$J_v(x) = \sum_{k=0}^{\infty} \frac{(-1)^k}{k!\Gamma(v+k+1)} \left(\frac{x}{2}\right)^{v+2k} \quad \textbf{(5.7)}$$

$$J_{-v}(x) = \sum_{k=0}^{\infty} \frac{(-1)^k}{k!\Gamma(-v+k+1)} \left(\frac{x}{2}\right)^{-v+2k} \quad \textbf{(5.8)}$$

Since $J_v(x)$ and $J_{-v}(x)$ are linearly independent when $v \neq$ integer, the general solution is given by

$$y(x) = AJ_v(x) + \beta J_{-v}(x) \qquad v \neq \text{integer}$$

Although we have defined Bessel functions for any real order, we will discover that developing some of their elementary properties will be tedious because the definition of the Bessel function is in the form of a series. Like trigonometric and hyperbolic functions, large numbers of identities, differentiation formulas, integral formulas, tables, and so on have been developed for Bessel functions.

We list a few of the more well-known properties concerning Bessel functions in Table 2 and later justify some of them.

TABLE 2 Properties of Bessel Functions

$J_0(0) = 1$
$J_v(0) = 0$
$J_n(x)$ is an even function if n is even.
$J_n(x)$ is an odd function if n is odd.

$$J_{-n}(x) = (-1)^n J_n(x) \quad \textbf{(5.9)}$$

$$\frac{d}{dx}[x^{-v}J_v(x)] = -x^{-v}J_{v+1}(x) \quad \textbf{(5.10)}$$

$$\frac{d}{dx}[x^{v}J_v(x)] = x^{v}J_{v-1}(x)$$

$$\frac{d}{dx}J_v(x) = \frac{1}{2}[J_{v-1}(x) - J_{v+1}(x)]$$

$$xJ_{v+1}(x) = 2vJ_v(x) - xJ_{v-1}(x) \quad \textbf{(5.11)}$$

$$\int x^{-v}J_{v+1}(x)\,dx = -x^{-v}J_v(x) + C \quad \textbf{(5.12)}$$

$$\int x^{v}J_{v-1}(x)\,dx = x^{v}J_v(x) + C$$

EXAMPLE 5. Prove the formula in Equation (5.9).

Solution. We know

$$J_{-n}(x) = \sum_{k=0}^{\infty} \frac{(-1)^k}{k!\Gamma(-n+k+1)} \left(\frac{x}{2}\right)^{-n+2k}$$

But since $\Gamma(-n+k+1) = \pm\infty$ for $k = 0, 1, \ldots, n-1$, we take

$$\frac{1}{\Gamma(-n+k+1)} = 0$$

Using this fact we see that

$$J_{-n}(x) = \sum_{k=n}^{\infty} \frac{(-1)^k}{k!\Gamma(-n+k+1)} \left(\frac{x}{2}\right)^{-n+2k}$$

The next step is to reindex this series by letting $k = \ell + n$, from which it follows that

$$\begin{aligned} J_{-n}(x) &= \sum_{\ell=0}^{\infty} \frac{(-1)^{\ell+n}}{(\ell+n)!\Gamma(\ell+1)} \left(\frac{x}{2}\right)^{-n+2(\ell+n)} \\ &= (-1)^n \sum_{\ell=0}^{\infty} \frac{(-1)^{\ell}}{(\ell+n)!\Gamma(\ell+1)} \left(\frac{x}{2}\right)^{n+2\ell} \end{aligned} \tag{5.13}$$

The denominator of the coefficients is

$$(\ell+n)!\Gamma(\ell+1) = (\ell+n)(\ell+n-1)\ldots(\ell+1)\Gamma(\ell+1)\ell(\ell-1)\ldots 1$$

Using the identity for (5.4) in Section 4, we can write

$$(\ell+n)!\Gamma(\ell+1) = \Gamma(n+\ell+1)\ell!$$

Returning to Equation (5.13), we have

$$\begin{aligned} J_{-n}(x) &= (-1)^n \sum_{\ell=0}^{\infty} \frac{(-1)^{\ell}}{\ell!\Gamma(n+\ell+1)} \left(\frac{x}{2}\right)^{n+2\ell} \\ &= (-1)^n J_n(x) \end{aligned}$$

■

In Example 6 we prove a differentiation formula.

EXAMPLE 6. Prove the differentiation formula in Equation (5.10).

Solution. Multiplying both sides of Equation (5.7) by $x^{-\nu}$, we can write

$$x^{-\nu}J_\nu(x) = \sum_{k=0}^{\infty} \frac{(-1)^k}{k!\Gamma(\nu+k+1)} \frac{x^{2k}}{2^{\nu+2k}}$$

Differentiating both sides with respect to x and observing that the first term in the differentiated series for $k = 0$ is zero, we can write

$$\begin{aligned} \frac{d}{dx}[x^{-\nu}J_\nu(x)] &= \sum_{k=1}^{\infty} \frac{(-1)^k 2k}{k!\Gamma(\nu+k+1)} \frac{x^{2k-1}}{2^{\nu+2k}} \\ &= x^{-\nu} \sum_{k=1}^{\infty} \frac{(-1)^k}{(k-1)!\Gamma(\nu+k+1)} \left(\frac{x}{2}\right)^{\nu+2k-1} \end{aligned}$$

Let $\ell = k - 1$. Then

$$\frac{d}{dx}[x^{-\nu}J_\nu(x)] = x^{-\nu}\sum_{\ell=0}^{\infty}\frac{(-1)^{\ell+1}}{\ell!\Gamma(\nu+\ell+2)}\left(\frac{x}{2}\right)^{\nu+1+2\ell} = x^{-\nu}J_{\nu+1}(x) \quad \textbf{(5.14)}$$

In a similar way we can show that

$$\frac{d}{dx}[x^{\nu}J_\nu(x)] = x^{\nu}J_{\nu-1}(x) \quad \textbf{(5.15)}$$

Applying the product rule to Equations (5.14) and (5.15), we have

$$x^{-\nu}\frac{d}{dx}J_\nu(x) - \nu x^{-\nu-1}J_\nu(x) = -x^{-\nu}J_{\nu+1}(x)$$

$$x^{\nu}\frac{d}{dx}J_\nu(x) + \nu x^{\nu-1}J_\nu(x) = x^{\nu}J_{\nu-1}(x)$$

Multiplying the top equation by x^ν and the bottom equation by $x^{-\nu}$ and subtracting one equation from the other, we are led to the recurrence relationship

$$xJ_{\nu+1}(x) = 2\nu J_\nu(x) - xJ_{\nu-1}(x)$$ ■

EXAMPLE 7. What is the value of $J_2(3)$?

Solution. Using Equation (5.9) we have

$$3J_2(3) = 2 \cdot 1J_1(3) - 3J_0(3)$$

or

$$J_3(3) = \tfrac{2}{3}J_1(3) - J_0(3)$$

By using Table 3 in the appendix, we find

$$\begin{aligned} J_3(3) &= \tfrac{2}{3}(0.3391) - (-0.2601) \\ &= 0.4862 \end{aligned}$$

Finally, using our derivative formulas it is easy to derive the following integrals:

$$\int x^{-\nu}J_{\nu+1}(x)\,dx = -x^{-\nu}J_\nu(x) + C$$

$$\int x^{\nu}J_{\nu-1}(x)\,dx = x^{\nu}J_\nu(x) + C$$

In particular, we can write

$$\int_0^x sJ_0(s)\,ds = xJ_1(x)$$ ■

EXAMPLE 8. Find the indefinite integral of $\int J_3(x)\,dx$.

Solution. Notice from our integral forms that

$$\int x^{-2} J_3(x)\,dx = -x^{-2} J_2(x)$$

If we multiply the integrand of our given integral by $x^2 x^{-2} = 1$, we see that

$$\int J_3(x)\,dx = \int x^2 x^{-2} J_3(x)\,dx$$

Using integration by parts the integral on the right-hand side yields

$$\begin{aligned}\int J_3(x)\,dx &= -J_2(x) + 2\int x^{-1} J_2(x)\,dx \\ &= -J_2(x) - 2x^{-1} J_1(x) + C \\ &= -J_2(x) - \frac{2J_1(x)}{x} + C\end{aligned}$$

■

SECTION 5 EXERCISES

18. By using the transformation $u = x^{1/2} y$ in the Bessel equation, show that this equation reduces to

$$u'' + \left[1 + \frac{\frac{1}{4} - n^2}{x^2}\right] u = 0 \qquad n = \text{integer}$$

19. By direct substitution in the Bessel's equation, show that

(a) $J_0(x) = \dfrac{2}{\pi}\displaystyle\int_0^{\pi/2} \cos(x \sin\theta)\,d\theta$

(b) $J_1(x) = \dfrac{2}{\pi}\displaystyle\int_0^{\pi/2} \sin(x \sin\theta)\sin\theta\,d\theta$

20. Use the series representation of Bessel's function to show that

(a) $\dfrac{d}{dx}[x^{\nu} J_{\nu}(\lambda x)] = \lambda x^{\nu} J_{\nu-1}(\lambda x)$

(b) $\dfrac{d}{dx}[x^{-\nu} J_{\nu}(\lambda x)] = -\lambda x^{-\nu} J_{\nu+1}(\lambda x)$

21. Evaluate the following limit:

$$\lim_{x\to 0} \frac{J_n(x)}{x^n}$$

22. Show that

(a) $J_{1/2}(x) = \left(\dfrac{2}{\pi x}\right)^{1/2} \sin x$

☐ ***Hints.*** $J_{1/2}(x) = \displaystyle\sum_{k=0}^{\infty} \frac{(-1)^k}{k!\,\Gamma\left(k + \frac{3}{2}\right)} \left(\frac{x}{2}\right)^{\frac{1}{2}+2k}$

Write $k!$ as a Γ function; show that $\Gamma(k+1)\Gamma[(k+3)/2] = \Gamma(2k+2)$. □

(b) Show that

$$J_{-1/2}(x) = \left[\frac{2}{\pi x}\right]^{1/2} \cos x$$

23. Obtain an explicit formula for $J_{3/2}(x)$.

24. (a) Show that $J_{n-1}(x) - J_{n+1}(x) = 2J_n'(x)$, $n = 0, 1, 2, \ldots$.
(b) What happens in part (a) for $n = 0$?

25. Prove

(a) $\displaystyle\int x^\nu J_{\nu-1}(x)\,dx = x^\nu J_\nu(x) + C$

(b) $\displaystyle\int x^{-\nu} J_{\nu+1}(x)\,dx = -x^{-\nu} J_\nu(x) + C$

(c) $\displaystyle\int_0^x sJ_0(s)\,ds = xJ_1(x)$

26. Find the integrals of

(a) $\displaystyle\int x^{10} J_9(x)\,dx$

(b) $\displaystyle\int x^{-3/2} J_{5/2}(x)\,dx$

(c) $\displaystyle\int x^5 J_2(x)\,dx$

□ ***Hint.*** Use integration by parts. □

(d) $\displaystyle\int x^{2-\nu} J_{\nu+1}(x)\,dx$

(e) $\displaystyle\int J_1(x)\,dx$

(f) $\displaystyle\int J_2(x)\,dx$

□ ***Hint.*** Multiply by $x^{-k}x^k = 1$. □

27. Find the integral

(a) $\displaystyle\int [J_5(x) - J_7(x)]\,dx$

(b) $\displaystyle\int_0^x s^4 J_1(s)\,ds$

(c) $\displaystyle\int x^6 J_3(x)\,dx$

28. Show that

$$\int_0^5 x^3 J_2(\alpha_j x)\,dx = \frac{125}{\alpha_j} J_3(5\alpha_j)$$

29. Use integration by parts to prove

$$\int_0^x s^3 J_0(s)\,ds = x^3 J_1(x) - 2x^2 J_2(x)$$

30. Establish the following recursion formula:

$$J'_\nu(x) = J_{\nu-1}(x) - \frac{\nu}{x} J_\nu(x)$$

31. Prove that

$$x^2 J''_n(x) = n(n-1)J_n(x) - (2n+1)xJ_{n+1}(x) + x^2 J_{n+2}(x)$$

32. Show that

$$\frac{d}{dr} J_n(\alpha r) = \alpha J'_n(\alpha r)$$

□ ***Note.***

$$J'_n(x) = \frac{d}{dx} J_n(x) \quad \square$$

SECTION 6

FURTHER PROPERTIES OF BESSEL FUNCTIONS

Bessel's equation can be written in an alternate way called the **self-adjoint form**:

$$(xy')' + \left(x - \frac{\nu^2}{x}\right)y = 0 \qquad x \neq 0$$

Theorem 1 states an oscillation theorem from ordinary differential equations.

THEOREM 1. Given the differential equation $[r(x)y']' + p(x)y = 0$ where $r(x) > 0$ and $r(x)$ and $p(x)$ are continuous on the interval $0 < x < +\infty$, if the two integrals

$$\int_1^\infty \frac{dx}{r(x)} = +\infty$$

and

$$\int_1^\infty p(x)\,dx = +\infty$$

then every solution $y(x)$ has an infinite number of zeros [i.e., the graph of $y(x)$ crosses the x-axis infinitely often] on the interval $(1, +\infty)$. ■

Now in the case of Bessel's equation,

$$r(x) = x \qquad \text{and} \qquad p(x) = x - \frac{v^2}{x}$$

The integral

$$\int_1^{+\infty} \frac{dx}{x} = \lim_{b \to +\infty} \ln x \Big|_1^b = +\infty$$

and the integral

$$\int_1^{\infty} p(x)\, dx = \int_1^{\infty} \left[x - \frac{v^2}{x}\right] dx \geq \int_1^{\infty} (x - v^2)\, dx = \lim_{b \to +\infty} x\left[\frac{x}{2} - v^2\right]\Big|_1^b = +\infty$$

The solutions of any Bessel's equation have an infinite number of zeros on $(0, +\infty)$. A section of the graphs of $J_0(x)$, $J_1(x)$, $Y_0(x)$, and $Y_1(x)$ are shown in Figure 2.

TABLE 3 **Some Zeros of Bessel Functions for $x > 0$**

	First	Second	Third	Fourth
$J_0(x)$	2.41	5.52	8.65	11.79
$J_1(x)$	3.83	7.02	10.17	13.32
$J_{1/2}(x)$	3.14	6.28	9.42	12.57
$J_8(x)$	12.20	16.04	19.60	22.90

Table 3 lists some of the values of the zeros of certain Bessel functions. It is important to identify these zeros of a Bessel function $J_v(x)$ because if we let $x_j =$ the jth positive root of $J_v(x) = 0$, then we can show that the infinite set of functions

$$\left\{J_v\left(\frac{x_j x}{c}\right)\right\} \qquad j = 1, 2, \ldots, c = \text{positive constant}$$

is particularly useful. Recall that the way we solved problems by separation of variables depended a great deal on the boundary conditions. The boundary conditions are equally important when we attempt to solve boundary value problems involving Bessel functions. We shall look at some examples in the next sections.

SECTION 6 EXERCISES

33. (a) Show by solving $y'' = 0$ that $y(x)$ crosses the x-axis at most once if the solution is nontrivial.

(b) Show that $\int_1^{+\infty} p(x)\, dx = 0$.

34. (a) Using Theorem 1, show that $y'' + 4y = 0$ has an infinite number of zeros.
(b) Solve $y'' + 4y = 0$. Show that a particular solution has an infinite number of zeros.

35. Use Theorem 1 to prove that solutions of

$$xy'' + y' + e^x y = 0$$

have an infinite number of zeros.

36. Use Theorem 1 to prove that solutions of

$$[(x+1)y']' + (x^2+1)y = 0$$

have an infinite number of zeros.

37. Does the modified Bessel's equation

$$(xy')' - (x^2 + \nu^2)y = 0$$

meet the hypothesis of Theorem 1?

38. Use Theorem 1 to prove that the solutions of

$$x^3 y'' + x^2 y' + (x^2 - 1)y = 0$$

have an infinite number of zeros.

SECTION 7
TEMPERATURE IN A DISK CONTINUED

We are now in a position to solve the differential equations in Equations (5.1) and (5.2) from Section 2; that is,

$$r^2 R'' + rR' + \lambda r^2 R = 0 \qquad R(c) = 0 \tag{5.16}$$

$$T' + \lambda \kappa T = 0 \tag{5.17}$$

Unfortunately, the first equation is "not quite" Bessel's equation, but if we let $\lambda = \alpha^2 > 0$, we are able to convert it into Bessel's equation by using the transformation $s = \alpha r$. If we write

$$R(r) = R\left(\frac{s}{\alpha}\right) = \bar{R}(s)$$

then

$$\frac{dR}{dr} = \frac{d\bar{R}}{ds}\frac{ds}{dr} = \alpha\,\frac{d\bar{R}}{ds} \qquad \text{and} \qquad \frac{d^2R}{dr^2} = \alpha^2\,\frac{d^2\bar{R}}{ds^2}$$

Equation (5.16) then becomes

$$s^2 \bar{R}''(s) + s\bar{R}'(s) + s^2 \bar{R}(s) = 0 \tag{5.18}$$

This is now a Bessel's equation of order 0. The solution to Equation (5.18) is

$$\bar{R}(s) = AJ_0(s) + BY_0(s)$$

Or converting back to r, we have the general solution

$$R(r) = AJ_0(\alpha r) + BY_0(\alpha r)$$

Now since $r = 0$ is in the domain of $R(r)$, we must set $B = 0$; otherwise, the solution would be unbounded contrary to a physical interpretation of the problem. [Remember $\lim_{r \to 0^+} Y_0(\alpha r) = -\infty$]. Therefore,

$$R(r) = AJ_0(\alpha r)$$

We must also meet the boundary condition $R(c) = 0$. Therefore, it follows that

$$R(c) = 0 = AJ_0(\alpha c)$$

The A cannot be zero; otherwise, we are led to the trivial solution. It follows that

$$J_0(\alpha c) = 0$$

Let $x_1, x_2, \ldots$ be the positive values of x that make $J_0(x) = 0$. Then

$$\alpha c = x_j \qquad j = 1, 2, \ldots$$

or

$$\alpha_j = \frac{x_j}{c}$$

(Notice the subscript added to α to match the root x_j.) The positive eigenvalues are

$$\lambda_j = \alpha_j^2 = \left(\frac{x_j}{c}\right)^2 \qquad j = 1, 2, \ldots$$

and the corresponding eigenfunctions are

$$R_j(r) = J_0(\alpha_j r) = J_0\left(\frac{x_j r}{c}\right)$$

Is it possible for λ to be nonpositive? To show this is not possible we must look at two situations:

$$\lambda = 0, \qquad \lambda < 0$$

If $\lambda = 0$, Bessel's equation becomes

$$r^2 R'' + rR' = 0$$

This equation is easily solved because it can be converted to a first-order equation. Its general solution is

$$R(r) = A + B \ln r$$

As before, B must equal zero; otherwise, $R(r)$ would be unbounded as $r \to 0^+$. Therefore, we have

$$R(r) = A$$

but

$$R(c) = 0 = A$$

$R(r) \equiv 0$ and therefore $\lambda = 0$ cannot be an eigenvalue.

When $\lambda < 0$, we let $\lambda = -\alpha^2$, and Equation (5.16) becomes

$$r^2R'' + rR' - \alpha^2 r^2 R = 0 \tag{5.19}$$

If we let $s = \alpha r$, as before, we can transform Equation (5.19) into the equation

$$s^2\bar{R}''(s) + s\bar{R}'(s) - s^2\bar{R}(s) = 0$$

This equation is known as the **modified Bessel's equation** and has the general solution

$$R(s) = AI_0(s) + BK_0(s)$$

or

$$R(r) = AI_0(\alpha r) + BK_0(\alpha r)$$

(see Figure 5). $I_0(x)$ and $K_0(x)$ are never zero, and $K_0(x)$ approaches $+\infty$ as $x \to 0^+$. Once again since $r = 0$ is in the domain of R, $B = 0$; otherwise, R will be unbounded. Thus we have

$$R(r) = AI_0(\alpha r)$$

But

$$R(c) = 0 = AI_0(\alpha c)$$

Since $I_0(\alpha c) \neq 0$, we must choose $A = 0$. It follows that

$$R(r) \equiv 0$$

and therefore $\lambda < 0$ has no eigenvalues.

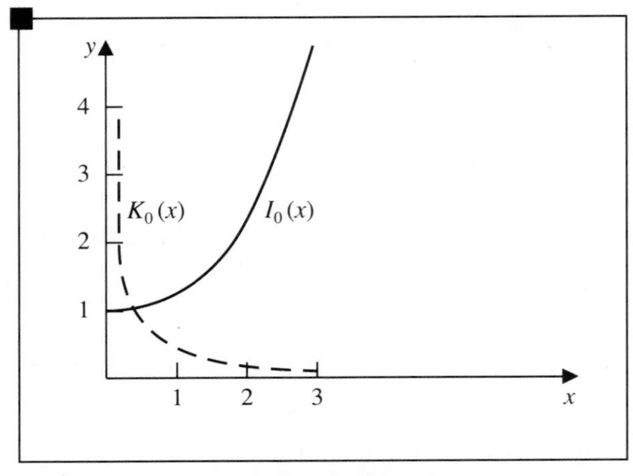

FIGURE 5 **Graph of $I_0(x)$, $K_0(x)$**

We have now completed the eigenvalue portion of the problem and turn to the other ordinary differential Equation (5.17)

$$T' + \alpha_j^2 kT = 0$$

whose general solution is

$$T_j = C_j e^{-\alpha_j^2 kt}$$

We can write

$$u_j(r, t) = e^{-\alpha_j^2 kt} J_0(\alpha_j r) \qquad j = 1, 2, \ldots$$

Forming a linear combination of these solutions, we have

$$u(r, t) = \sum_{j=1}^{\infty} A_j e^{-\alpha_j^2 kt} J_0(\alpha_j r)$$

There is still one condition to be met; that is

$$u(r, 0) = f(r) = \sum_{j=1}^{\infty} A_j J_0(\alpha_j r)$$

The series on the right-hand side is known as a Fourier–Bessel series if we compute the A_j's according to a particular formula, which we will discuss in Section 8. That formula is

$$A_j = \frac{2}{c^2[J_1(\alpha_j c)]^2} \int_0^c rf(r)J_0(\alpha_j r)\, dr$$

The final answer to our temperature problem is

$$u(r, t) = \frac{2}{c^2} \sum_{j=1}^{\infty} \frac{J_0(\alpha_j r)}{[J_1(\alpha_j c)]^2} e^{-\alpha_j^2 kt} \left[\int_0^c wf(w)J_0(\alpha_j w)\, dw \right]$$

SECTION 8
THE FOURIER–BESSEL SERIES

In order to complete the preceding problem it was necessary to find the coefficients of a series of the form

$$f(r) = \sum_{j=1}^{\infty} A_j J_\nu(\alpha_j r)$$

We already know from our discussion of the Sturm–Liouville problem that

$$\int_0^c rJ_\nu(\alpha_i r)J_\nu(\alpha_j r)\, dr = 0 \qquad i \neq j$$

where $\alpha_j = x_j/c$ and x_j are the roots of the boundary condition

$$b_1 J_\nu(x) + b_2 J_\nu'(x) = 0$$

b_1, b_2 not both zero. In other words, we say $J_\nu(\alpha_i r)$ is orthogonal to $J_\nu(\alpha_j r)$ with respect to the weight function r over the interval $[0, c]$.

We run into a problem when we wish to evaluate

$$\int_0^c rJ_\nu^2(\alpha_j r)\,dr$$

because the value of this integral depends on the particular form of the boundary condition chosen. The three forms we will look at follow:

Case 1. $J_\nu(\alpha_j c) = 0$

Case 2. $b_1 J_\nu(\alpha_j c) + b_2 J_\nu'(\alpha_j c) = 0 \qquad b_1 \neq 0,\, b_2 \neq 0$

Case 3. $J_\nu'(\alpha_j c) = 0$

In any case, however, we start with Bessel's equation written in self-adjoint form:

$$(rR')' + \left(\alpha^2 r - \frac{\nu^2}{r}\right)R = 0 \tag{5.20}$$

If we multiply Equation (5.20) by $2rR'$ and integrate from 0 to c, we have

$$\int_0^c \frac{d}{dr}(rR')^2\,dr = 2\nu^2 \int_0^c RR'\,dr - 2\alpha^2 \int_0^c r^2 RR'\,dr$$

The integral on the left and the first integral on the right-hand side can be evaluated directly, so we can write

$$(rR')^2\Big|_0^c = \nu^2 R^2\Big|_0^c - 2\alpha^2 \int_0^c r^2 RR'\,dr$$

Applying integration by parts to the remaining integral, it follows that

$$2\alpha^2 \int_0^c rR^2\,dr = (\alpha^2 r^2 - \nu^2)R^2 + (rR')^2\Big|_0^c$$

or

$$\int_0^c rR^2\,dr = \frac{1}{2\alpha^2}\left[(\alpha^2 r^2 - \nu^2)R^2 + (rR')^2\right]\Big|_0^c$$

If we replace $R(r)$ by $J_\nu(\alpha_j r)$ and $R'(r)$ by $\alpha_j J_\nu'(\alpha_j r)$ where

$$J_\nu'(\alpha_j r) = \frac{d}{du} J_\nu(u)\Big|_{u=\alpha_j r}$$

we can write

$$\begin{aligned}\int_0^c rJ_\nu^2(\alpha_j r)\,dr &= \frac{1}{2\alpha_j^2}\left[(\alpha_j^2 r^2 - \nu^2)J_\nu^2(\alpha_j r) + \alpha_j^2 r^2 J_\nu'^2(\alpha_j r)\right]\Big|_0^c \\ &= \frac{1}{2\alpha_j^2}\left[(\alpha_j^2 c^2 - \nu^2)J_\nu^2(\alpha_j c) + \alpha_j^2 c^2 J_\nu'^2(\alpha_j c)\right]\end{aligned} \tag{5.21}$$

Case 1. If $J_\nu(\alpha_j c) = 0$, then

$$\int_0^c rJ_\nu^2(\alpha_j r)\,dr = \frac{c^2}{2} J_\nu'^2(\alpha_j c) = \frac{c^2}{2} J_{\nu+1}^2(\alpha_j c)$$

The latter form follows from the identity

$$J_\nu'(\alpha_j c) = \frac{\nu}{\alpha_j c} J_\nu(\alpha_j c) - J_{\nu+1}(\alpha_j c)$$

Case 2. $b_1 J_\nu(\alpha_j c) + b_2 J_\nu'(\alpha_j c) = 0, \quad b_1 \neq 0, b_2 \neq 0$. From this equation we see that

$$J_\nu'(\alpha_j c) = -\frac{b_1}{b_2} J_\nu(\alpha_j c)$$

When we substitute this result into Equation (5.21), we find that

$$\int_0^c rJ_\nu^2(\alpha_j r)\,dr = \frac{1}{2\alpha_j^2}\left[(\alpha_j^2 c^2 - \nu^2)J_\nu^2(\alpha_j c) + \frac{b_1^1 \alpha_j^2 c^2}{b_2^2} J_\nu^2(\alpha_j c)\right] \tag{5.22}$$

For convenience it is customary to let the ratio

$$\frac{b_1}{b_2} = \frac{h}{\alpha_j c} \qquad h > 0$$

which implies that the boundary condition should be written in the form

$$hJ_\nu(\alpha_j c) + \alpha_j c J_\nu'(\alpha_j c) = 0$$

Under these circumstances Equation (5.22) can be written as

$$\int_0^c rJ_\nu^2(\alpha_j r)\,dr = \frac{1}{2\alpha_j^2}[\alpha_j^2 c^2 - \nu^2 + h^2]J_\nu^2(\alpha_j c)$$

Case 3. $J_\nu'(\alpha_j c) = 0$. When we consider this boundary condition, two situations arise, depending on whether or not $\nu = 0$.

If $\nu \neq 0$, then Equation (5.19) can be written as

$$\int_0^c rJ_\nu^2(\alpha_j r)\,dr = \frac{1}{2\alpha_j^2}[\alpha_j^2 c^2 - \nu^2]J_\nu^2(\alpha_j c)$$

However, when $\upsilon = 0$, we see that the condition

$$J_0'(\alpha_j c) = 0$$

is satisfied when

$$\alpha_j c = 0 \qquad x_1, \ldots, x_j, \ldots \qquad \text{or} \qquad \alpha_j = \frac{x_j}{c} \qquad j = 0, 1, \ldots$$

where $x_0 = 0$ and $x_1, \ldots, x_j$ are the positive roots of

$$J_0'(x) = 0.$$

For these positive roots Equation (5.21) yields

$$\int_0^c rJ_0^2(\alpha_j r)\,dr = \frac{c^2}{2}\,J_0^2(\alpha_j c)$$

and for the root $\alpha_0 = 0$ yields

$$\int_0^c rJ_0^2(0)\,dr = \int_0^c r\,dr = \frac{c^2}{2}$$

Suppose we want to represent $f(r)$, a function defined on $[0, c]$, by a series of Bessel functions; that is,

$$f(r) \sim \sum_{j=1}^{\infty} A_j J_\nu(\alpha_j r) \tag{5.23}$$

subject to the condition that $J_0(\alpha_j c) = 0$. As in the case of the Fourier series (see Section 4 in Chapter 3), we assume that the series converges to $f(r)$ and the operations used below are permissible.

Multiply both sides of Equation (5.23) by $rJ_\nu(\alpha_j r)$ and integrate from 0 to c. It follows that

$$\int_0^c rf(r)J_\nu(\alpha_i r)\,dr = \sum_{j=1}^{\infty} A_j \int_0^c rJ_\nu(\alpha_j r)J_\nu(\alpha_i r)\,dr$$

But when $i \neq j$, the integral on the right-hand side is zero by the orthogonal property. All that remains on the right-hand side is

$$A_i \int_0^c rJ_\nu^2(\alpha_i r)\,dr = \int_0^c rf(r)J_\nu(\alpha_i r)\,dr$$

Solving for A_i and using the properties above,

$$A_i = \frac{2}{c^2[J_{\nu+1}(\alpha_i c)]^2} \int_0^c rf(r)J_\nu(\alpha_i r)\,dr$$

Or since i is a dummy variable,

$$A_j = \frac{2}{c^2[J_{\nu+1}(\alpha_j c)]^2} \int_0^c rf(r)J_\nu(\alpha_j r)\,dr \tag{5.24}$$

A_j is called a **Fourier–Bessel coefficient**, and the series

$$\sum_{j=1}^{\infty} A_j J_\nu(\alpha_j r)$$

is called a **Fourier–Bessel series**.

EXAMPLE 9. Write the Fourier–Bessel series of

$$f(x) = x^3 = \sum_{j=1}^{\infty} A_j J_3(\alpha_j x)$$

subject to the condition $J_3(10\alpha_j) = 0$, where $c = 10$, $\alpha_j = x_j/10$, and $J_3(x_j) = 0$.

Solution. From Equation (5.24) we see that

$$A_j = \frac{2}{10^2[J_4(10\alpha_j)]^2} \int_0^{10} x^3 x J_3(\alpha_j x)\,dx$$
$$= \frac{1}{50[J_4(10\alpha_j)]^2} \int_0^{10} x^4 J_3(\alpha_j x)\,dx$$

To carry out the integration, we multiply the right-hand side by $[\alpha_j/\alpha_j]^5$ and rewrite it as

$$A_j = \frac{1}{50[J_4(10\alpha_j)]^2} \frac{1}{\alpha_j^5} \int_0^{10} (\alpha_j x)^4 J_3(\alpha_j x) d(\alpha_j x)$$

Let $s = \alpha_j x$. Then

$$A_j = \frac{1}{50[J_4(10\alpha_j)]^2} \frac{1}{\alpha_j^5} \int_0^{10\alpha_j} s^4 J_3(s)\,ds$$
$$= \frac{1}{50[J_4(10\alpha_j)]^2} \frac{(10\alpha_j)^4 J_4(10\alpha_j)}{\alpha_j^5}$$
$$= \frac{200}{\alpha_j J_4(10\alpha_j)}$$

The Fourier–Bessel series becomes

$$x^3 = 200 \sum_{j=1}^{\infty} \frac{J_3(\alpha_j x)}{\alpha_j J_4(10\alpha_j)}$$ ■

□ ***Warning.*** When dealing with a Fourier–Bessel series, it is necessary to know which boundary conditions are to be applied in order to use the correct formula for evaluating A_j. In Example 9 the integral

$$\int_0^c r J_\nu^2(\alpha_j r)\,dr \tag{5.25}$$

was evaluated using the results of Case 1. If other boundary conditions had been used, the appropriate value for the integral in Equation (5.25) must be used. The whole idea of the Fourier–Bessel series is summarized in Table 4. □

■ **EXAMPLE 10.** Write the Fourier–Bessel series of

$$f(x) = x = \sum_{j=0}^{\infty} A_j J_0(\alpha_j x)$$

subject to the condition $J_0'(x_j) = 0$, $\alpha_j = x_j/c$, $j = 0, 1, \ldots$.

Solution. First we see from Case 3 in Table 4 that

$$A_0 = \frac{2}{c^2} \int_0^c x[x J_0(\alpha_0 x)]\,dx$$
$$= \frac{2}{c^2} \int_0^c x^2 J_0(0)\,dx$$
$$= \frac{2}{c^2} \int_0^c x^2\,dx = \frac{2}{c^2}\left[\frac{x^3}{3}\right]\Bigg|_0^c = \frac{2c}{3}$$

TABLE 4 **Table for Finding Fourier–Bessel Coefficients**

	$f(x) = \sum_{j=1}^{\infty} A_j J_\nu(\alpha_j x)$		$f(x) = \sum_{j=0}^{\infty} A_j J_0(\alpha_j x)$
	Case 1	*Case 2*	*Case 3*
	$\nu \geq 0$	$h \geq 0 \quad \nu \geq 0$ (not both zero)	$h = 0 \quad \nu = 0$
Boundary conditions	$J_\nu(\alpha_j c) = 0$	$hJ_\nu(\alpha_j c) + \alpha_j c J'_\nu(\alpha_j c) = 0$	$J'_0(\alpha_j c) = 0$
x_j are roots of:	$J_\nu(x) = 0$	$hJ_\nu(x) + xJ'_\nu(x) = 0$	$J'_0(x) = 0$
$j =$	$1, 2, 3, \ldots$	$1, 2, 3, \ldots$	$0, 1, 2, \ldots$
$\alpha_j =$	$\dfrac{x_j}{c}$	$\dfrac{x_j}{c}$	$\alpha_0 = 0, \quad \alpha_j = \dfrac{x_j}{c} \quad j = 1, 2, \ldots$
Eigenvalues	$\lambda_j = \dfrac{x_j^2}{c^2}$	$\lambda_j = \dfrac{x_j^2}{c^2}$	$\lambda_0 = 0, \quad \lambda_j = \dfrac{x_j^2}{c^2} \quad j = 1, 2, \ldots$
Eigenfunctions	$[J_\nu(\alpha_j x)]$	$[J_\nu(\alpha_j x)]$	$J_0(0) = 1, \quad J_0(\alpha_j x) \quad j = 1, 2, \ldots$
Fourier–Bessel coefficients	$A_j = \dfrac{2}{c^2[J_{\nu+1}(\alpha_j c)]^2} \times \int_0^c xf(x)J_\nu(\alpha_j x)\,dx$	$A_j = \dfrac{2\alpha_j^2}{(\alpha_j^2 c^2 - \nu^2 + h^2)[J_\nu(\alpha_j c)]^2} \times \int_0^c xf(x)J_\nu(\alpha_j x)\,dx$	$A_0 = \dfrac{2}{c^2}\int_0^c xf(x)\,dx$; $A_j = \dfrac{2}{c^2[J_0(\alpha_j c)]^2} \times \int_0^c xf(x)J_0(\alpha_j x)\,dx \quad j = 1, 2, 3, \ldots$

To evaluate the remaining coefficients, we continue using information in Case 3 and write

$$A_j = \frac{2}{c^2[J_0(\alpha_j c)]^2} \int_0^c x^2 J_0(\alpha_j x)\, dx \qquad j = 1, 2, \ldots$$

$$= \frac{2}{c^2[J_0(\alpha_j c)]^2} \frac{1}{\alpha_j^3} \int_0^{\alpha_j c} s^2 J_0(s)\, ds$$

$$= \frac{2}{c^2[J_0(\alpha_j c)]^2} \frac{1}{\alpha_j^3} \left\{ \alpha_j^2 c^2 J_1(\alpha_j c) - \alpha_j c J_0(\alpha_j c) + \int_0^{\alpha_j c} J_0(s)\, ds \right\}$$

Even if the value of c is given, considerable work is required to find the actual value of A_j. However, with tables, a calculator, and time, it is possible to evaluate some A_j's. Substituting these values back into the series $\sum_{j=0}^{\infty} A_j J_0(\alpha_j x)$, we will have our Fourier–Bessel series expansion for $f(x) = x$. ■

SECTION 8 EXERCISES

Find the Fourier–Bessel series in Exercises 39–46.

39. $100 = \sum_{j=1}^{\infty} A_j J_0(\alpha_j x) \qquad \alpha_j = \frac{x_j}{c},\ J_0(x_j) = 0$

40. $x = \sum_{j=1}^{\infty} A_j J_1(\alpha_j x) \qquad \alpha_j = \frac{x_j}{5},\ c = 5,\ J_1(x_j) = 0$

41. $10 = \sum_{j=1}^{\infty} A_j J_2(\alpha_j x) \qquad \alpha_j = \frac{x_j}{c},\ J_2(x_j) = 0$

42. $x^5 = \sum_{j=1}^{\infty} A_j J_3(\alpha_j x) \qquad \alpha_j = \frac{x_j}{20},\ c = 20,\ J_3(x_j) = 0$

Leave A_j in integral form.

43. $x^2 = \sum_{j=1}^{\infty} A_j J_0(\alpha_j x) \qquad \alpha_j = \frac{x_j}{c},\ J_0(x_j) = 0$

44. $ax^2 + b = \sum_{j=1}^{\infty} A_j J_0(\alpha_j x) \qquad \alpha_j = \frac{x_j}{50},\ c = 50,\ J_0(x_j) = 0$

45. $x^2 = \sum_{j=1}^{\infty} A_j J_3(\alpha_j x) \qquad \alpha_j = \frac{x_j}{c},\ J_3(x_j) = 0$

Leave A_j in integral form.

46. $f(x) = \begin{cases} 0 & 0 < x < 5 \\ 1 & 5 < x < 10 \end{cases}$

$f(x) = \sum_{j=1}^{\infty} A_j J_0(\alpha_j x) \qquad \alpha_j = \frac{x_j}{10},\ c = 10,\ J_0(x_j) = 0$

Find the Fourier–Bessel series in Exercises 47–49.

47. $3 = \sum_{j=1}^{\infty} A_j J_0(\alpha_j x) \qquad \alpha_j = \frac{x_j}{c},\ J_0(x_j) + J_0'(x_j) = 0$

48. $2x = \sum_{j=1}^{\infty} A_j J_1(\alpha_j x) \qquad \alpha_j = \frac{x_j}{10},\ c = 10,\ J_1(x_j) - J_1'(x_j) = 0$

49. $x^2 = \sum_{j=1}^{\infty} A_j J_2(\alpha_j x) \qquad \alpha_j = \frac{x_j}{2},\ c = 2,\ 3J_2(x_j) + 4J_2'(x_j) = 0$

Find the Fourier–Bessel series in Exercises 50–53.

50. $50 = \sum_{j=1}^{\infty} A_j J_1(\alpha_j x) \qquad \alpha_j = \frac{x_j}{c},\ J_1'(x_j) = 0$

Leave A_j in integral form.

51. $x^2 = \sum_{j=0}^{\infty} A_j J_0(\alpha_j x) \qquad \alpha_j = x_j,\ c = 1,\ x_0 = 0,\ J_0'(x_j) = 0$

52. $f(x) = |x - 1| \qquad 0 \le x \le 2$

$f(x) = \sum_{j=0}^{\infty} A_j J_0(\alpha_j x) \qquad \alpha_j = \frac{x_j}{2},\ c = 2,\ x_0 = 0,\ J_0'(x_j) = 0$

Leave A_j in integral form.

53. $x^3 = \sum_{j=1}^{\infty} A_j J_2(\alpha_j x) \qquad \alpha_j = \frac{x_j}{4},\ c = 4,\ J_2'(x_j) = 0$

SECTION 9*

PROBLEM INVOLVING BESSEL FUNCTIONS OF FIRST AND SECOND KIND

Suppose we design an unusual drumhead. The head is made of parchment in the form of a ring whose inner radius is a centimeters and whose outer radius is b centimeters. Both edges are fixed. We strike the drumhead, which is at rest, in such a way as to impart a velocity of $g(r)$. We will assume the displacement of the drumhead depends only on r and t. Formally stated, this boundary value problems looks like

$$\frac{1}{r}\frac{\partial}{\partial r}\left(r\frac{\partial u}{\partial r}\right) = \frac{1}{c^2}\frac{\partial^2 u}{\partial t^2}$$

where the solutions are of the form $u(r, t)$

$$u(a, t) = 0$$
$$u(b, t) = 0$$
$$u(r, 0) = 0$$
$$\frac{\partial u}{\partial t}(r, 0) = g(r)$$

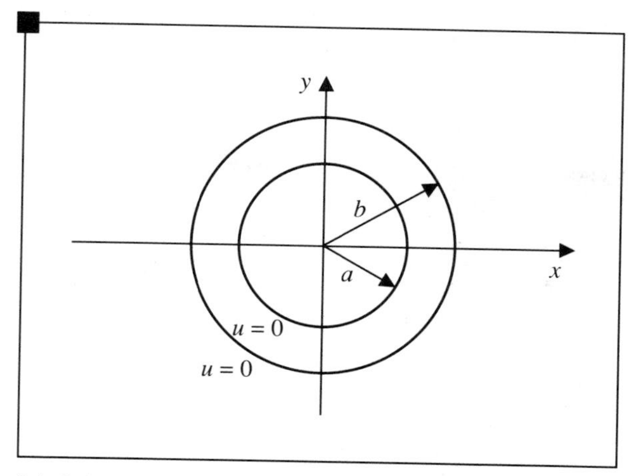

FIGURE 6 **Unusual Drumhead**

(see Figure 6). Let $u(r, t) = R(r)T(t)$ and substitute into the differential equation. It follows that

$$\frac{1}{r}\frac{\partial}{\partial r}(rR'T) = \frac{1}{c^2}RT''$$

where upon separating variables we have

$$\frac{R''}{R} + \frac{1}{r}\frac{R'}{R} = \frac{1}{c^2}\frac{T''}{T} = -\lambda \tag{5.26}$$

From Equation (5.26) we arrive at the two ordinary differential equations with their conditions:

$$r^2R'' + rR' + \lambda r^2 R = 0 \qquad R(a) = 0,\ R(b) = 0 \tag{5.27}$$

$$T'' + \lambda c^2 T = 0 \qquad T(0) = 0 \tag{5.28}$$

The general solution of Equation (5.27) with $\lambda = \alpha^2 > 0$ is

$$R(r) = AJ_0(\alpha r) + BY_0(\alpha r)$$

Using the boundary conditions we find that

$$R(a) = 0 = AJ_0(\alpha a) + BY_0(\alpha a)$$

and

$$R(b) = 0 = AJ_0(\alpha b) + BY_0(\alpha b) \tag{5.29}$$

The only way Equation (5.27) can have a nontrivial solution is if the determinant of the coefficients is zero; therefore,

$$J_0(\alpha a)Y_0(\alpha b) - J_0(\alpha b)Y_0(\alpha a) = 0$$

Let α_j be the jth root of this cross-product. There are tables that give some of these roots (see the *Handbook of Mathematical Functions* published by the U.S.

Department of Commerce). Therefore, the eigenvalues are $\lambda = \alpha_j^2$, and

$$R_j(r) = Y_0(\alpha_j a)J_0(\alpha_j r) - J_0(\alpha_j a)Y_0(\alpha_j r) \qquad j = 1, 2, \ldots$$

are eigenfunctions of the boundary value problems in Equation (5.27) for $\lambda > 0$. It can be shown that these are the only eigenvalues and eigenfunctions.

Next we solve the differential equation (5.28),

$$T'' + \alpha_j^2 c^2 T = 0$$

which yields

$$T_j(t) = C_j \cos \alpha_j ct + D \sin \alpha_j ct$$

Applying the initial condition, we can write

$$T_j(0) = 0 = C_j$$

which implies that

$$T_j(t) = \sin \alpha_j ct$$

We form the infinite linear combination

$$u(r, t) = \sum_{j=1}^{\infty} A_j[Y_0(\alpha_j a)J_0(\alpha_j r) - J_0(\alpha_j a)Y_0(\alpha_j r)] \sin \alpha_j ct \tag{5.30}$$

Using our final condition we have

$$\begin{aligned} \frac{\partial u}{\partial t}(r, 0) = g(r) &= \sum_{j=1}^{\infty} A_j \alpha_j c[Y_0(\alpha_j a)J_0(\alpha_j r) - J_0(\alpha_j a)Y_0(\alpha_j r)] \\ &= \sum_{j=1}^{\infty} A_j \alpha_j c R_j(r) \end{aligned} \tag{5.31}$$

Now $\int_a^b rR_i(r)R_j(r)\, dr = 0$, $\lambda_i \neq \lambda_j$; that is, $\{R_j(r)\}$ is an orthogonal set.

We can evaluate A_j by multiplying both sides of Equation (5.31) by $rR_i(r)$ and integrating from a to b; thus,

$$\int_a^b rg(r)R_i(r)\, dr = A_i \alpha_i c \int_a^b rR_i^2(r)\, dr$$

from which it follows that

$$A_i = \frac{\displaystyle\int_a^b rg(r)R_i(r)\, dr}{\alpha_i c \displaystyle\int_a^b rR_i^2(r)\, dr}$$

And since i is a dummy variable,

$$A_j = \frac{\displaystyle\int_a^b rg(r)R_j(r)\, dr}{a_j c \displaystyle\int_a^b rR_j^2(r)\, dr}$$

Substituting these values of A_j back onto the right side of Equation (5.30), we have the solution to our problem, which is

$$u(r, t) = \sum_{j=1}^{\infty} \left[\frac{\displaystyle\int_a^b sg(s)R_j(s)\, ds}{\alpha_j c \displaystyle\int_a^b sR_j^2(s)\, ds} \right] [Y_0(\alpha_j a)J_0(\alpha_j r) - J_0(\alpha_j a)Y_0(\alpha_j r)] \sin \alpha_j ct$$

SECTION 9 EXERCISES

54. Solve the following boundary value problems in terms of the variables given:

(a) $u_{rr} + \dfrac{1}{r} u_r = \dfrac{1}{k} u_t$, $\quad u(r, t);\ u(a, t) = 0,\ u(r, 0) = 10$

(b) $u_{rr} + \dfrac{1}{r} u_r = \dfrac{1}{c^2} u_{tt}$, $\quad u(r, t);\ u(20, t) = 0,\ u(r, 0) = 20 - r,\ u_t(r, 0) = 0$

(c) $u_{rr} + \dfrac{1}{r} u_r + u_{zz} = 0$, $\quad u(r, z);\ u(r, 0) = 0,\ u(r, L) = 5,\ u(a, z) = 0$

(d) $u_{rr} + \dfrac{1}{r} u_r - \dfrac{9u}{r^2} = \dfrac{1}{k} u_t$, $\quad u(r, t);\ u(a, t) = 0,\ u(r, 0) = f(r)$

55. A circular plate of radius a is insulated laterally. The temperature on its circumference is 0. Find the temperature u in the plate if it depends only on r and t, and the initial temperature is r^2.

56. A customer purchases a round hamburger 10 centimeters in diameter. Consider the roll to be a perfect insulator. When the hamburger is inserted in the roll, it is 200 °C, much too hot to eat. In order to speed up the cooling, the customer leaves the diner and walks into the parking lot, where the temperature is 0° C.
(a) Find the temperature of the hamburger as a function of r and t.
(b) How long will it take for the hamburger to cool below 50° C? Use only the first term in the series.

57. A circular plate of radius 20 is insulated laterally and insulated perfectly on its circumference. Assuming the temperature depends only on r and t,

(a) write the partial differential equation and boundary and initial conditions.
(b) use separation of variables to yield two ordinary differential equations.
(c) solve for all eigenvalues and eigenfunctions.
(d) solve the equation that depends on t.
(e) write the solution as an infinite linear combination of solutions of ordinary differential equations.
(f) *stop* — you cannot go further without more information.

58. Consider the problem of finding the steady-state temperature $u(r, z)$ in a right circular cylinder of length 15 centimeters and radius 3 centimeters. Assume the temperature is independent of θ. If the edge at $z = 0$ is perfectly insulated, the edge at $z = 15$ has a temperature of 100, and the curved lateral edge is 0, find the temperature in the cylinder.

59. Solve Exercise 58 except change the condition at edge $z = 0$ to temperature equals zero.

60. Consider a circular plate insulated laterally whose radius is a. The temperature $u(r, t)$ in the plate satisfies the differential equation

$$\frac{1}{r}\frac{\partial}{\partial r}\left(r\frac{\partial u}{\partial r}\right) - \frac{9}{r^2}u = \frac{1}{k}\frac{\partial u}{\partial t}$$

which indicates that heat is generated within the plate. Solve for u if the temperature on the boundary is 0 and the initial temperature is $f(r)$.

61. A circular plate insulated laterally has a radius of 30. Find the temperature $u(r, t)$ in the plate if the temperature on the circumference is 70 and the initial temperature is 100.

62. Consider a circular drumhead of radius 10 that is fixed on the circumference. If the drumhead is at rest at $t = 0$ and is given a velocity of $g(r)$ at $t = 0$, find the solution for the displacement $u(r, t)$.

63. Solve Exercise 62 except replace the initial conditions by $u(r, 0) = 10 - r^2$ and $(\partial u/\partial t)(r, 0) = 0$.

64. A circular trampoline is 5 meters in diameter and is fixed to a frame. An acrobat has jumped into the center of the trampoline. The canvas has returned to the equilibrium position and has an upward velocity of 50 when the acrobat leaves the trampoline. Find the displacement $u(r, t)$ of the trampoline.

65. A circular drumhead with damping has a radius of 40 centimeters and satisfies the partial differential equation

$$\frac{1}{r}\frac{\partial}{\partial r}\left(r\frac{\partial u}{\partial r}\right) - \frac{25}{r^2}u = \frac{1}{c^2}\frac{\partial^2 u}{\partial t^2}$$

The membrane is fixed on the circumference, and while the membrane is at rest the drummer imparts a velocity of $f(r)$ to the head. Find the displacement $u(r, t)$.

66. A circular rubber diaphragm whose radius is 2 centimeters is damped such that its equation is

$$u_{rr} + \frac{1}{r}u_r = \frac{1}{c^2}u_{tt} + 100$$

If $u(2, t) = 0$ and $u(r, 0) = g(r)$, find the displacement $u(r, t)$.

□ ***Hint.*** Let $u(r, t) = w(r, t) + \varphi(r)$. □

67. A heavy chain of length L and weight w newton/meters hangs from a fixed hook. If the initial displacement $u(x, 0)$ of the chain is $f(x)$ and the initial velocity is zero, find the displacement $u(x, t)$. The equation that u must satisfy is

$$xu_{xx} + u_x = \frac{1}{g}u_{tt}$$

where g is the gravitational constant (see Figure 7).

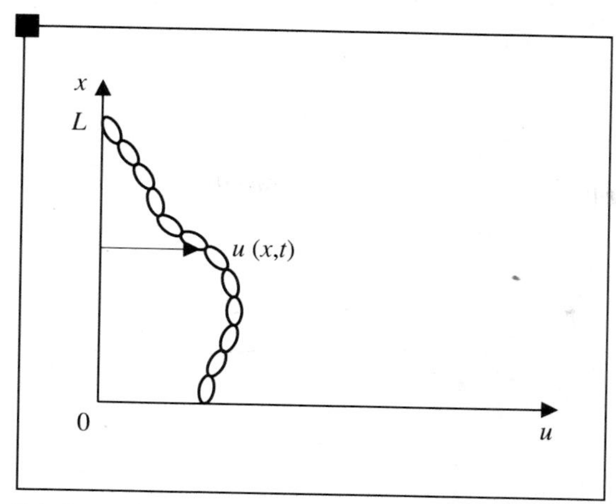

FIGURE 7 **Hanging Chain**

68. If a drumhead is not perfectly flexible, it is necessary to take into account the stiffness when deriving the equation. The equation for finding the displacement $u(r, t)$ of a drumhead under such circumstances is

$$c^2\left(u_{rr} + \frac{1}{r}\,u_r\right) = u_{tt} + w_c\,u$$

where c^2 and w_c are constants. If $u(a, t) = 0$, $u(r, 0) = 0$, and $u_t(r, 0) = 1$, find the displacement $u(r, t)$.

69. A steam pipe whose inside radius is 2 centimeters and whose outside radius is 3 centimeters carries steam whose temperature is 100° C. If the temperature outside the pipe is 25° C, find the steady-state temperature in the wall of the pipe. (This exercise is more difficult than the others.)

70. Consider a semi-infinite solid of radius a whose base is in the xy plane. Find the temperature $u(r, z)$ in the solid if $u(r, 0) = 100$ and $u(a, z) = 0$.

71. Consider a circular plate whose radius is 15 meters. Find the radial vibrations in the plate if the circumference is fixed and the initial displacement is zero, and the initial velocity is 10 m/sec. Assume the radial vibrations $u(r, t)$ depend only on r and t, and satisfy the partial differential equation

$$u_{rr} + \frac{1}{r}\,u_r = \frac{1}{c^2}\,u_{tt}$$

CHAPTER SIX

Legendre Polynomials

SECTION 1
PROBLEMS LEADING TO LEGENDRE'S EQUATION

Boundary value problems on spherical domains in R^3 often lead to an ordinary differential equation known as Legendre's differential equation, some of whose solutions are polynomials. To motivate our study of those polynomials, consider the following example.

Suppose a mine, spherical in shape, is floating so that one-fourth of its diameter of 160 centimeters is above the water. If the temperature of the air and sea are 0° and 10°, respectively, what is the steady-state temperature T at any point inside the mine? Assume the mine has uniform density.

We recall from Chapter 4, Example 4 that the steady-state temperature T in R^2 must satisfy Laplace's equation $\nabla^2 T = 0$. This property also holds in R^3. Because of the shape of our region, it is most convenient to solve the problem in terms of the spherical coordinates, which are shown in Figure 1. Therefore, we see that

$$\nabla^2 T(\rho, \varphi, \theta) = \frac{1}{\rho^2}\left\{\frac{\partial}{\partial \rho}\left(\rho^2 \frac{\partial T}{\partial \rho}\right) + \frac{1}{\sin\theta}\frac{\partial}{\partial \theta}\left(\sin\theta \frac{\partial T}{\partial \theta}\right) + \frac{1}{\sin^2\theta}\frac{\partial^2 T}{\partial \varphi^2}\right\} = 0$$

The boundary conditions are

$$T(80, \varphi, \theta) = \begin{cases} 0^\circ & 0 \le \theta \le 2\pi \quad 0 \le \theta < \dfrac{\pi}{3} \\ 10^\circ & 0 \le \theta \le 2\pi \quad \dfrac{\pi}{3} < \theta \le \pi \end{cases}$$

(see Figure 2). If you look at Figure 2 you will see that the temperature T on the surface is independent of φ. It is natural, therefore, to assume that the

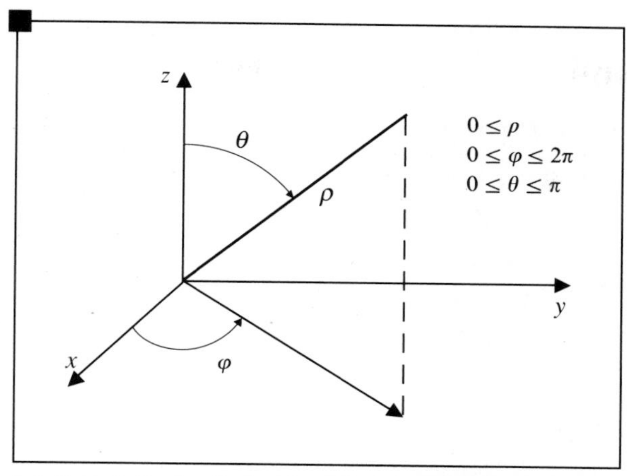

FIGURE 1 **Spherical Coordinates**

temperature throughout the mine is also independent of φ. Our boundary value problem can be rewritten as

$$\frac{1}{\rho^2}\left\{\frac{\partial}{\partial \rho}\left(\rho^2 \frac{\partial T}{\partial \rho}\right) + \frac{1}{\sin\theta}\frac{\partial}{\partial \theta}\left(\sin\theta \frac{\partial T}{\partial \theta}\right)\right\} = 0$$

$$T(80, \theta) = \begin{cases} 0 & 0 \le \theta < \dfrac{\pi}{3} \\ 10 & \dfrac{\pi}{3} < \theta \le \pi \end{cases}$$

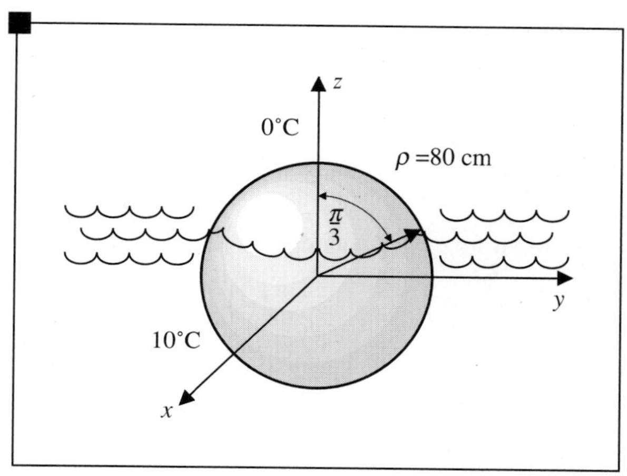

FIGURE 2 **Floating Mine**

To begin, we seek a solution of this equation in the form $T(\rho, \theta) = R(\rho)\Theta(\theta)$. Substituting into the differential equation, we have

$$\frac{1}{\rho^2}\left\{(\rho^2 R')'\Theta + \frac{1}{\sin\theta}(\sin\theta\Theta')'R\right\} = 0$$

or

$$\frac{(\rho^2 R')'}{R} = -\frac{(\Theta' \sin\theta)'}{\Theta \sin\theta} = \lambda$$

which leads to the two ordinary differential equations,

$$(\rho^2 R')' - \lambda R = 0 \tag{6.1}$$

$$[(\sin\theta)\Theta']' + \lambda(\sin\theta)\Theta = 0 \tag{6.2}$$

Notice that there are no "obvious" homogeneous boundary conditions in this problem. Equation (6.1) is a Cauchy–Euler equation, which we solved in Chapter 4, Section 3; therefore, we will concentrate on Equation (6.2), which is equivalent to **Legendre's equation**.

This equation can be handled more easily if we make the transformation $x = \cos\theta, 0 \le \theta \le \pi$, from which it follows that

$$\sqrt{1 - x^2} = \sin\theta$$

Let

$$\Theta(\theta) = \Theta(\cos^{-1} x) = y(x)$$

Then

$$\Theta'(\theta) = \frac{dy}{dx}\frac{dx}{d\theta} = -\sin\theta\frac{dy}{dx}$$

Transforming Equation (6.2), it follows that

$$\frac{d}{dx}\left(-\sin^2\theta\frac{dy}{dx}\right)(-\sin\theta) + \lambda(\sin\theta)y = 0$$

or

$$\frac{d}{dx}\left[(1 - x^2)\frac{dy}{dx}\right] + \lambda y = 0 \tag{6.3}$$

Equation (6.3) is the well-known form of Legendre's equation. It can be solved using a Taylor series,

$$y = \sum_{j=0}^{\infty} a_j x^j$$

which leads us to the recursion formula

$$a_{j+2} = \frac{j(j+1) - \lambda}{(j+1)(j+2)} a_j \qquad j = 0, 1, 2, \ldots$$

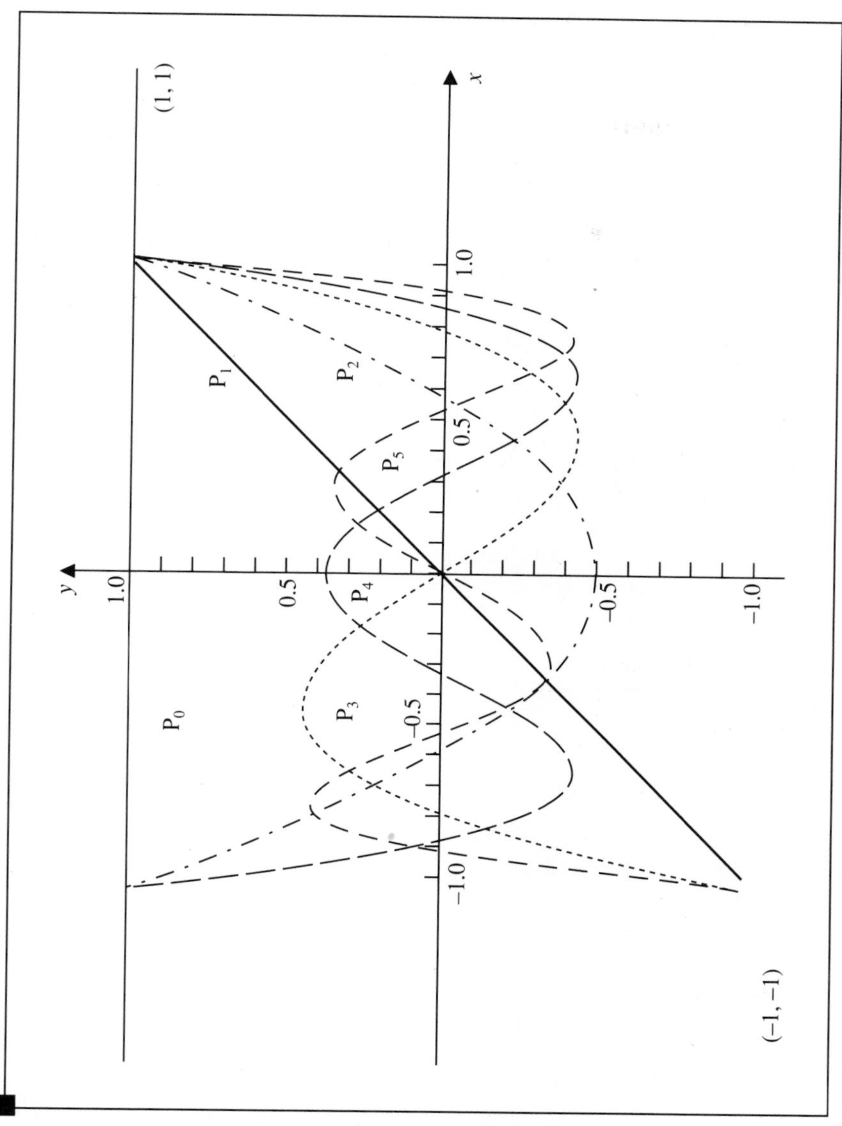

FIGURE 3 Legendre Polynomials

First notice that if $\lambda = n(n+1)$, $n = 0, 1, \ldots$, the recursion formula will yield all zeros after $j = n$. Hence, one of the solutions of Legendre's differential equation becomes a *polynomial* whereas the other solution remains an infinite series. Second, if $\lambda \neq n(n+1)$, then both linearly independent solutions are infinite series. In all cases whether $\lambda = n(n+1)$ or not, there are points on $[-1, 1]$ at which the series solutions are unbounded so that the only bounded solutions of Legendre's differential equation are the polynomials

$$\begin{aligned} y_1 &= a_0 + a_2 x^2 + \cdots + a_n x^n \qquad a_n \neq 0 \qquad n \text{ even} \\ y_2 &= a_1 x + a_3 x^3 + \cdots + a_n x^n \qquad a_n \neq 0 \qquad n \text{ odd} \end{aligned} \tag{6.4}$$

For convenience it is customary to choose the coefficient

$$a_n = \frac{(2n)!}{2^n (n!)^2}$$

When this is done the polynomials are called **Legendre polynomials** and are symbolized by $P_n(x)$. The reason for the special choice of a_n is to force every Legendre polynomial to have the value 1 when $x = 1$. A few Legendre polynomials are shown in Table 1, and their graphs are shown in Figure 3.

TABLE 1 **Some Legendre Polynomials**

$P_0(x) = 1$
$P_1(x) = x$
$P_2(x) = \frac{1}{2}(3x^2 - 1)$
$P_3(x) = \frac{1}{2}(5x^3 - 3x)$
$P_4(x) = \frac{1}{8}(35x^4 - 30x^2 + 3)$
$P_5(x) = \frac{1}{8}(63x^5 - 70x^3 + 15x)$

SECTION 1 EXERCISES

1. Write Laplace's equation if the dependent variable $u(r, \varphi, \theta)$ is independent of θ.
2. Write Laplace's equation if the dependent variable $u(r, \varphi, \theta)$ is independent of φ and θ.
3. Write boundary conditions to go along with the partial differential equation

$$\frac{1}{\rho^2}\left\{\frac{\partial}{\partial \rho}\left(\rho^2 \frac{\partial u}{\partial \rho}\right) + \frac{1}{\sin\theta}\frac{\partial}{\partial \theta}(\sin\theta)\frac{\partial u}{\partial \theta}\right\} = 0$$

where u is a function of ρ and θ and the shape is

(a) a sphere, $\rho = a$, $u = 100°$ on the surface.

(b) a sphere, $\rho = 20$, $u = \begin{cases} 80 \text{ on the upper half} \\ 0 \text{ on the lower half} \end{cases}$ of the sphere.

(c) a hemisphere, $\rho = L$, $u = 0$ on the spherical surface and $u = 100$ on the bounding plane through the center.

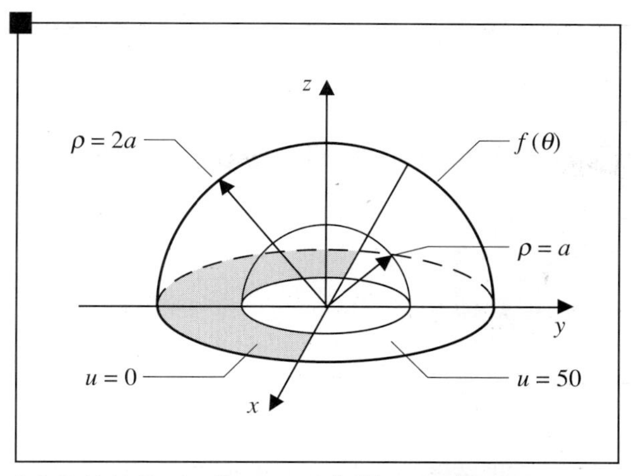

FIGURE 4 **Concentric Hemispheres**

(d) concentric hemispheres as shown in Figure 4.
(e) a sphere, $\rho = 50$, u on the upper half of the sphere is 0; the lower half is perfectly insulated.
(f) a sphere, $\rho = a$. The total sphere is insulated with inferior insulation such that the temperature u equals the rate of change of u in the direction of the radius.
(g) a sphere of radius L. Let u be the electrostatic potential on the sphere. The potential on the surface is zero.
(h) a hemisphere of radius 16. Let u be the electrostatic potential on the hemisphere. The potential on the spherical surface is zero, and the potential on the bounding plane through the center is 100.

SECTION 2
PROPERTIES OF LEGENDRE POLYNOMIALS

In this section we will show that Legendre polynomials can be expressed in two other ways that are useful in developing some of the properties of these polynomials.

Let $F(x, t)$ be a function of two variables. If F can be represented as a power series in t whose coefficients are functions of x, F is called a **generating function** for the set of coefficients.

EXAMPLE 1. $F(x, t) = 1/(1 - xt)$ is a generating function for the set of functions $\{x^n\}$ since

$$F(x, t) = \frac{1}{1 - xt} = \sum_{n=0}^{\infty} x^n t^n \qquad \text{provided } |xt| < 1$$

The given series is just the binomial series for $(1 - xt)^{-1}$. ■

Functions like Bessel functions and orthogonal polynomials can also be represented by generating functions. We state just a few in Example 2.

EXAMPLE 2. The generating function for Bessel functions is

$$\exp\left[\frac{tx - x}{2t}\right] = \sum_{n=-\infty}^{\infty} J_n(x)t^n \qquad t \neq 0$$

For Legendre polynomials the generating function is

$$\frac{1}{\sqrt{1 - 2tx + t^2}} = \sum_{n=0}^{\infty} P_n(x)t^n \qquad -1 < x < 1, |t| < 1 \tag{6.5}$$

And for Hermite polynomials the generating function is

$$\exp(2tx - t^2) = \sum_{n=0}^{\infty} \frac{H_n(x)}{n!} t^n$$ ■

The following equation, called **Rodrigue's formula**, expresses individual Legendre polynomials:

$$P_n(x) = \frac{1}{2^n n!} \frac{d^n}{dx^n} (x^2 - 1)^n \qquad n = 0, 1, 2, \ldots \tag{6.6}$$

Some of the more well-known properties of Legendre polynomials are listed in Table 2.

TABLE 2 **Some Well-Known Properties of Legendre Polynomials**

$$P_n(x) = \begin{cases} \text{even function} & n \text{ even} \\ \text{odd function} & n \text{ odd} \end{cases} \tag{6.7}$$

$$P_n(1) = 1$$

$$P_n(-1) = (-1)^n$$

$$(2n + 1)P_n(x) = P'_{n+1}(x) - P'_{n-1}(x) \qquad n = 1, 2, \ldots \tag{6.8}$$

$$(2n + 1)P_n(x) = (n + 1)P_{n+1}(x) + nP_{n-1}(x) \qquad n = 1, 2, \ldots \tag{6.9}$$

$$\int_{-1}^{x} P_n(s)\, ds = \frac{1}{2n + 1} [P_{n+1}(x) - P_{n-1}(x)]$$

The proof of Equation (6.7) follows immediately from Equation (6.4). Identities such as Equation (6.9) are derived using the generating function, as shown in Example 3.

EXAMPLE 3. Derive the identity for Equation (6.9).

Solution. Differentiating both sides of Equation (6.5) with respect to λ yields

$$\frac{x-\lambda}{(1-2\lambda x+\lambda^2)^{3/2}} = \sum_{n=0}^{\infty} n\lambda^{n-1}P_n(x)$$

or

$$\frac{x-\lambda}{1-2\lambda x+\lambda^2}\sum_{n=0}^{\infty}\lambda^n P_n(x) = \sum_{n=0}^{\infty} n\lambda^{n-1}P_n(x) \tag{6.10}$$

If we multiply both sides of Equation (6.10) by $1 - 2\lambda x + \lambda^2$, it follows that

$$\sum_{n=0}^{\infty} x\lambda^n P_n(x) - \sum_{n=0}^{\infty}\lambda^{n+1}P_n(x) = \sum_{n=0}^{\infty} n\lambda^{n-1}P_n(x) - \sum_{n=0}^{\infty} 2nx\lambda^n P_n(x) + \sum_{n=0}^{\infty} n\lambda^{n+1}P_n(x)$$

Combining terms with the same powers of λ, this equation can be written as

$$\sum_{n=0}^{\infty}(2n+1)x\lambda^n P_n(x) = \sum_{n=0}^{\infty}(n+1)\lambda^{n+1}P_n(x) + \sum_{n=1}^{\infty} n\lambda^{n-1}P_n(x)$$

If the two series on the right-hand side are reindexed, we see that

$$\sum_{n=0}^{\infty}(2n+1)x\lambda^n P_n(x) = \sum_{n=0}^{\infty} n\lambda^n P_{n-1}(x) + \sum_{n=0}^{\infty}(n+1)\lambda^n P_{n+1}(x)$$

The two sides of the series are equal if they are equal term by term, which gives us the result we are seeking; that is,

$$(2n+1)xP_n(x) = nP_{n-1}(x) + (n+1)P_{n+1}(x)$$ ■

SECTION 2 EXERCISES

4. Using the series solution of Legendre's equation in Equation (6.4), show that $P_n(x)$ is an even function if n is even and an odd function if n is odd.

5. Using the series representations in Equation (6.4), find $P_n(x)$ for
   (a) $P_5(x)$
   (b) $P_8(x)$

6. Use the generating function to find P_0, P_1, P_2, and P_3.

7. (a) Show that $P_n(1) = 1$, $n = 0, 1, 2, \ldots$.
   (b) Show that $P_n(-1) = (-1)^n$, $n = 0, 1, 2, \ldots$.

   □ ***Hint.*** Use the generating function. □

8. Use the generating function to prove

$$P_n(0) = \begin{cases} 0 & n \text{ odd} \\ \dfrac{(-1)^n(2n)!}{2^{2n}(n!)^2} & n \text{ even} \end{cases} \qquad n = 0, 1, 2, \ldots$$

9. Prove $(2n + 1)P_n(x) = P'_{n+1}(x) - P'_{n-1}(x)$, $n = 1, 2, \ldots$.

10. Prove

$$\int_{-1}^{x} P_n(s)\, ds = \frac{1}{2n+1}\,[P_{n+1}(x) - P_{n-1}(x)]$$

11. Prove

$$\int_{x}^{1} P_n(s)\, ds = \frac{1}{2n+1}\,[P_{n-1}(x) - P_{n+1}(x)]$$

12. Use Rodrigue's formula to compute P_2, P_5.

13. Find $P'_2(x)$ directly from Rodrigue's formula.

14. If n is odd, show that $\int_{-1}^{1} P_n(x)\, dx = 0$.

15. (a) Show that

$$\int_a^b f(x)P_n(x)\, dx = \frac{(-1)^n}{2^n n!} \int_a^b (x^2 - 1)^n \frac{d^n}{dx^n} f(x)$$

☐ ***Hint.*** Use integration by parts. ☐

(b) Using 17(a) show $\int_{-1}^{1} P_n(x)\, dx = 0, \qquad n = 0, 1, 2, \ldots$.
(c) Show $\int_{-1}^{1} P_0(x)\, dx = 2 \int_0^1 P_0(x)\, dx = 2$.

16. If $f(x)$ is a polynomial of degree $<(n - 1)$, show that

$$\int_a^b f(x)P_n(x)\, dx = 0$$

☐ ***Hint.*** Use Exercise 15. ☐

17. Suppose $f(x) = x^{n+1}$. Show that

$$\int f(x)P_n(x)\, dx = \frac{(-1)^n}{2^n} \frac{(x^2 - 1)^{n+1}}{n+1} + C$$

SECTION 3
FOURIER–LEGENDRE SERIES

From the Sturm–Liouville theory (see Section 6 in Chapter 4), it follows directly that the infinite set of polynomials $P_n(x)$ are eigenfunctions corresponding to the eigenvalues $\lambda_n = n(n + 1)$, $n = 0, 1, 2, \ldots$, and that, therefore, these polynomials form an orthogonal set over the interval $(-1, 1)$. It is worthwhile, however, to show this property directly from Legendre's equation.

Let P_m and P_n be solutions, respectively, of the equations

$$[(1-x^2)P'_m]' + m(m+1)P_m = 0 \tag{6.11}$$

$$[(1-x^2)P'_n]' + n(n+1)P_n = 0 \tag{6.12}$$

If we multiply Equation (6.11) by P_n and Equation (6.12) by P_m, subtract one equation from the other, and then integrate from -1 to 1, we have

$$[m(m+1) - n(n+1)] \int_{-1}^{1} P_m(x)P_n(x)\,dx = \int_{-1}^{1} \{[(1-x^2)P'_m(x)]'P_n(x) - [(1-x^2)P'_n(x)]'P_m(x)\}\,dx \tag{6.13}$$

Applying integration by parts to the integral on the right-hand side, it can be written as

$$(1-x^2)[P_n(x)P'_m(x) - P_m(x)P'_n(x)]_{-1}^{1} - \int_{-1}^{1} (1-x^2)[P'_m(x)P'_n(x) - P'_n(x)P'_m(x)]\,dx = 0$$

If $m \neq n$, then it follows from Equation (6.8) that

$$\int_{-1}^{1} P_m(x)P_n(x)\,dx = 0 \tag{6.14}$$

In order to evaluate

$$\int_{-1}^{1} P_n^2(x)\,dx$$

we replace $P_n(x)$ by

$$P_n(x) = \frac{(2n-1)xP_{n-1}(x) - (n-1)P_{n-2}(x)}{n} \qquad 2 \leq n$$

which follows directly from Equation (6.9). Therefore,

$$\begin{aligned}\int_{-1}^{1} P_n^2(x)\,dx &= \frac{2n-1}{n}\int_{-1}^{1} xP_{n-1}(x)P_n(x)\,dx - \frac{n-1}{n}\int_{-1}^{1} P_{n-2}(x)P_n(x)\,dx \\ &= \frac{2n-1}{n}\int_{-1}^{1} xP_{n-1}(x)P_n(x)\,dx\end{aligned} \tag{6.15}$$

since the second integral on the right-hand side is zero by using the orthogonal property in Equation (6.14). From Equation (6.9) we can write

$$xP_n(x) = \frac{(n+1)P_{n+1}(x) + nP_{n-1}(x)}{2n+1} \tag{6.16}$$

Substituting the right-hand side of Equation (6.16) into Equation (6.15), we have

$$\int_{-1}^{1} P_n^2(x)\,dx = \frac{n+1}{n}\frac{2n-1}{2n+1}\int_{-1}^{1} P_{n+1}(x)P_{n-1}(x)\,dx + \frac{2n-1}{2n+1}\int_{-1}^{1} P_{n-1}^2(x)\,dx$$

$$= \frac{2n-1}{2n+1}\int_{-1}^{1} P_{n-1}^2(x)\,dx \qquad n = 1, 2, \ldots \tag{6.17}$$

since the first integral on the right-hand side is zero by Equation (6.14). The relation in Equation (6.17) is also true for $n = 1$ by direct substitution.

By repeated use of Equation (6.17) we see that

$$\int_{-1}^{1} P_n^2(x)\,dx = \frac{2n-1}{2n+1}\cdot\frac{2n-3}{2n-1}\cdots\frac{3}{5}\cdot\frac{1}{3}\int_{-1}^{1} P_0^2(x)\,dx$$

$$= \frac{1}{2n+1}\int_{-1}^{1} 1^2 dx = \frac{2}{2n+1} \tag{6.18}$$

SECTION 4
FORMULA FOR COEFFICIENTS OF FOURIER–LEGENDRE SERIES

Since the set of functions $\{P_n(x)\}$ is orthogonal, it is natural to attempt to describe functions f defined over the interval $[-1, 1]$ in terms of a series of Legendre polynomials:

$$\sum_{n=0}^{\infty} A_n P_n(x) \tag{6.19}$$

As in the case of determining the Fourier coefficients for a Fourier series (see Section 4 in Chapter 3), we do not know a priori whether the series in Equation (6.19) so obtained converges; and provided it does, if it converges to the function $f(x)$.

If we assume that the series of Legendre polynomials converges to $f(x)$, we can write

$$f(x) = \sum_{n=0}^{\infty} A_n P_n(x) \tag{6.20}$$

And if the operations used below are permissible, we can derive the formula for evaluating the coefficients A_n.

We start by multiplying both sides of Equation (6.20) by $P_m(x)$ and integrating from -1 to 1. These operations give us

$$\int_{-1}^{1} f(x)P_m(x)\,dx = \sum_{i=1}^{\infty} A_n \int_{-1}^{1} P_n(x)P_m(x)\,dx$$

from which it follows using the orthogonal property of Legendre polynomials.

$$\int_{-1}^{1} f(x)P_m(x)\,dx = \frac{2}{2m+1} A_m$$

Since m is a dummy variable, we can write

$$A_n = \frac{2n+1}{2} \int_{-1}^{1} f(x)P_n(x)\,dx \tag{6.21}$$

The coefficients A_n are called **Legendre–Fourier coefficients.**

Summarizing these ideas, we can state the definition of a **Fourier–Legendre series.**

Definition 1. Let f be a function defined on $(-1, 1)$; then the Fourier–Legendre series of f is given by

$$f(x) = \sum_{n=0}^{\infty} A_n P_n(x)$$

provided the coefficients

$$A_n = \frac{2n+1}{2} \int_{-1}^{1} f(x)P_n(x)\,dx$$

all exist.

A theorem similar to the Fourier integral theorem involves Legendre polynomials.

THEOREM 1. If f is a piecewise smooth function in the interval $[-1, 1]$, the series

$$f(x) = \sum_{i=1}^{\infty} A_n P_n(x)$$

where A_n is given by Equation (6.21), will converge to

$$\frac{f(x^+) + f(x^-)}{2}$$

for every value of x in $(-1, 1)$. ■

EXAMPLE 4. Find the Fourier–Legendre series for

$$f(x) = \begin{cases} 0 & -1 < x < 0 \\ 1 & 0 < x < 1 \end{cases}$$

Solution. Using the formula in Equation (6.21), it follows that

$$A_0 = \frac{1}{2}\int_0^1 1 \cdot P_0\,dx = \frac{1}{2}\int_0^1 dx = \frac{1}{2}$$

$$A_1 = \frac{3}{2}\int_0^1 1 \cdot P_1\,dx = \frac{1}{2}\int_0^1 x\,dx = \frac{x^2}{4}\bigg|_0^1 = \frac{1}{4}$$

$$A_2 = \frac{5}{2}\int_0^1 1 \cdot P_2\,dx = \frac{5}{2}\int_0^1 \frac{1}{2}(3x^2 - 1)\,dx = \frac{5}{4}[x^3 - x]_0^1 = 0$$

$$A_3 = \frac{7}{2}\int_0^1 1 \cdot P_3\,dx = \frac{7}{4}\int_0^1 (5x^3 - 3x)\,dx = \frac{7}{4}\left[\frac{5}{4}x^4 - \frac{3}{2}x^2\right]_0^1 = -\frac{7}{16}$$

and the Fourier–Legendre series can be written as

$$f(x) = \frac{1}{2}P_0(x) + \frac{1}{4}P_1(x) - \frac{7}{16}P_3(x) + \cdots$$ ■

If we know beforehand whether $f(x)$ is even or odd over $(-1, 1)$, we can simplify the work required in computing the A_n much in the same way we did in the Fourier cosine and sine series. These results are stated in Definitions 2 and 3.

Definition 2. If $f(x)$ is even over $(-1, 1)$, then

$$f(x) = \sum_{n=0}^{\infty} A_{2n}P_{2n}(x)$$

where

$$A_{2n} = (4n + 1)\int_0^1 f(x)P_{2n}(x)\,dx \qquad \text{if they exist, } n = 0, 1, 2, \ldots$$

Definition 3. If $f(x)$ is odd over $(-1, 1)$, then

$$f(x) = \sum_{n=0}^{\infty} A_{2n+1} P_{2n+1}(x)$$

where

$$A_{2n+1} = (4n+3) \int_0^1 f(x) P_{2n+1}(x)\, dx \qquad \text{if they exist, } n = 0, 1, 2, \ldots$$

SECTION 5

CONCLUSION TO PROBLEM INVOLVING LEGENDRE'S EQUATION

Let us return to Section 6.1 and our problem concerning the temperature inside a mine. From our discussions, λ must equal $n(n+1)$ where $n = 0, 1, 2, \ldots$ in order that all solutions be bounded. The solution to Legendre's equation

$$\frac{d}{dx}\left[(1-x^2)\frac{dy}{dx}\right] + n(n+1)y = 0$$

is

$$y(x) = P_n(x) \qquad n = 0, 1, 2, \ldots$$

or

$$\Theta(\theta) = y(\cos\theta) = P_n(\cos\theta) \tag{6.22}$$

is the solution to Equation (6.2).

Going back to the Cauchy–Euler equation, we know its solution is

$$R_n(\rho) = C_n \rho^n + D_n \rho^{-n-1}$$

Since we are looking for the temperature inside the mine, it follows that the point $\rho = 0$ must be considered. It is necessary to set $D_n = 0$ since $\lim_{\rho \to 0^+} \rho^{-n-1} = +\infty$. Our solution of the Cauchy–Euler equation is

$$R(\rho) = \rho^n \tag{6.23}$$

Combining the solution in Equations (6.22) and (6.23), we find

$$T_n(\rho, \theta) = \rho^n P_n(\cos\theta)$$

and we assume the solution to our partial differential equation is

$$T(\rho, \theta) = \sum_{n=0}^{\infty} A_n \rho^n P_n(\cos\theta)$$

One of the boundary conditions still remains to be satisfied:

$$T(80, \theta) = \begin{cases} 0 & 0 \le \theta < \dfrac{\pi}{3} \\ 10 & \dfrac{\pi}{3} < \theta < \pi \end{cases} = \sum_{n=0}^{\infty} A_n(80)^n P_n(\cos\theta)$$

For convenience, let $x = \cos\theta$; then

$$T(80, x) = \begin{cases} 0 & \frac{1}{2} < x \le 1 \\ 10 & -1 \le x < \frac{1}{2} \end{cases}$$

From our study of the Fourier–Legendre series, we can write

$$\begin{aligned} A_n(80)^n &= \frac{2n+1}{2} \int_{-1}^{1/2} 10 \cdot P_n(x)\, dx + \frac{2n+1}{2} \int_{1/2}^{1} 0 \cdot P_n(x)\, dx \\ &= (10n+5) \int_{-1}^{1/2} P_n(x)\, dx \end{aligned}$$

Therefore,

$$A_n = \frac{10n+5}{80^n} \int_{-1}^{1/2} P_n(x)\, dx$$

The solution to our problem is given by

$$T(\rho, \theta) = \sum_{n=0}^{\infty} (10n+5) \left[\frac{\rho}{80}\right]^n \left[\int_{-1}^{1/2} P_n(s)\, ds\right] P_n(\cos\theta)$$

SECTIONS 3 - 5 EXERCISES

18. Show that the following functions are orthogonal over $[-1, 1]$.

(a) $P_1(x), P_3(x)$
(b) $P_0(x), P_4(x)$
(c) $x + 1, P_2(x)$

19. Let $f_1(x) = x + B, f_2(x) = \frac{1}{2}x^2 + Dx + E$. If $f_1(x)$ and $f_2(x)$ are orthogonal over $[-1, 1]$ and $\int_{-1}^{1} f_1^2(x)\, dx = \frac{2}{3}$ and $\int_{-1}^{1} f_1^2(x)\, dx = \frac{2}{5}$,

show that

$$\begin{aligned} f_1(x) &= x = P_1(x) \\ f_2(x) &= \tfrac{1}{2}(3x^2 - 1) = P_2(x) \end{aligned}$$

20. Prove $A_{2n} = (4n + 1) \displaystyle\int_0^1 f(x) P_{2n}(x)\, dx$ if f is even.

21. Prove $A_{2n+1} = (4n + 3) \displaystyle\int_0^1 f(x) P_{2n+1}(x)\, dx$ if f is odd.

22. Prove by integrating, $\int_{-1}^{1} P_3^2(x)\,dx = \frac{2}{7}$.

23. Write the Fourier–Legendre series

$$f(x) = \sum_{n=0}^{\infty} A_n P_n(x)$$

over the interval $[-1, 1]$ for the functions listed below. Find at least three nonzero terms if possible.

(a) $f(x) = 100 + x$
(b) $f(x) = a_0 x^2 + a_1 x + a_2$
(c) $f(x) = x$
(d) $f(x) = x^2$
(e) $f(x) = \sin x$ (leave in integral form)
(f) $f(x) = 100 + 50 \cos x$
(leave in integral form)

(g) $f(x) = \begin{cases} 1 & 0 \le x \le 1 \\ -1 & -1 \le x < 0 \end{cases}$

(h) $f(x) = |x|$

(i) $f(x) = \begin{cases} x & 0 \le x \le 1 \\ 0 & -1 \le x < 0 \end{cases}$

(j) $f(x) = \begin{cases} 2x & 0 < x < 1 \\ 0 & -1 < x < 0 \end{cases}$

(k) $f(x) = \begin{cases} 2 & -1 \le x < 0 \\ 2(1 - x) & 0 < x \le 1 \end{cases}$

(l) $f(x) = \begin{cases} 0 & -1 \le x \le \frac{1}{2} \\ 4 & \frac{1}{2} < x \le 1 \end{cases}$

24. Show that if $x = \cos \theta$,

$$\int_{-1}^{1} f(x) P_n(x)\,dx = \int_0^{\pi} [f(\cos\theta) P_n(\cos\theta)] \sin\theta\,d\theta$$

25. Compute the Fourier–Legendre series

$$f(\cos\theta) = \sum_{n=0}^{\infty} A_n P_n(\cos\theta)$$

over the interval $[0, \pi]$ for the functions listed below. Compute at least three nonzero terms if possible

(a) $f(\cos\theta) = \cos\theta$
(b) $f(\cos\theta) = \sin^2\theta$

(c) $f(\cos\theta) = \begin{cases} 1 & 0 \le x \le \dfrac{\pi}{2} \\ \cos\theta & \dfrac{\pi}{2} < x \le \pi \end{cases}$

(d) $f(\theta) = \theta$

□ ***Hint.*** Write θ in terms of $\cos\theta$. □

26. Find the Fourier–Legendre series

$$f(x) = \sum_{n=0}^{\infty} A_n P_n\left[\frac{x}{4}\right]$$

if

$$f(x) = \begin{cases} 4 - x & 0 \le x \le 4 \\ 0 & -4 \le x < 0 \end{cases}$$

27. Find the Fourier–Legendre series

$$f(x) = \sum_{n=0}^{\infty} A_n P_n(x)$$

over [1, 1] by extending $f(x)$ in (0, 1) as (1) an even function and (2) an odd function into (−1, 0) for the functions listed below.

(a) $f(x) = x^2$
(b) $f(x) = 1 - x^2$
(c) $f(x) = 1$
(d) $f(x) = \sin x$ (2 terms only)
(e) $f(x) = \begin{cases} x & 0 \le x \le \frac{1}{2} \\ 1 & \frac{1}{2} \le x \le 1 \end{cases}$ (3 terms only)

SECTION 6
PROBLEMS ON A HEMISPHERE

Very often the region in which our problem is set is not the whole sphere but just part of it, as for example, a hemisphere. We will show in Example 5 the added steps needed to solve this type of problem.

EXAMPLE 5. Consider a hemisphere of radius 0.1 meter on whose surface is a charge of uniform density. If the potential at any point on the surface is 100 and the potential on the plane $\theta = \pi/2$ is 0, find the electrostatic potential u at any point outside the hemisphere.

Solution. The electrostatic potential $u(\rho, \theta)$ satisfies Laplace's equation $\nabla^2 u = 0$ and on the surface u is given by

$$u(0.1, \theta) = 100 \qquad 0 \le \theta < \frac{\pi}{2}$$

$$u\left(\rho, \frac{\pi}{2}\right) = 0 \qquad 0 \le \rho < 0.1$$

As before we use Laplace's equation in spherical coordinates:

$$\nabla^2 u = \frac{1}{\rho^2} \frac{\partial}{\partial \rho}\left[\rho^2 \frac{\partial u}{\partial \rho}\right] + \frac{1}{\sin\theta} \frac{\partial}{\partial \theta}\left[\sin\theta \frac{\partial u}{\partial \theta}\right]$$

Letting $u(\rho, \theta) = R(\rho)\Theta(\theta)$, we are led to the ordinary differential equations and conditions

$$(\rho^2 R')' - \lambda R = 0$$

$$(\sin\theta\Theta')' + \lambda \sin\theta\Theta = 0 \qquad \Theta\left(\frac{\pi}{2}\right) = 0$$

Now we recall that $\lambda = n(n + 1)$ in order that all solutions be bounded.

The solution of the Legendre's equation is

$$\Theta_n(\theta) = P_n(\cos\theta) \qquad n = 0, 1, 2, \ldots$$

From the homogeneous boundary condition, we have

$$\Theta_n\left(\frac{\pi}{2}\right) = 0 = P_n\left(\cos\frac{\pi}{2}\right) = P_n(0)$$

Since $P_n(0) \neq 0$ for n even and $P_n(0) = 0$ for n odd,

$$\Theta_n(\theta) = P_n(\cos\theta) \qquad n = 1, 3, 5, \ldots$$

The other differential equation is a Cauchy–Euler equation of the form

$$\rho^2 R'' + 2\rho R' - n(n+1)R = 0$$

whose solution is

$$R_n(\rho) = C_n \rho^n + D_n \rho^{-n-1} \qquad n = 0, 1, 2, \ldots$$

Inasmuch as we are looking for the potential u outside of the hemisphere, we must set $C_n = 0$ because $\lim_{\rho \to +\infty} \rho^n = +\infty$. Therefore,

$$R_n(\rho) = \rho^{-n-1}$$

are solutions of the Cauchy–Euler equation.

Using the solutions of the differential equations, we can write

$$u_n(\rho, \theta) = \rho^{-n-1} P_n(\cos\theta) \qquad n = 1, 3, 5, \ldots$$

or

$$u_n(\rho, \theta) = \rho^{-2(n+1)} P_{2n+1}(\cos\theta) \qquad n = 0, 1, 2, \ldots$$

which are solutions of the partial differential equation satisfying the boundary condition $u(\rho, \pi/2) = 0$. Thus by the superposition principle, the most general solution to the differential equation takes the form

$$u(\rho, \theta) = \sum_{n=0}^{\infty} A_{2n+1} \rho^{-2(n+1)} P_{2n+1}(\cos\theta)$$

To find the A_{2n+1}'s, we use the boundary condition

$$u(0.1, \theta) = 100 = \sum_{n=0}^{\infty} A_{2n+1}(0.1)^{-2(n+1)} P_{2n+1}(\cos\theta) \tag{6.24}$$

To simplify our work let $x = \cos\theta$; then Equation (6.24) becomes

$$100 = \sum_{n=0}^{\infty} A_{2n+1}(0.1)^{-2(n+1)} P_{2n+1}(x)$$

Therefore, since we have only odd Legendre polynomials,

$$A_{2n+1}(0.1)^{-2(n+1)} = (4n+3) \int_0^1 100 P_{2n+1}(x)\, dx$$

from which we find

$$A_{2n+1} = (4n + 3)(0.1)^{2(n+1)} \int_0^1 100P_{2n+1}(x)\, dx$$

Using the result of Exercise 11, we have

$$A_{2n+1} = \frac{(4n + 3)(0.1)^{2(n+1)}100}{4n + 3} [P_{2n}(0) - P_{2n+2}(0)]$$

or

$$A_{2n+1} = 100(0.1)^{2(n+1)}[P_{2n}(0) - P_{2n+2}(0)]$$

The solution to our problem is

$$u(\rho, \theta) = 100 \sum_{n=0}^{\infty} \left(\frac{0.1}{\rho}\right)^{2(n+1)} [P_{2n}(0) - P_{2n+2}(0)]P_{2n+1}(\cos \theta)$$

■

SECTION 6 EXERCISES

28. The diameter of the sun is 1.4×10^6 kilometers. If the temperature at the surface is 20,000° C, find the temperature in space as a function of ρ and θ. What is the approximate temperature at earth if the distance from the sun to the earth is 1.5×10^8 kilometers?

29. The electrostatic potential on the surface of a sphere of radius 10 centimeters is $100 - \cos^2 \theta$. Find the potential outside of the sphere.

30. A hemisphere of radius 30 centimeters has a temperature on the curved surface of zero; the temperature on the bounding surface is 100. Find the temperature inside the hemisphere.

31. The radius of the earth is 6400 kilometers. If the gravitational potential on the surface is

$$f(\theta) = \begin{cases} 200 - \cos \theta & 0 \le \theta < \dfrac{\pi}{2} \\ 200 & \dfrac{\pi}{2} \le \theta \le \pi \end{cases}$$

find the potential at any point in space.

32. Repeat Exercise 31 but find the potential inside the earth.

33. A hemisphere of radius c has an electrostatic potential equal to $\cos \theta$ on the curved surface; the potential on the bounding plane through the "center" is zero. Find the potential outside of the hemisphere (see Figure 5).

34. If a sphere of radius c is placed in a stream whose velocity is $\bar{v}$, the velocity potential $u(r, \theta)$ satisfies Laplace's equation. Find u if the velocity potential on the circumference is $f(\theta)$.

35. A sphere of radius 10 has an electrostatic potential on the upper half of the circumference equal to 40; the rate of change of potential in the ρ direction on the lower half is zero. Find the potential $u(\rho, \theta)$ outside of the sphere.

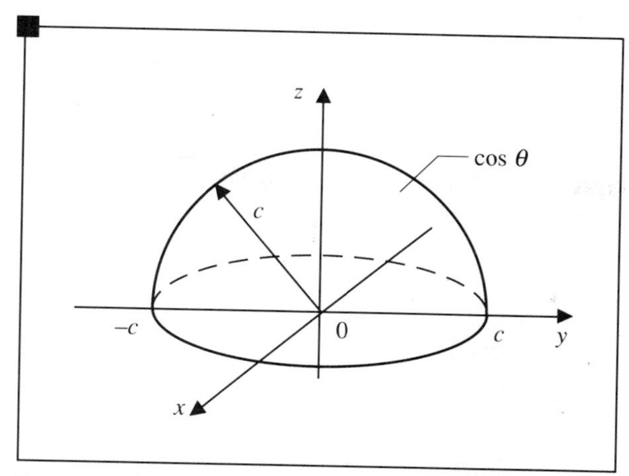

FIGURE 5 **Potential on Hemisphere**

36. A hemisphere of radius c has a temperature $f(\theta) = \cos^2 \theta$ on the curved portion of the hemisphere. The bounding plane $z = 0$ is perfectly insulated. Find the temperature inside the hemisphere.

37. The flat side of a hemisphere sets on a plate that is 50°; the curved surface is 100°. Find the temperature $u(r, \theta)$ inside the hemisphere.

□ ***Hint.*** Let $u(r, \theta) = w(r, \theta) + S(\theta)$. □

38. The electrostatic potential $u(r, \theta)$ on a hemisphere of radius 5 satisfies the boundary value problem

$$\nabla^2 u = 0$$

$$\frac{\partial u}{\partial \theta}\left(\rho, \frac{\pi}{2}\right) = 16$$

$$u(5, \theta) = 4\theta$$

Find u outside the hemisphere. Leave your answer in integral form.

CHAPTER SEVEN

Fourier Integral

SECTION 1

INTRODUCTION

In the examples of Chapter 4 we discussed a method for finding the solution of a boundary value problem on a finite region. By applying the technique of separation of variables, we were led to the problem of finding the coefficients of a Fourier series that satisfied a given nonhomogeneous condition over a finite region. These coefficients allowed us to write the solution of the boundary value problem.

In some situations a solution is sought for a boundary value problem set on a region where dimensions may be infinite. For example, consider the problem of finding the magnitude of an electric field produced by a horizontally polarized antenna when a voltage is impressed upon its length. In analyzing such a problem, we notice that the electric field starts at the antenna but radiates infinitely far in many directions (see Figure 1).

There are also boundary value problems that are set on regions where the dimensions are finite but very large. It may be convenient in such circumstances to view these large dimensions as infinite in length. Suppose we wish to find the voltage or current in a transatlantic cable. Although this cable is finite in length (4000 kilometers), it may be easier to consider a similar boundary value problem where the length of the cable is infinite.

In this chapter we will extend the technique of separation of variables to find the solutions of boundary value problems over infinite intervals. We will see that under these circumstances the Fourier series step will no longer work but must be replaced by an improper integral known as the Fourier integral.

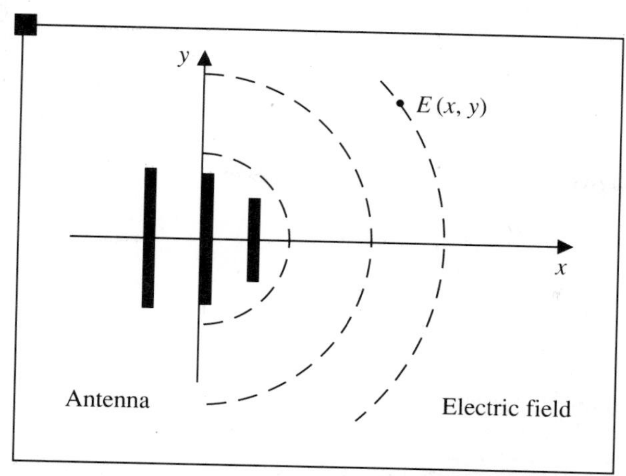

FIGURE 1 **Electric Field Produced by Antenna**

SECTION 2
THE FOURIER INTEGRAL

We will consider the solution of a boundary value problem over the two types of intervals defined in Definition 1.

Definition 1

1. A **semi-infinite** interval begins at a fixed point c and extends to infinity.
2. An **infinite** interval extends from $-\infty$ to $+\infty$ (see Figure 2).

Suppose a function f is defined on one of these intervals. It might be in the form of a periodic function with period $2p$ but it does not have to be. The function f might not repeat itself over the whole semi-infinite or infinite interval, or if it does repeat itself it may not be periodic (see Figure 3).

If the function is not periodic, then it *cannot* have a Fourier series representation. We might ask then, "Is there some way to express f in terms of sine

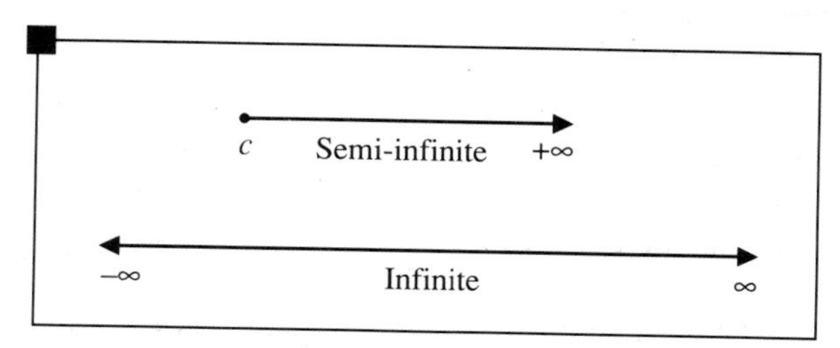

FIGURE 2 **Semi-Infinite and Infinite Intervals**

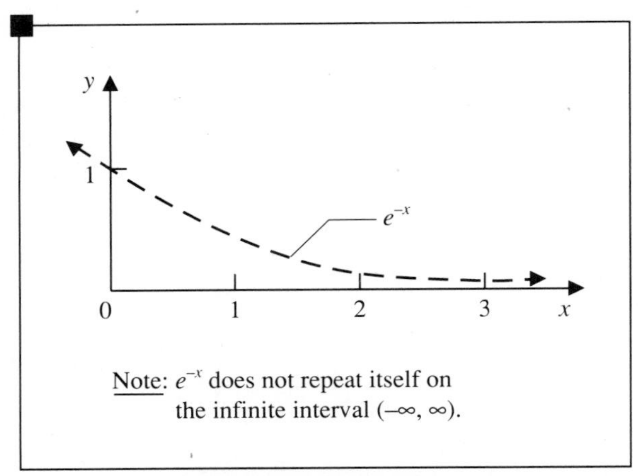

FIGURE 3 **Nonperiodic Function**

or cosine functions when f is not periodic?" The answer is yes; it is called the Fourier integral.

The actual derivation of the Fourier integral is quite involved and beyond the scope of this book. However, it is possible through a heuristic argument to make the result seem plausible.

Suppose we have a function $f(t)$ integrable on the infinite interval $(-\infty, \infty)$. Let f_p be the restriction of f to $(-p, p)$; that is,

$$f_p(t) = f(t) \qquad -p < t < p$$

as shown in Figure 4.

After a proper extension of f_p to a periodic function F_p on $(-\infty, \infty)$, we can write the Fourier series of $F_p(t)$, which takes the form

$$F_p(t) = \frac{a_0}{2} + \sum_{n=1}^{\infty} a_n \cos \frac{n\pi}{p} t + b_n \sin \frac{n\pi}{p} t$$

where

$$a_n = \frac{1}{p} \int_{-p}^{p} f_p(t) \cos \frac{n\pi}{p} t \, dt \qquad n = 0, 1, 2, \ldots$$

$$b_n = \frac{1}{p} \int_{-p}^{p} f_p(t) \sin \frac{n\pi}{p} t \, dt \qquad n = 1, 2, \ldots$$

Let

$$\omega_n = \frac{n\pi}{p}$$

and the difference equation

$$\Delta\omega_n = \omega_{n+1} - \omega_n = \frac{\pi}{p} \qquad \textbf{(7.1)}$$

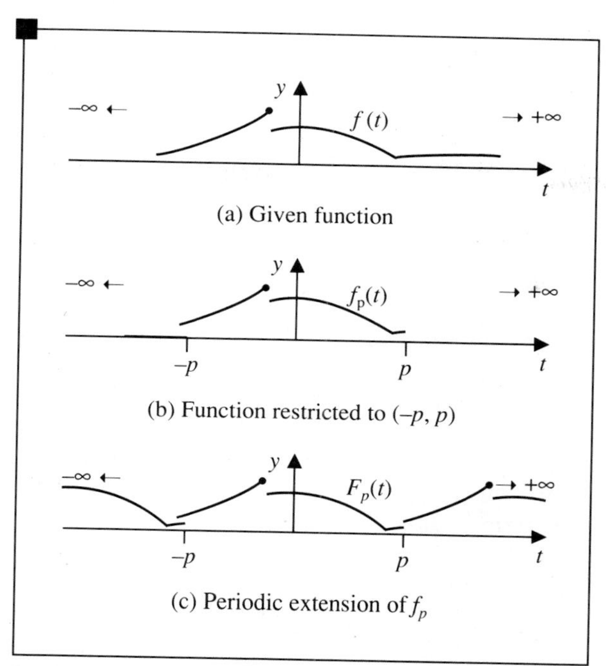

FIGURE 4 **Development of Periodic Function from Nonperiodic Function**

Using the notation above, the Fourier series for $F_p(t)$ can be written

$$F_p(t) = \frac{a_0}{2} + \sum_{n=1}^{\infty} \left\{ \left[\frac{1}{p} \int_{-p}^{p} f_p(s) \cos \omega_n s \, ds \right] \cos \omega_n t + \left[\frac{1}{p} \int_{-p}^{p} f_p(s) \sin \omega_n s \, ds \right] \sin \omega_n t \right\} \tag{7.2}$$

Using Equation (7.1) we see that

$$\frac{1}{p} = \frac{\Delta\omega_n}{\pi}$$

and replacing $1/p$ in Equation (7.2) we have

$$F_p(t) = \frac{a_0}{2} + \sum_{n=1}^{\infty} \left\{ \left[\frac{1}{\pi} \int_{-p}^{p} f_p(s) \cos \omega_n s \, ds \right] \cos \omega_n t \, \Delta\omega_n + \left[\frac{1}{\pi} \int_{-p}^{p} f_p(s) \sin \omega_n s \, ds \right] \sin \omega_n t \, \Delta\omega_n \right\} \tag{7.3}$$

If we let

$$A_n = \frac{1}{\pi} \int_{-p}^{p} f_p(s) \cos \omega_n s \, ds = \frac{p}{\pi} a_n \qquad n = 0, 1, 2, \ldots$$

and

$$B_n = \frac{1}{\pi}\int_{-p}^{p} f_p(s)\sin\omega_n s\,ds = \frac{p}{\pi}b_n \qquad n = 1, 2, \ldots$$

then Equation (7.3) can be written more simply as

$$F_p(t) = \frac{\pi}{2p}A_0 + \sum_{n=1}^{\infty}\{A_n(\omega_n)\cos\omega_n t\,\Delta\omega_n + B_n(\omega_n)\sin\omega_n t\,\Delta\omega_n\} \qquad (7.4)$$

Adding and subtracting $(\pi/p)A_0$ to Equation (7.4),

$$F_p(t) = -\frac{\pi}{2p}A_0 + \sum_{n=0}^{\infty}\{A_n(\omega_n)\cos\omega_n t\,\Delta\omega_n + B_n(\omega_n)\sin\omega_n t\,\Delta\omega_n\} \qquad (7.5)$$

Notice that the term $(\pi/p)A_0$ is taken inside the summation sign and the lower index changed from $n = 1$ to $n = 0$.

The next step in our argument is to let $p \to +\infty$ so that $\lim_{p\to\infty} f_p(t) = f(t)$, which is the function for which we are seeking a Fourier integral representation. First notice that the term $(\pi/2p)A_0$ goes to zero as $p \to +\infty$.

Now let us examine the terms inside the summation sign. The sum $\sum_0^{n-1} A_n(\omega_n)\cos\omega_n t\,\Delta\omega_n$ is similar to the Riemann sum for the function $A_n(\omega) \cdot \cos\omega t$ (t fixed). If we let $n \to +\infty$, it would appear that the Riemann sum approaches the area inside the rectangles from 0 to ∞ as shown in Figure 5.

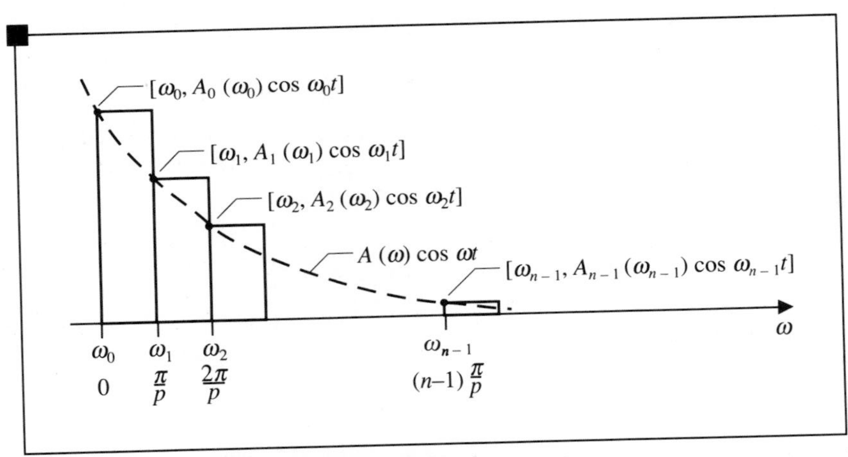

FIGURE 5 **Riemann Sum of $A(\omega)\cos\omega t$**

Suppose $A(\omega)$ is a continuous function such that $A(\omega_n) = A_n(\omega_n)$ for all n. Let us compare the area under the curve $A(\omega)\cos\omega t$, which is given by

$$\int_0^{\infty} A(\omega)\cos\omega t\,d\omega$$

within the area under the rectangles, given by

$$\sum_{n=0}^{\infty} A_n(\omega_n)\cos\omega_n t\,\Delta\omega_n$$

When p is relatively small, then the area inside the rectangles is not a good approximation to the area under the curve given by $A(\omega)\cos\omega t$ as shown in Figure 5. However, as p gets larger, the width of the rectangles gets smaller and the area inside the rectangles becomes a better approximation to the area given by the improper integral (see Figure 6). Therefore, it would appear that as $p \to +\infty$,

$$\int_0^{\infty} A(\omega)\cos\omega t\, d\omega = \sum_{n=0}^{\infty} A_n(\omega_n)\cos\omega_n t\,\Delta\omega_n$$

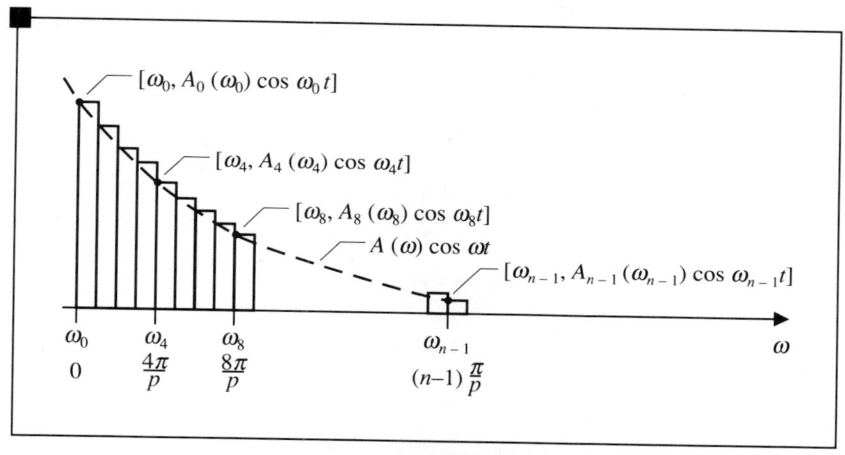

FIGURE 6 **Finer Partition for Riemann Sum of $A(\omega)$ cos ωt**

A similar argument would apply to the $B_n(\omega_n)\sin\omega_n t\,\Delta\omega_n$ terms. If we return to Equation (7.5), we see from the previous discussion that as $p \to +\infty$, $f_p(t) \to f(t)$ and

$$f(t) = \int_0^{\infty} [A(\omega)\cos\omega t + B(\omega)\sin\omega t]\, d\omega \tag{7.6}$$

where

$$A(\omega) = \frac{1}{\pi}\int_{-\infty}^{\infty} f(t)\cos\omega t\, dt \tag{7.7}$$

$$B(\omega) = \frac{1}{\pi}\int_{-\infty}^{\infty} f(t)\sin\omega t\, dt \tag{7.8}$$

The integral representation of $f(t)$ is called the **Fourier integral**, and $A(\omega)$ and $B(\omega)$ are called the **Fourier cosine** and **sine transform**, respectively, of the function $f(t)$. In certain situations $A(\omega)$ and $B(\omega)$ are called the **spectrum of $f(t)$**.

Remember that although the formulas above are correct under certain restrictions on f, their derivation is not rigorous. Theorem 1, called the **Fourier integral theorem**, puts this whole idea on a firm mathematical foundation.

THEOREM 1. Let f be a function that is piecewise smooth on every bounded interval and suppose that $\int_{-\infty}^{\infty} |f(t)|\, dt$ converges; that is, it is absolutely integrable over the interval $(-\infty, \infty)$. Then the Fourier integral

$$\int_0^{\infty} [A(\omega) \cos \omega t + B(\omega) \sin \omega t]\, d\omega$$

where $A(\omega)$ and $B(\omega)$ are given by Equations (7.7) and (7.8) converges to the value

$$\frac{f(t^+) + f(t^-)}{2}$$

at each point t belonging to $(-\infty, \infty)$.

If we substitute $A(\omega)$ and $B(\omega)$ [i.e., Equations (7.7) and (7.8)] into Equation (7.6), we have

$$f(t) = \frac{1}{\pi} \int_0^{\infty} \int_{-\infty}^{\infty} f(s)[\cos \omega s \cos \omega t + \sin \omega s \sin \omega t]\, ds\, d\omega$$

The factor in the brackets can be simplified using a well-known trigonometric identity, and therefore,

$$f(t) = \frac{1}{\pi} \int_0^{\infty} \int_{-\infty}^{\infty} f(s) \cos \omega(s - t)\, ds\, d\omega$$

which is an alternate form for the Fourier integral. ■

EXAMPLE 1. Given the function

$$f(t) = \begin{cases} t & -1 \le t \le 1 \\ 0 & \text{elsewhere} \end{cases}$$

which is shown in Figure 7, (a) find the Fourier cosine transform; (b) find the Fourier sine transform; and (c) find the Fourier integral.

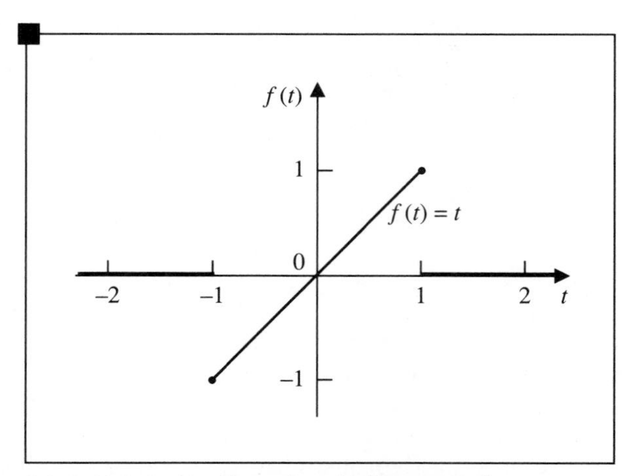

FIGURE 7 **Function in Example 1**

Solution

(a) Using Equation (7.7) we see that

$$A(\omega) = \frac{1}{\pi} \int_{-1}^{1} t \cos \omega t \, dt = 0$$

(b) Similarly from Equation (7.8) it follows that

$$\begin{aligned} B(\omega) &= \frac{1}{\pi} \int_{-1}^{1} t \sin \omega t \, dt \\ &= \frac{1}{\pi\omega^2} \int_{-1}^{1} \omega t \sin \omega t \, d(\omega t) = \frac{1}{\pi\omega^2} [\sin \omega t - \omega t \cos \omega t]_{-1}^{1} \\ &= \frac{2}{\pi\omega^2} [\sin \omega - \omega \cos \omega] \end{aligned}$$

(c) The Fourier integral in Equation (7.6) becomes

$$f(t) = \frac{2}{\pi} \int_{0}^{\infty} \frac{\sin \omega - \omega \cos \omega}{\omega^2} \sin \omega t \, d\omega$$

■

When we are dealing with the function f defined over a semi-infinite interval $(c, +\infty)$, it will be necessary to extend the given function into the remainder of the interval $(-\infty, \infty)$. This is similar to the extensions done in Section 5 of Chapter 3 to the functions defined on the half Fourier interval when we wished to write its Fourier series.

Various options are available when we do this extension. We can extend f into $(-\infty, c)$ by defining any "reasonable" bounded function over $(-\infty, c)$. In general, this approach leads to a lengthy solution.

Perhaps the best way to carry out this extension is to extend F in $(-\infty, c)$ as an even or odd function. Very often the boundary conditions force this option upon us. If the function f is extended in this way, then the work involved in computing its Fourier integral is greatly reduced.

THEOREM 2. If the function f is defined on the interval $(0, \infty)$ and is extended into $(-\infty, 0)$ as an even function f_e over $(-\infty, \infty)$, which meets the hypothesis of Theorem 1, then

$$f_e(t) = \int_{0}^{\infty} A(\omega) \cos \omega t \, d\omega$$

where

$$A(\omega) = \frac{2}{\pi} \int_{0}^{\infty} f(t) \cos \omega t \, dt$$

Proof. Using the formula in Equation (7.7), we have

$$\begin{aligned} A(\omega) = \frac{1}{\pi} \int_{-\infty}^{\infty} f_e(t) \cos \omega t \, dt &= \frac{1}{\pi} \int_{-\infty}^{0} f_e(t) \cos \omega t \, dt \\ &+ \frac{1}{\pi} \int_{0}^{\infty} f_e(t) \cos \omega t \, dt \end{aligned} \quad \textbf{(7.9)}$$

If in the first integral in the right-hand side we replace t by $-s$, it can be written as

$$\int_{-\infty}^{0} f_e(t) \cos \omega t \, dt = \int_{\infty}^{0} f_e(-s) \cos \omega(-s) \, d(-s) = \int_{0}^{\infty} f_e(t) \cos \omega t \, dt$$

Combining these results with the integral in Equation (7.9) we see that

$$A(\omega) = \frac{2}{\pi} \int_0^{\infty} f_e(t) \cos \omega t \, dt = \frac{2}{\pi} \int_0^{\infty} f(t) \cos \omega t \, dt \qquad \blacksquare$$

In a similar way we can show that $B(\omega) = 0$ if f is extended as an even function.

THEOREM 3. If the function f is defined on the interval $(0, \infty)$ and it is extended into $(-\infty, 0)$ as an odd function f_0 over $(-\infty, \infty)$, which needs the hypothesis of Theorem 1, then

$$f_0(t) = \int_0^{\pi} B(\omega) \sin \omega t \, d\omega$$

where

$$B(\omega) = \frac{2}{\pi} \int_0^{\infty} f(t) \sin \omega t \, dt$$

Proof. The proof for Theorem 3 is similar to Theorem 2. You are asked to prove Theorem 3 in Exercise 16. $\blacksquare$

SECTION 2 EXERCISES

1. Find the Fourier cosine transform and the Fourier sine transform for

(a) $f(t) = \begin{cases} 1 & -4 \le t \le 4 \\ 0 & \text{elsewhere} \end{cases}$

(b) $f(t) = \begin{cases} t & 0 \le t \le 1 \\ 0 & \text{elsewhere} \end{cases}$

(c) $f(t) = \begin{cases} t^2 & -1000 \le t \le 1000 \\ 0 & \text{elsewhere} \end{cases}$

(d) $f(t) = \begin{cases} -h & -1 < t < 0 \\ h & 0 < t < 1 \\ 0 & \text{elsewhere} \end{cases}$

(e) $f(t) = \begin{cases} e^{-2t} & 0 \le t \\ 0 & \text{elsewhere} \end{cases}$

(f) $f(t) = \begin{cases} e^{-t} & 0 \le t \\ e^{t} & t < 0 \end{cases}$

2. Find the Fourier cosine transform and Fourier sine transform of $u(t)$ where

$$u(t) = \begin{cases} 0 & t < 0 \\ 1 & 0 < t \end{cases}$$

(i.e., a unit step function).

☐ ***Hint.*** Let $f(t) = u(t)e^{-bt}$, $b > 0$. ☐

3. Find the Fourier sine transform of $f(t) = u(t) \sin at$.

☐ ***Hint.*** See Exercise 2.

4. Find the Fourier cosine transform of

$$f(t) = \begin{cases} |t| & t^2 < 4 \\ 0 & \text{elsewhere} \end{cases}$$

5. If $f(t)$ is an even function, show that

(a) $A(\omega) = \dfrac{2}{\pi} \displaystyle\int_0^{\infty} f(t) \cos \omega t \, dt$

(b) $B(\omega) = 0$

6. If $f(t)$ is an odd function, show that

(a) $A(\omega) = 0$

(b) $B(\omega) = \dfrac{2}{\pi} \displaystyle\int_0^{\infty} f(t) \sin \omega t \, dt$

Find the Fourier integral in Exercises 7–11.

7. $f(t) = \begin{cases} h & -1 \le t \le 1 \\ 0 & \text{elsewhere} \end{cases}$

8. $f(t) = \begin{cases} 1 - t & 0 \le t \le 1 \\ 0 & \text{elsewhere} \end{cases}$

9. $f(t) = \begin{cases} 1 - t & 0 < t < 1 \\ 1 + t & -1 < t < 0 \\ 0 & \text{elsewhere} \end{cases}$

10. $f(t) = \begin{cases} e^{-t} & 0 < t \\ e^{t} & t < 0 \end{cases}$

11. $f(t) = \begin{cases} \sin t & -\pi < t < \pi \\ 0 & \text{elsewhere} \end{cases}$

12. Write the Fourier integral in Exercises 7–11 in alternate form.

SECTION 3
AN APPLICATION TO A PHYSICAL PROBLEM

EXAMPLE 2. We wish to find the temperature in an infinite rod that is insulated laterally. If the initial temperature $f(x)$ is 400° K, $-4 \leq x \leq 4$, and 0° K outside of this interval, find the temperature $u(x, t)$ in the rod. We are working under the assumption that $u(x, t)$ is bounded on the rod for all $t \geq 0$.

Solution. Formally this problem can be stated as

$$\frac{\partial^2 u}{\partial x^2} = \frac{1}{k}\frac{\partial u}{\partial t} \qquad -\infty < x < \infty,\ 0 < t$$

$$u(x, 0) = \begin{cases} 400 & -4 \leq x \leq 4 \\ 0 & 4 < |x| < \infty \end{cases}$$

We use separation of variables as before; that is, we assume that $u(x, t) = X(x)T(t)$. Then after substituting into the partial differential equation, we have

$$X''T = \frac{1}{k}XT'$$

This equation can be separated as follows:

$$\frac{X''}{X} = \frac{1}{k}\frac{T'}{T} = -\lambda$$

from which we get the two ordinary differential equations

$$X'' + \lambda X = 0$$

$$T' + k\lambda T = 0$$

The general solution of the equation $X'' + \lambda X = 0$ where $\lambda = \alpha^2 > 0$ is

$$X(x) = \bar{A}(\alpha)\cos \alpha x + \bar{B}(\alpha)\sin \alpha x$$

Notice that this solution is bounded for all x and since we have no further condition to restrict α, any $\alpha \geq 0$ is acceptable. (We need not consider solutions for $\alpha < 0$ because they can be combined, using trigonometry, with the corresponding solutions for $\alpha > 0$.)

Are there any solutions when $\lambda = 0$ or $\lambda < 0$? We investigate the answer to this question by noting that the general solution of the differential equation for $\lambda = 0$ is

$$X(x) = \bar{A} + \bar{B}x$$

We must set $\bar{B} = 0$; otherwise, $\lim_{x \to +\infty} X(x)$ would be unbounded. Therefore, $X(x) = \bar{A} = \text{constant}$ is a bounded solution. This case can be included under the case $\cos \alpha x$, $\alpha = 0$.

When $\lambda < 0$, the general solution can be written

$$X(x) = \bar{A}(\alpha)\cosh \alpha x + \bar{B}(\alpha)\sinh \alpha x$$

Unless both $\bar{A}(\alpha) = \bar{B}(\alpha) = 0$, the previous solution will be unbounded on $(-\infty, \infty)$.

We now solve the other differential equation, which yields

$$T(t) = \bar{C}(\alpha)e^{-k\alpha^2 t}$$

Combining these two solutions, we write

$$u_\alpha(x, t) = [A(\alpha) \cos \alpha x + B(\alpha) \sin \alpha x]e^{-k\alpha^2 t} \tag{7.10}$$

where $A(\alpha) = \bar{A}(\alpha)\bar{C}(\alpha)$ and $B(\alpha) = \bar{B}(\alpha)\bar{C}(\alpha)$. The expression in Equation (7.10) satisfies the partial differential equation and boundedness condition.

We continue by trying a solution of the form

$$u(x, t) = \int_0^\infty [A(\alpha) \cos \alpha x + B(\alpha) \sin \alpha x]e^{-k\alpha^2 t}\, d\alpha$$

from which it follows that

$$u(x, 0) = f(x) = \int_0^\infty [A(\alpha) \cos \alpha x + B(\alpha) \sin \alpha x]\, d\alpha$$

where

$$A(\alpha) = \frac{1}{\pi}\int_{-\infty}^{\infty} f(x) \cos \alpha x\, dx = \frac{400}{\pi}\int_{-4}^{4} \cos \alpha x\, dx = \frac{800}{\pi\alpha} \sin 4\alpha$$

$$B(\alpha) = \frac{1}{\pi}\int_{-\infty}^{\infty} f(x) \sin \alpha x\, dx = \frac{400}{\pi}\int_{-4}^{4} \sin \alpha x\, dx = 0$$

The solution to our problem is

$$u(x, t) = \frac{800}{\pi}\int_0^\infty \frac{\sin 4\alpha}{\alpha} e^{-k\alpha^2 t} \cos \alpha x\, d\alpha$$

■

We next examine a problem defined over a semi-infinite interval.

EXAMPLE 3. Consider a long wire fixed at $x = 0$ and stretching to infinity. If, while at rest, the wire is struck with a felt mallet so as to impose the initial velocity $x/100$ m/s over $0 \le x < 1$ and zero elsewhere, find the displacement $u(x, t)$.

Solution. Formally this boundary value problem can be stated

$$c^2 \frac{\partial^2 u}{\partial x^2} = \frac{\partial^2 u}{\partial t^2} \qquad 0 < x,\ 0 < t,$$

$$u(0, t) = 0 \qquad 0 < t$$

$$u(x, t) \text{ is bounded} \qquad \text{for } 0 < x < \infty,\ 0 \le t$$

$$u(x, 0) = 0 \qquad 0 < x$$

$$\frac{\partial u}{\partial t}(x, 0) = f(x) = \begin{cases} \dfrac{x}{100} & 0 \le x < 1 \\ 0 & 1 \le x < \infty \end{cases}$$

Using the technique of separation of variables by setting $u(x, t) = X(t)T(t)$, we arrive at the ordinary differential equations and conditions

$$X'' + \lambda X = 0 \qquad X(0) = 0$$
$$T'' + \lambda c^2 T = 0 \qquad T(0) = 0$$

The general solution of the first equation for $\lambda = \alpha^2 > 0$ is

$$X(x) = A(\alpha) \cos \alpha x + B(\alpha) \sin \alpha x$$

Applying the boundary condition $X(0) = 0$, we find $A(\alpha) = 0$ and

$$X(x) = B(\alpha) \sin \alpha x$$

It follows from the boundary condition and boundedness that there are no nontrivial solutions corresponding to $\lambda = 0$ or $\lambda < 0$. Since there are no further boundary conditions, α can be any number $\alpha > 0$.

Solving the other differential equation and its initial condition, we find

$$T(t) = \sin \alpha ct$$

Combining these results we assume the solution to the problem is

$$u(x, t) = \int_0^\infty B(\alpha) \sin \alpha x \sin \alpha ct \, d\alpha$$

To meet the remaining initial condition, we see that

$$\frac{\partial u}{\partial t}(x, 0) = f(x) = \int_0^\infty \alpha c B(\alpha) \sin \alpha x \, d\alpha$$

This indefinite integral is of the type found in Theorem 3, and therefore,

$$\begin{aligned} \alpha c B(\alpha) &= \frac{2}{\pi} \int_0^\infty f(x) \sin \alpha x \, dx = \frac{2}{\pi} \int_0^1 \frac{x}{100} \sin \alpha x \, dx \\ &= \frac{\pi}{50\alpha^2} [\sin \alpha - \alpha \cos \alpha] \end{aligned}$$

The final answer to the boundary value problem is

$$u(x, t) = \frac{\pi}{50c} \int_0^\infty \frac{\sin \alpha - \alpha \cos \alpha}{\alpha^3} \sin \alpha x \sin \alpha ct \, d\alpha$$

■

SECTION 3 EXERCISES

Solve the boundary value problems in Exercises 15–18 using separation of variables and the Fourier integral.

13. $u_{xx} = \frac{1}{k} u_t \qquad -\infty < x < \infty, \ 0 < t$

$$u(x, 0) = \begin{cases} 1 & 0 < x < 1 \\ 0 & \text{elsewhere} \end{cases}$$

14. $c^2u_{xx} = u_{tt} \qquad -\infty < x < \infty,\ 0 < t$

$$u(x, 0) = \begin{cases} 4 + x & -4 < x < 0 \\ 4 - x & 0 < x < 4 \\ 0 & \text{elsewhere} \end{cases}$$

$$\frac{\partial u}{\partial t}(x, 0) = 0 \qquad -\infty < x < \infty$$

15. $e_{xx} = RCe_t \qquad 0 < x < \infty,\ 0 < t$

$e(0, t) = 0 \qquad 0 < t$

$$e(x, 0) = \begin{cases} 5 & 0 < x < 10 \\ 0 & \text{elsewhere} \end{cases}$$

16. $c^2u_{xx} = u_{tt} \qquad 0 < x < \infty$

$$\frac{\partial u}{\partial x}(0, t) = 0 \qquad 0 < t$$

$u(x, 0) = e^{-x} \qquad 0 \le x$

$$\frac{\partial u}{\partial t}(x, 0) = 0 \qquad 0 \le x$$

17. A submarine cable is so long it can be considered semi-infinite in length. The voltage $e(x, t)$ satisfies the differential equation

$$\frac{\partial^2 e}{\partial x^2} = RC\frac{\partial e}{\partial t}$$

If the voltage at $x = 0$ is zero and the initial voltage is $f(x) = x$, $0 < x < 4$; $f(x) = 0$, $4 < x$, find the voltage in the cable.

18. Consider the temperature in an infinite rod that is insulated laterally. If the initial temperature is $f(x)$ where

$$f(x) = \begin{cases} e^{-2x} & 0 < x \\ e^{2x} & x < 0 \end{cases}$$

find the temperature $u(x, t)$ in the rod.

19. A semi-infinite string is fixed at $x = 0$. If the initial displacement is $f(x) = \sin x$, $0 < x < \pi$; $f(x) = 0$, $\pi < x$, and the initial velocity is $g(x) = 0$, $0 < x < \infty$, find the displacement $u(x, t)$ of the string.

20. A high-frequency line is semi-infinite in length. Suppose that originally the current on the line is $i(x, 0) = I_0$ and $i_t(x, 0) = 0$. Then the end at $x = 0$ is opened so that no current flows. Find the current in the line as a function of x and t.

21. An infinite string has the initial conditions $u(x, 0) = 0$ and

$$\frac{\partial u}{\partial t}(x, 0) = g(x)$$

Find the displacement of the string as a function of x and t.

22. Consider a semi-infinite bar whose longitudinal displacement is 50 millimeters at $x = 0$. The initial displacement is zero and the initial velocity is $g(x)$. Let $u(x, t) = v(x, t) + \varphi(x)$.

(a) Find $\varphi(x)$ so that $v(x, t)$ satisfies a homogeneous boundary problem.
(b) Write out but do not solve the boundary value problem $v(x, t)$ must satisfy.

23. A semi-infinite rod insulated laterally satisfies the heat equation

$$u_{xx} + \frac{1}{k} u_t = Ae^{-x}$$

and conditions

$$u(0, t) = 0 \qquad 0 < t$$
$$u(x, 0) = f(x) \qquad 0 < x < \infty$$

(a) Let $u(x, t) = v(x, t) + \varphi(x)$. Find $\varphi(x)$ so that $v(x, t)$ satisfies a homogeneous boundary value problem.
(b) Write but do not solve the boundary value problem $v(x, t)$ must satisfy.

24. The telegraph equation is given by

$$e_{xx} = LCe_{tt} + (LG + RC)e_t + RGe$$

Suppose the voltage $e(x, t)$ on a semi-infinite line satisfies the telegraph equation and

$$\frac{\partial e}{\partial x}(0, t) = 0 \qquad 0 < t$$

If the initial voltage is $f(x)$ where

$$f(x) = \begin{cases} 5 & 0 < x < 4 \\ 0 & 4 < x \end{cases}$$

and

$$\frac{\partial e}{\partial t}(x, 0) = 0 \qquad 0 < x < t$$

find the voltage e as a function of x and t along the line.

SECTION 4
THE COMPLEX FORM OF FOURIER INTEGRAL

We have already seen that

$$f(t) = \int_0^{\infty} [A(\omega) \cos \omega t + B(\omega) \sin \omega t] \, d\omega \tag{7.11}$$

where

$$A(\omega) = \frac{1}{\pi} \int_{-\infty}^{\infty} f(s) \cos \omega s \, ds$$

$$B(\omega) = \frac{1}{\pi} \int_{-\infty}^{\infty} f(s) \sin \omega s \, ds$$

Using Euler's formula $e^{i\theta} = \cos \theta + i \sin \theta$, we can write Equation (7.11) in the form

$$f(t) = \int_0^{\infty} \left[A(\omega) \frac{e^{i\omega t} + e^{-i\omega t}}{2} + B(\omega) \frac{e^{i\omega t} - e^{-i\omega t}}{2i} \right] d\omega$$

The expression on the right-hand side can be rewritten as

$$f(t) = \frac{1}{2}\int_0^{\infty} \{[A(\omega) - iB(\omega)]e^{i\omega t} + [A(\omega) + iB(\omega)]e^{-i\omega t}\}\, d\omega \tag{7.12}$$

Let

$$F(\omega) = \frac{1}{2}[A(\omega) - iB(\omega)] = \frac{1}{2\pi}\int_{-\infty}^{\infty} f(s)e^{-i\omega s}\, ds$$

and

$$\bar{F}(\omega) = \frac{1}{2}[A(\omega) + iB(\omega)] = \frac{1}{2\pi}\int_{-\infty}^{\infty} f(s)e^{i\omega s}\, ds$$

Now

$$\bar{F}(-\omega) = \frac{1}{2}[A(-\omega) + iB(-\omega)] = \frac{1}{2}[A(\omega) - iB(\omega)] = F(\omega)$$

Rewriting Equation (7.12) we see that

$$f(t) = \int_0^{\infty} F(\omega)e^{i\omega t}\, d\omega + \int_0^{\infty} \bar{F}(\omega)e^{-i\omega t}\, d\omega$$

If in the second integral, ω is replaced by $-u$, we have

$$\begin{aligned}\int_0^{\infty} \bar{F}(\omega)e^{-i\omega t}\, d\omega &= \int_0^{-\infty} \bar{F}(-u)e^{iut}\, d(-u)\\ &= \int_{-\infty}^{0} F(u)e^{iut}\, du = \int_{-\infty}^{0} F(\omega)e^{i\omega t}\, d\omega\end{aligned}$$

Therefore,

$$f(t) = \int_0^{\infty} F(\omega)e^{i\omega t}\, d\omega + \int_{-\infty}^{0} F(\omega)e^{i\omega t}\, d\omega$$

or

$$f(t) = \int_{-\infty}^{\infty} F(\omega)e^{i\omega t}\, d\omega \tag{7.13}$$

where

$$F(\omega) = \frac{1}{2\pi}\int_{-\infty}^{\infty} f(t)e^{-i\omega t}\, dt \tag{7.14}$$

The right-hand sides of Equations (7.14) and (7.13) are called the **Fourier transform of $f(t)$** and the **inverse Fourier transform of $F(\omega)$**, respectively. Sometimes these two equations are spoken of as **Fourier transform pairs**. At times we will write $\mathscr{F}[f(t)] = F(\omega)$ and $\mathscr{F}^{-1}[F(\omega)] = f(t)$.

EXAMPLE 4.

(a) What is the Fourier transform of a rectangular pulse given by

$$f(t) = \begin{cases} h & |t| < a \\ 0 & \text{elsewhere} \end{cases}$$

(see Figure 8.)

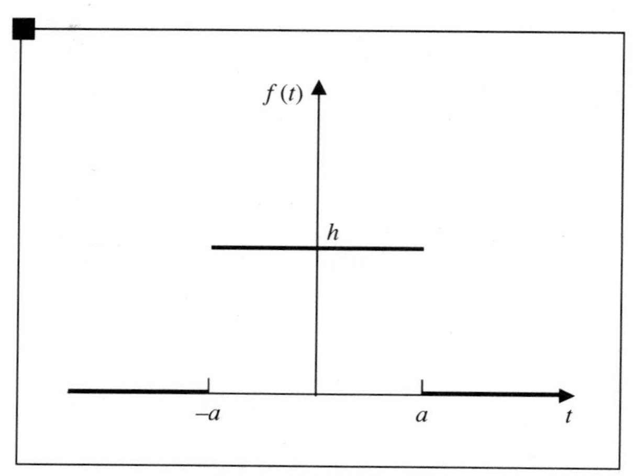

FIGURE 8 **Function in Example 4(a)**

(b) What is the Fourier transform of

$$f(t) = \begin{cases} e^{-bt} & 0 < t \\ 0 & t < 0 \end{cases}$$

(See Figure 9.)

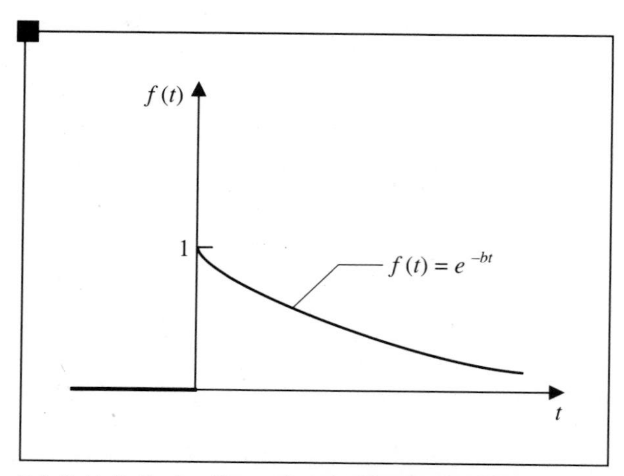

FIGURE 9 **Function in Example 4(b)**

Solution

(a) From Equation (7.14) we see that

$$\begin{aligned} F(\omega) &= \frac{1}{2\pi}\int_{-a}^{a} he^{-i\omega t}\,dt \\ &= \frac{h}{-i2\pi\omega}\int_{-a}^{a} e^{-i\omega t}\,d(-i\omega t) \\ &= \frac{h}{-i2\pi\omega}(e^{-i\omega a} - e^{i\omega a}) \\ &= \frac{2h}{2\pi\omega}\,\frac{e^{i\omega a} - e^{-i\omega a}}{2i} = \frac{h}{\omega\pi}\sin\omega a \end{aligned}$$

(b) Using (7.14) it follows that

$$\begin{aligned} F(\omega) &= \frac{1}{2\pi}\int_{0}^{\infty} e^{-bt}e^{-i\omega t}\,dt = \frac{1}{2\pi}\int_{0}^{\infty} e^{-(b+i\omega)t}\,dt \\ &= \frac{1}{-2\pi(b+i\omega)}\,e^{-(b+i\omega)t}\Big|_{0}^{\infty} \\ &= \frac{1}{2\pi}\left[\frac{1}{b+i\omega} - \lim_{d\to+\infty}\frac{1}{b+i\omega}\,e^{-bd}(\cos\omega d - i\sin\omega d)\right] \\ &= \frac{1}{2\pi(b+i\omega)} \quad \text{provided } b > 0 \end{aligned}$$

■

Although we cannot directly compute $\mathscr{F}[u(t)]$ where

$$u(t) = \begin{cases} 0 & t < 0 \\ 1 & 0 < t \end{cases}$$

we can find it by using the result from Example 4(b) and observing that $\lim_{b\to 0^+} e^{-bt} = 1$ (see Figure 10). Then

$$\mathscr{F}[u(t)] = \lim_{b\to 0^+}\mathscr{F}(e^{-bt}) = \frac{1}{i2\pi\omega}$$

As in the case with Laplace transforms, we build a table of Fourier transforms of well-known functions. Table 1 lists some of these Fourier transforms. In this table we will assume the functions listed in the left-hand column and their derivatives are piecewise smooth on every finite interval.

We will now justify some of the entries in Table 1.

Transform 2. Using Equation (7.14) we write the integral

$$\mathscr{F}[kf(t)] = \frac{1}{2\pi}\int_{-\infty}^{\infty} kf(t)e^{-i\omega t}\,dt = k\int_{-\infty}^{\infty} f(t)e^{-i\omega t}\,dt = kF(\omega)$$

Transform 3. From Equation (7.14) we have

$$\mathscr{F}[f(at)] = \frac{1}{2\pi}\int_{-\infty}^{\infty} f(at)e^{-i\omega t}\,dt$$

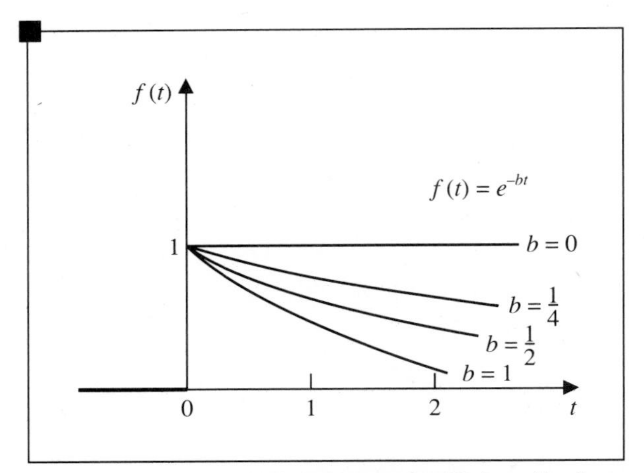

FIGURE 10 **Using Example 4(b) to Evaluate $\mathscr{F}[u(t)]$**

Let $\tau = at$, and the integral above becomes

$$\mathscr{F}[f(at)] = \frac{1}{2\pi}\int_{-\infty}^{\infty} f(\tau)e^{-i(\omega/a)\tau}\,\frac{d\tau}{a} = \frac{1}{a}F\left(\frac{\omega}{a}\right)$$

Transform 4. Starting with Equation (7.14) we write

$$\begin{aligned}\mathscr{F}[e^{-bt}f(t)] &= \frac{1}{2\pi}\int_{-\infty}^{\infty} e^{-bt}f(t)e^{-i\omega t}\,dt\\ &= \frac{1}{2\pi}\int_{-\infty}^{\infty} f(t)e^{-i(\omega-ib)t}\,dt\\ &= F(\omega - ib)\end{aligned}$$

Transform 5. To find the Fourier transform of the derivative of a function, we again start with Equation (7.14), which yields

$$\begin{aligned}\mathscr{F}[f'(t)] = \frac{1}{2\pi}\int_{-\infty}^{\infty} f'(t)e^{-i\omega t}\,dt &= \lim_{c\to-\infty}\frac{1}{2\pi}\int_{c}^{0} f'(t)e^{-i\omega t}\,dt\\ &+ \lim_{b\to+\infty}\frac{1}{2\pi}\int_{0}^{b} f'(t)e^{-i\omega t}\,dt\end{aligned}$$

Using integration by parts on both integrals on the right-hand side, we see that

$$\begin{aligned}\mathscr{F}[f'(t)] &= \lim_{c\to-\infty}\frac{1}{2\pi}\left[f(t)e^{-i\omega t}\Big|_{c}^{0} + \frac{i\omega}{2\pi}\int_{c}^{0} f(t)e^{-i\omega t}\,dt\right]\\ &+ \lim_{b\to+\infty}\frac{1}{2\pi}\left[f(t)e^{-i\omega t}\Big|_{0}^{b} + \frac{i\omega}{2\pi}\int_{0}^{b} f(t)e^{-i\omega t}\,dt\right]\\ &= -\frac{f(0)}{2\pi} + \frac{f(0)}{2\pi} + \frac{i\omega}{2\pi}\int_{-\infty}^{+\infty} f(t)e^{-i\omega t}\,dt\\ &= i\omega F(\omega)\end{aligned}$$

TABLE 1 **Some Fourier Transforms**

	$f(t)$	$F(\omega) = \mathscr{F}[f(t)]$	
1.	$g(t) \pm h(t)$	$G(\omega) \pm H(\omega)$	
2.	$kf(t)$	$kF(\omega)$	
3.	$f(at)$	$\frac{1}{a} F\left(\frac{\omega}{a}\right)$	
4.	$e^{-bt}f(t)$	$F(\omega - ib)$	
5.	$f'(t)\left[\lim_{\lvert t\rvert \to +\infty} f'(t) = 0\right]$	$i\omega F(\omega)$	
6.	$f''(t)$	$-\omega^2 F(\omega)$	
7.	$f^{(n)}(t)$	$(i\omega)^n F(\omega)$	
8.	$\int_{-\infty}^{t} f(s)\, ds$	$\frac{F(\omega)}{i\omega}$	
9.	$u(t)$	$\frac{1}{i2\pi\omega}$	
10.	$u(t)e^{-bt}$	$\frac{1}{2\pi(b + i\omega)}$	$b > 0$
11.	$u(t)t$	$-\frac{1}{2\pi\omega^2}$	
12.	$u(t) \sin at$	$\frac{1}{2\pi} \frac{a}{a^2 - \omega^2}$	
13.	$u(t) \cos at$	$\frac{1}{2\pi} \frac{i\omega}{a^2 - \omega^2}$	
14.	$e^{-bt}u(t)t$	$\frac{1}{2\pi} \frac{1}{(b + i\omega)^2}$	$b > 0$
15.	$e^{-bt}u(t) \sin at$	$\frac{1}{2\pi} \frac{a}{(b + i\omega)^2 + a^2}$	$b > 0$
16.	$e^{-bt}u(t) \cos at$	$\frac{1}{2\pi} \frac{b + i\omega}{(b + i\omega)^2 + a^2}$	$b > 0$
17.	$u(t - \tau)$	$\frac{1}{2\pi} \frac{e^{-i\omega\tau}}{i\omega}$	

Transform 8. Let $q(t) = \int_{-\infty}^{t} f(s)\, ds$. Then

$$\mathscr{F}[q(t)] = \frac{1}{2\pi} \int_{-\infty}^{\infty} q(t)e^{i\omega t}\, dt = Q(\omega)$$

And using Transform 5,

$$\mathscr{F}[q'(t)] = i\omega Q(\omega)$$

But $q'(t) = f(t)$ and

$$\mathscr{F}[q'(t)] = \mathscr{F}[f(t)] = F(\omega) = i\omega Q(\omega)$$

$$\mathscr{F}\left[\int_{\infty}^{t} f(s)\, ds\right] = \mathscr{F}[q(t)] = \frac{F(\omega)}{i\omega}$$

Transform 12. $\mathscr{F}[u(t)\sin at]$ cannot be done directly but can be found by evaluating $\mathscr{F}[u(t)e^{-bt}\sin at]$ and then letting $b \to 0^+$. As before we can write

$$\begin{aligned}\mathscr{F}[u(t)e^{-bt}\sin at] &= \lim_{c\to+\infty} \frac{1}{2\pi}\int_0^c e^{-(b+i\omega)t}\sin at\,dt \\ &= \lim_{c\to+\infty} -\frac{1}{2\pi}\frac{e^{-(b+i\omega)t}[a\cos at + (b+i\omega)\sin at]}{a^2+(b+i\omega)^2}\Bigg|_0^c \\ &= \frac{1}{2\pi}\frac{a}{a^2+(b+i\omega)^2}\end{aligned}$$

To find our transform we see that

$$\begin{aligned}\mathscr{F}[u(t)\sin at] &= \lim_{b\to0^+} \mathscr{F}[u(t)e^{-bt}\sin at] \\ &= \lim_{b\to0^+} \frac{a}{2\pi[a^2+(b+i\omega)^2]} \\ &= \frac{1}{2\pi}\frac{a}{a^2-\omega^2}\end{aligned}$$

SECTION 4 EXERCISES

Find the Fourier transform of $f(t)$ in Exercises 27–37.

25. $f(t) = \begin{cases} 1 & 0 < t < 1 \\ 0 & \text{elsewhere} \end{cases}$

26. $f(t) = \begin{cases} h & 0 < t < a \\ -h & -a < t < 0 \\ 0 & \text{elsewhere} \end{cases}$

27. $f(t) = \begin{cases} 2+t & -2 < t < 0 \\ 2-t & 0 < t < 2 \\ 0 & \text{elsewhere} \end{cases}$

28. $f(t) = \begin{cases} |t| & t^2 < 3 \\ 0 & \text{elsewhere} \end{cases}$

29. $f(t) = \begin{cases} 1-e^{-t} & 0 < t < 1 \\ 0 & \text{elsewhere} \end{cases}$

30. $f(t) = g(t) \pm h(t)$

31. $f(t) = u(t)e^{-bt}\cos at \qquad b > 0,\ -\infty < t < \infty$

□ ***Note.*** $u(t) = \begin{cases} 1 & 0 < t \\ 0 & t < 0 \end{cases}$ □

32. $f(t) = u(t)te^{-bt} \qquad b > 0,\ -\infty < t < \infty$

33. $f(t) = u(t) \cos at$ (See Exercise 33.)

34. $f(t) = u(t)t$ (See Exercise 34.)

35. If $\int_{-\infty}^{\infty} |f(t)|\, dt$ converges, then $\lim_{t\to\pm\infty} |f(t)| = 0$. Prove $\lim_{t\to\pm\infty} |f(t)| = 0$ implies that $\lim_{t\to\pm\infty} f(t) = 0$.

36. Show that the Fourier transform of $f'(t)$ is given by

$$\mathscr{F}[f'(t)] = i\omega\mathscr{F}[f(t)]$$

Assume $f'(t)$ is piecewise smooth.

37. Show that the Fourier transform of $f^{(n)}(t)$ is given by

$$\mathscr{F}[f^{(n)}(t)] = (i\omega)^n F(\omega)$$

Assume $f^{(n)}(t)$ is piecewise smooth.

38. Show that the Fourier transform of $\int_{-\infty}^{t} f(s)\, ds$ is given by

$$\mathscr{F}\left[\int_{-\infty}^{t} f(s)\, ds\right] = \frac{\mathscr{F}[f(t)]}{i\omega}$$

Assume $\int_{-\infty}^{t} f(s)\, ds$ is piecewise smooth.

SECTION 5
SOLVING A BOUNDARY VALUE PROBLEM USING A FOURIER TRANSFORM

EXAMPLE 5. We will solve the temperature problem in an infinite rod which was given in Example 2. It was required to solve the problem

$$u_{xx} = \frac{1}{k} u_t$$

$$u(x, 0) = \begin{cases} 400^\circ\text{ K} & -4 \le x \le 4 \\ 0^\circ\text{ K} & 4 < |x| < +\infty \end{cases}$$

Solution. We start by taking the Fourier transform of both sides of the partial differential equation, treating x as the variable of integration. Thus we can write

$$\mathscr{F}[u_{xx}] = \frac{1}{k}\mathscr{F}[u_t] \tag{7.15}$$

From Table 1 we see that

$$\mathscr{F}[u_{xx}] = \frac{-\omega^2 U(\omega, t)}{2\pi}$$

Now

$$\mathscr{F}[u_t] = \frac{1}{2\pi}\int_{-\infty}^{\infty} u_t(x, t)e^{-i\omega x}\, dx$$

If we interchange integration with differentiation, we have

$$\mathscr{F}[u_t] = \frac{1}{2\pi}\frac{\partial}{\partial t}\int_{-\infty}^{\infty} u(x, t)e^{-i\omega x}\, dx = \frac{1}{2\pi}\frac{\partial}{\partial t}\, U(\omega, t)$$

Substituting these results back into Equation (7.15), we find

$$-\omega^2 U(\omega, t) = -\frac{1}{k}\frac{\partial}{\partial t}\, U(\omega, t)$$

or

$$\frac{\partial}{\partial t}\, U(\omega, t) + k\omega^2 U(\omega, t) = 0 \tag{7.16}$$

Since ω can be considered fixed at some value, Equation (7.16) can be written as the ordinary differential equation

$$\frac{d}{dt}\, U(\omega, t) + k\omega^2 U(\omega, t) = 0$$

The solution of this first-order equation is

$$U(\omega, t) = A(\omega)e^{-k\omega^2 t}$$

To evaluate $A(\omega)$ we take the Fourier transform of both sides of our initial condition, from which we get

$$U(\omega, 0) = \mathscr{F}[u(x, 0)] = \frac{1}{2\pi}\int_{-4}^{4} 400e^{-i\omega x}\, dx = \frac{400}{2\pi i\omega}\,[e^{i4\omega} - e^{-i4\omega}]$$

$$= 400\,\frac{\sin 4\omega}{\pi\omega}$$

Therefore

$$U(\omega, 0) = A(\omega) = 400\,\frac{\sin 4\omega}{\pi\omega}$$

Then

$$U(\omega, t) = 400\,\frac{\sin 4\omega}{\pi\omega}\, e^{-k\omega^2 t}$$

This is the answer to our problem in the transformed state.

To get back to $u(x, t)$, we must use the inverse Fourier transform, which is given by

$$u(x, t) = \frac{400}{\pi}\int_{-\infty}^{\infty} \frac{\sin 4\omega}{\omega}\, e^{-k\omega^2 t}e^{i\omega x}\, d\omega$$

To show this solution is the same as before, we write $u(x, t)$ in the form

$$u(x, t) = \frac{400}{\pi}\int_{-\infty}^{\infty} \frac{\sin 4\omega}{\omega}\, e^{-k\omega^2 t}\cos \omega x\, d\omega + \frac{i400}{\pi}\int_{-\infty}^{\infty} \frac{\sin 4\omega}{\omega}\, e^{-k\omega^2 t}\sin \omega x\, d\omega \tag{7.17}$$

Now sin $(4\omega)/\omega$ and $e^{-k\omega^2 t}$ are even functions with respect to ω, and the first integral on the right-hand side of Equation (7.17) can be written as twice the integral over the interval $(0, +\infty)$, whereas the second integral is zero. Therefore, our solution can be written

$$u(x, t) = \frac{800}{\pi} \int_0^{\infty} \frac{\sin 4\omega}{\omega} e^{-k\omega^2 t} \cos \omega x \, d\omega$$

which is the same as the answer in Example 2. ■

SECTION 5 EXERCISES

Use the Fourier transform to solve Exercises 41–43.

39. $u_{xx} = \frac{1}{k} u_t$ Semi-infinite rod

$u(0, t) = 0$

$u(x, 0) = \begin{cases} 1 & 0 < x < \pi \\ 0 & \text{elsewhere} \end{cases}$

40. $c^2 u_{xx} = u_{tt}$ Infinite string

$u(x, 0) = f(x)$

$u_t(x, 0) = 0$ $\quad -\infty < x < \infty$

41. $e_{xx} = RCe_t$ Semi-infinite cable

$e_x(0, t) = 0$

$e(x, 0) = \begin{cases} \sin 100x & 0 < x < \frac{\pi}{100} \\ 0 & \text{elsewhere} \end{cases}$

42. A high-frequency transmission line is infinite in length. If the initial current is zero and the rate of change of current with respect to time is

$$\begin{cases} 2 \text{ amps} & -4 < x < 4 \\ 0 & \text{elsewhere} \end{cases}$$

find the current $i(x, t)$ using the Fourier transform.

43. A semi-infinite rod insulated laterally has perfect insulation at its left end $x = 0$. If the initial temperature is e^{-x}, find the temperature $u(x, t)$ using the Fourier transform.

44. Find the temperature in a semi-infinite plate $-\infty < x < \infty$, $0 \le y$, that is insulated laterally. Assume the temperature is independent of x; that is $u(y, t)$. The edge $y = 0$ is perfectly insulated and the initial temperature is $f(y)$. Use the Fourier transform.

SECTION 6* FREQUENCY SPECTRUM*

Very often in the study of the communications branch of electrical engineering, it is useful to break down a given wave form into sinusoidal waves. When studying the effect of filters or reconstructing a wave form from a finite number of sinusoids, we can use our knowledge of Fourier series and Fourier transforms to great advantage.

Recall that in Chapter 3 we examined wave forms by plotting various sines and cosines against time. The electrical engineer, however, finds that in certain instances it is more advantageous to plot the amplitude of the wave against the frequency. This technique is useful because we can estimate how important the contribution of each harmonic is to the total wave form. We will examine these ideas as applied to different wave forms.

EXAMPLE 6. Let us look at the square wave that is the periodic extension of

$$f(t) = \begin{cases} 1 & -\dfrac{\pi}{2} < t < \dfrac{\pi}{2} \\ 0 & -\pi < t < -\dfrac{\pi}{2}, \dfrac{\pi}{2} < t < \pi \end{cases}$$

(see Figure 11.)

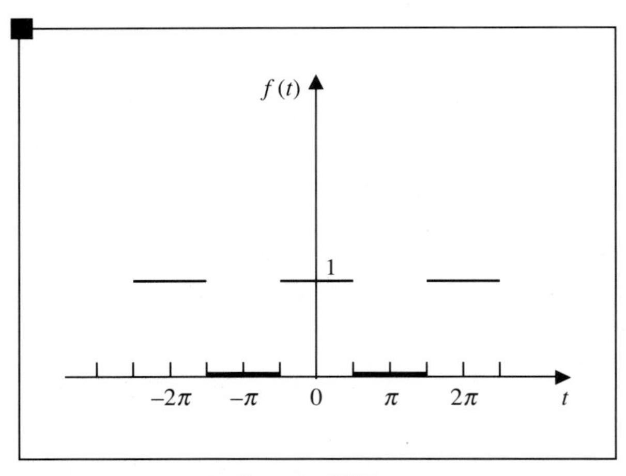

FIGURE 11 **Square Wave**

Solution. If we express this periodic function in the form of a complex Fourier series (see Section 6 in Chapter 3), we find the coefficients are given by

$$c_0 = \frac{1}{2\pi}\int_{-\pi/2}^{\pi/2} e^{-i0t}\,dt = \frac{1}{2\pi}\int_{-\pi/2}^{\pi/2} dt = \frac{1}{2}$$

and

$$c_n = \frac{1}{2\pi}\int_{-\pi/2}^{\pi/2} e^{-int}\,dt = \frac{\sin(n\pi/2)}{n\pi} \qquad n = \pm 1, \pm 2, \ldots \tag{7.18}$$

The series representation is

$$f(t) = \frac{1}{2} + \frac{1}{\pi} \sum_{n=-\infty}^{-1} \frac{\sin(n\pi/2)}{n} e^{int} + \sum_{n=1}^{\infty} \frac{\sin(n\pi/2)}{n} e^{int} \tag{7.19}$$

If we reindex the first series on the right-hand side by replacing n by $-n$, we arrive at the Fourier cosine series

$$f(t) = \frac{1}{2} + \frac{2}{\pi} \sum_{n=1}^{\infty} \frac{\sin(n\pi/2)}{n} \cos nt \tag{7.20}$$

In Chapter 3 we saw how the individual terms of the Fourier series add together to represent the given function $f(t)$. Thus a pictorial representation of Equation (7.20) is shown in Figure 12. Although this graphical representation is useful, it is very tedious to draw. There is, however, another graphical method to show the breakdown of a given wave form into its components along with their relative size.

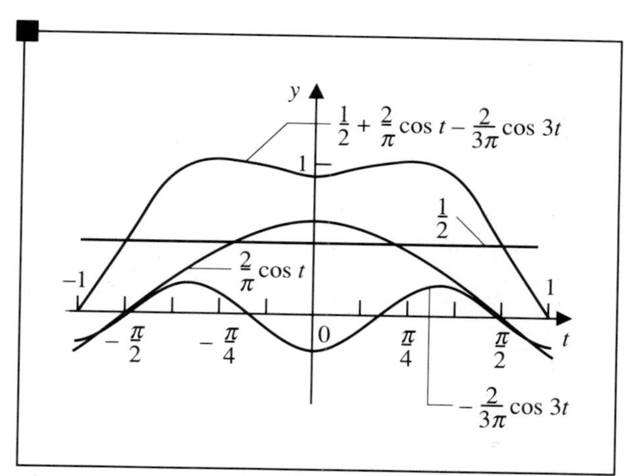

FIGURE 12 **Graphical Summation of Terms of a Fourier Series**

Using the complex Fourier series representation in Equation (7.19), we first notice that there is a constant equal to $\frac{1}{2}$. This represents the zero frequency term (or DC component in electrical engineering).

In order to find the frequency of the other terms, we recall that the period of the nth term ($n = \pm 1, \pm 2, \ldots$) is $2\pi/n$, from which it follows that

$$\text{Frequency} = \frac{1}{\text{period}} = \frac{n}{2\pi}$$

By repeated use of Equation (7.18) we can construct Table 2, which shows the frequency and amplitude of a representative sample of harmonics.

TABLE 2 **Frequency and Amplitude of Square Wave**

	c_0	c_1	c_{-1}	c_2	c_{-2}	c_3	c_{-3}	c_4	c_{-4}	c_5	c_{-5}
Frequency	0	$\frac{1}{2\pi}$	$\frac{1}{2\pi}$	$\frac{1}{\pi}$	$-\frac{1}{\pi}$	$\frac{3}{2\pi}$	$-\frac{3}{2\pi}$	$\frac{2}{\pi}$	$-\frac{2}{\pi}$	$\frac{5}{2\pi}$	$-\frac{5}{2\pi}$
Amplitude	$\frac{1}{2}$	$\frac{1}{\pi}$	$\frac{1}{\pi}$	0	0	$\frac{1}{3\pi}$	$-\frac{1}{3\pi}$	0	0	$\frac{1}{5\pi}$	$-\frac{1}{5\pi}$

Notice that in the complex Fourier series representation there are both positive and negative frequencies. If desired, it is possible to combine the corresponding positive and negative frequencies into a single term.

To plot the information in Table 2, we divide the abscissa, which is the frequency axis, into two parts. The positive frequencies are represented to the right of the division and the negative ones are to the left. The ordinate represents the amplitude. Such a graph is called the **frequency spectrum**. Figure 13 shows the frequency spectrum for $f(t)$. ■

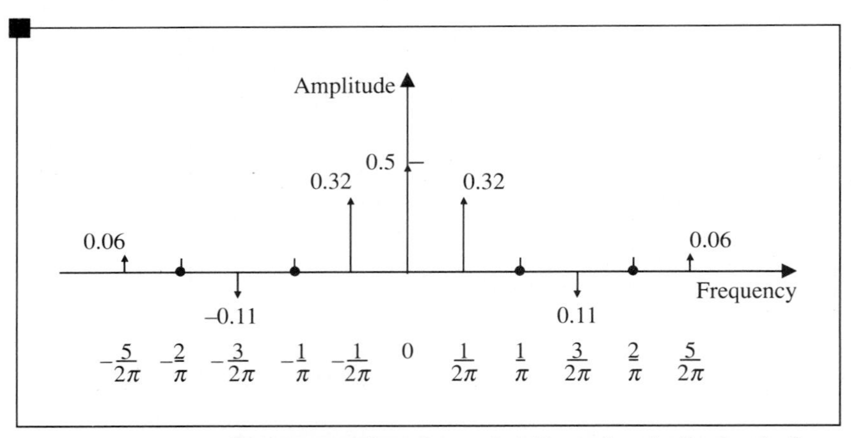

FIGURE 13 **Frequency Spectrum for Function in Example 6**

An interesting study can be made of the square wave given in Example 6 when the distance between pulses becomes greater and greater.

EXAMPLE 7. Let $2p$ be the period of a square wave whose pulse height is 1 and width is $2a$, $a \le p$. To be more specific, we represent the wave by the equation

$$f(t) = \begin{cases} 1 & 2mp - a < t < 2mp + a \qquad a \text{ fixed } 0 < a \le p \\ 0 & 2mp + a < t < 2(m+1)p - a \qquad m = 0, \pm 1, \pm 2, \ldots \end{cases}$$

(See Figure 14.)

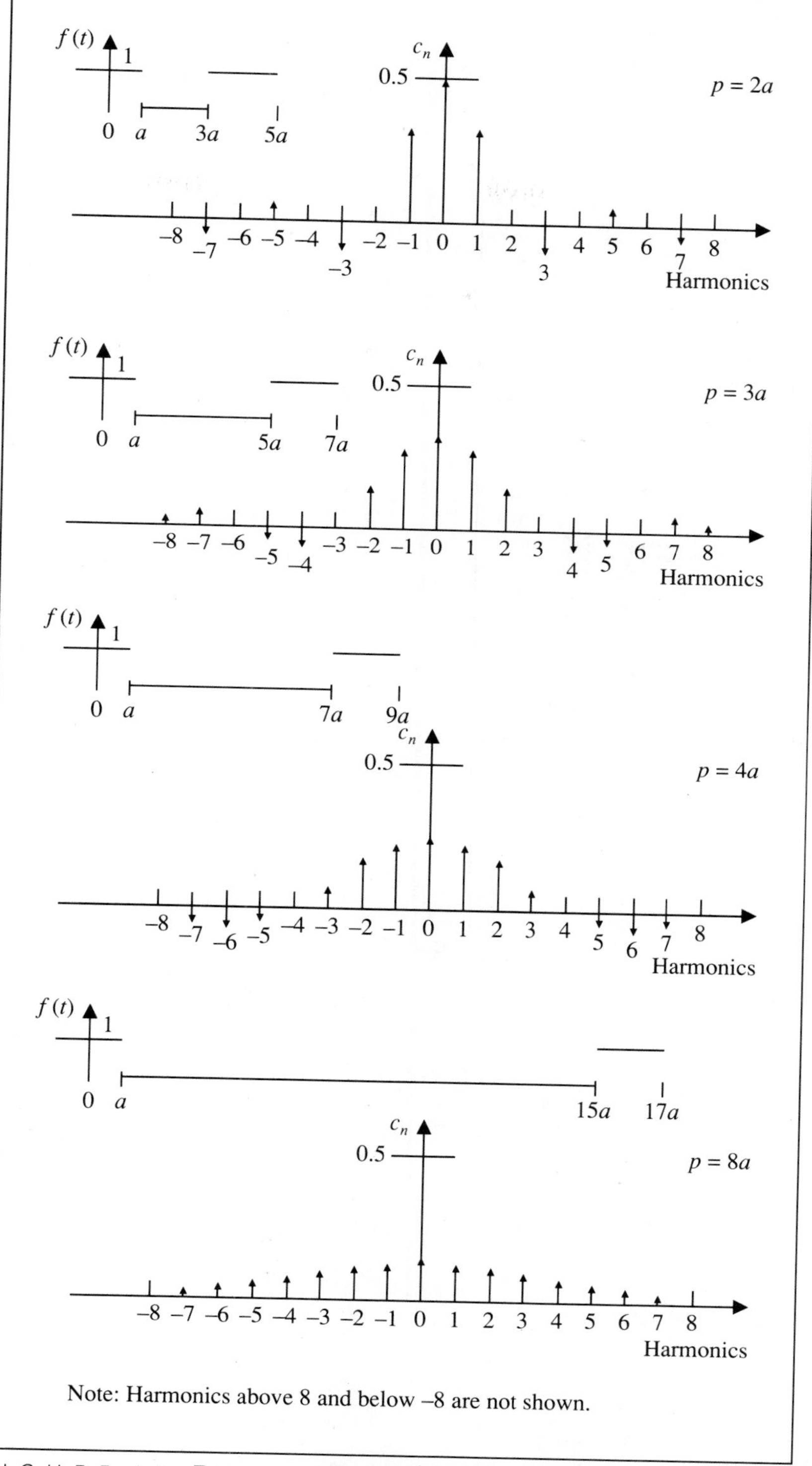

FIGURE 14 **Frequency Spectra for Example 7**

TABLE 3 **Values of c_0 and c_n**

Harmonic	0	1	2	3	4	5	6	7	8
Frequency	0	$\pm\frac{1}{2p}$	$\pm\frac{1}{p}$	$\pm\frac{3}{2p}$	$\pm\frac{2}{p}$	$\pm\frac{5}{2p}$	$\pm\frac{3}{p}$	$\pm\frac{7}{2p}$	$\pm\frac{4}{p}$
$p = 2a$	0.5	0.32	0	−0.11	0	0.06	0	0.045	0
$p = 3a$	0.33	0.28	0.14	0	−0.07	−0.055	0	0.040	0.034
$p = 4a$	0.25	0.23	0.16	0.075	0	−0.045	−0.053	−0.032	0
$p = 8a$	0.125	0.12	0.11	0.10	0.08	0.06	0.038	0.012	0

Solution. As p increases, the distance between the pulses increases, although the duration of the pulse remains fixed.

The value of c_n depends not only on the frequency but also on p and is given by

$$c_0 = \frac{1}{2p}\int_{-a}^{a} dt = \frac{a}{p}$$

and

$$\begin{aligned} c_n &= \frac{1}{2p}\int_{-a}^{a} e^{-in\pi t/p}\, dt \\ &= \frac{i}{2n\pi} e^{-in\pi t/p}\Big|_{-a}^{a} \\ &= \frac{\sin(n\pi a/p)}{n\pi} \qquad n = \pm 1, \pm 2, \ldots \end{aligned}$$

Let us look at four cases: $p = 2a$, $3a$, $4a$, and $8a$. The information is collected in Table 3. The frequency spectra are shown in Figure 14.

Finally, we notice by Example 4(a) that if we let $p \to +\infty$, the Fourier transform of our example is

$$F(\omega) = \frac{a}{\pi}\frac{\sin \omega a}{\omega a}$$

$F(\omega)$ can be interpreted as the amplitude of all frequencies from $-\infty$ to $+\infty$ and is essentially the envelope of the discrete Fourier transforms (see Figure 15). ■

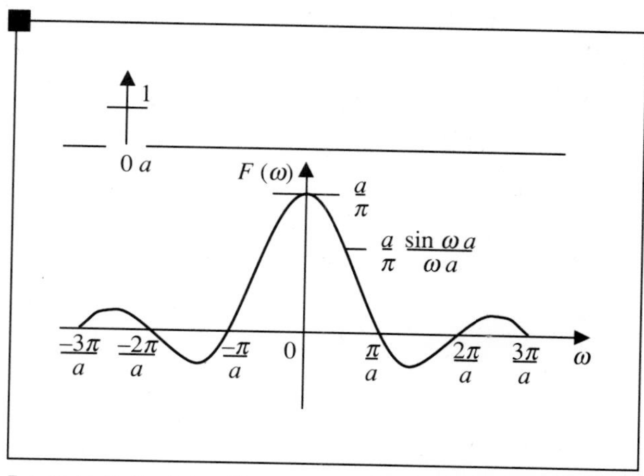

FIGURE 15 **Frequency Spectrum of Single Square Pulse**

SECTION 6 EXERCISE

45. Find the frequency spectrum for the periodic waves indicated in Figure 16. Calculate and plot through eight harmonics.

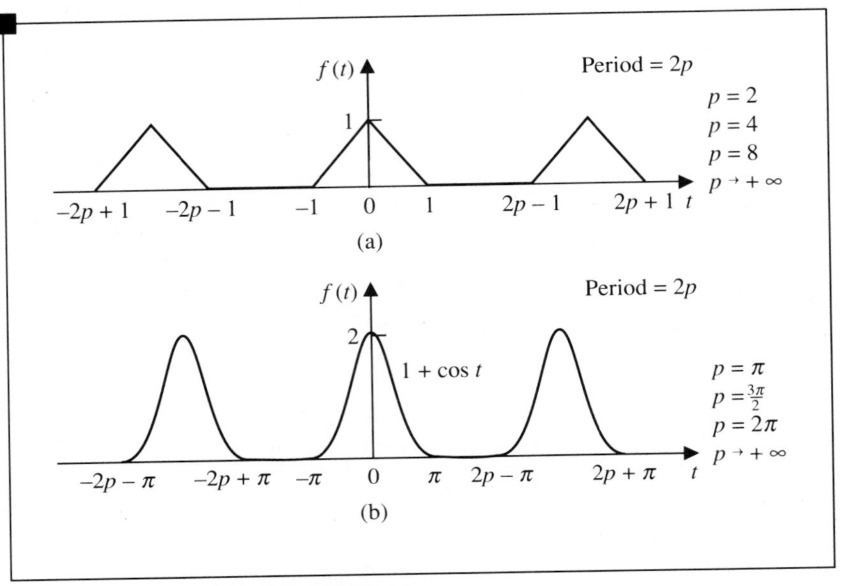

FIGURE 16 **Periodic Waves**

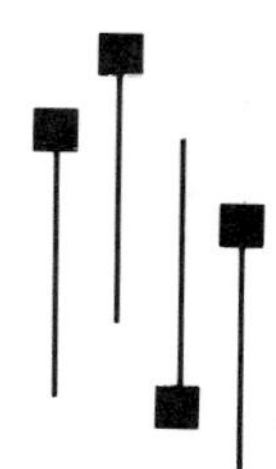

CHAPTER EIGHT

Double Fourier Series

SECTION 1
INTRODUCTION

In Chapter 1 we derived a number of partial differential equations but never attempted to solve them. We delayed solving these partial differential equations because before doing so it is necessary to understand Fourier series techniques for functions of two or more variables.

Suppose we have a periodic function $f(x, y)$ of two variables. The periods in x and y do not have to be the same, although in many problems they are; so let the period of $f(x, y)$ be $2a$ in x and $2b$ in y.

EXAMPLE 1. Find the period for the given functions.

(a) Let $g(x, y) = \sin x \sin 2y$. The smallest period in x is $2a = 2\pi$, whereas that in y is $2b = \pi$.

(b) Let

$$h(x, y) = \begin{cases} 0 & 0 < x < 1, \quad 0 < y < 1 \\ 1 & -1 < x < 0, \quad 0 < y < 1 \\ 0 & -1 < x < 0, \quad -1 < y < 0 \\ 1 & 0 < x < 1, \quad -1 < y < 0 \end{cases}$$

Then if $H(x, y)$ is the periodic extension of $h(x, y)$, a possible period of $H(x, y)$ in x and y is $2a = 2b = 2$ (see Figure 1). ■

Suppose it is possible to express the periodic function $f(x, y)$ as the series

$$f(x, y) = \sum_{m=0}^{\infty} \sum_{n=0}^{\infty} \left\{ a_{mn} \cos \frac{m\pi x}{a} \cos \frac{n\pi y}{b} + b_{mn} \sin \frac{m\pi x}{a} \sin \frac{n\pi y}{b} + c_{mn} \sin \frac{m\pi x}{a} \cos \frac{n\pi y}{b} + d_{mn} \cos \frac{m\pi x}{a} \sin \frac{n\pi y}{b} \right\}$$

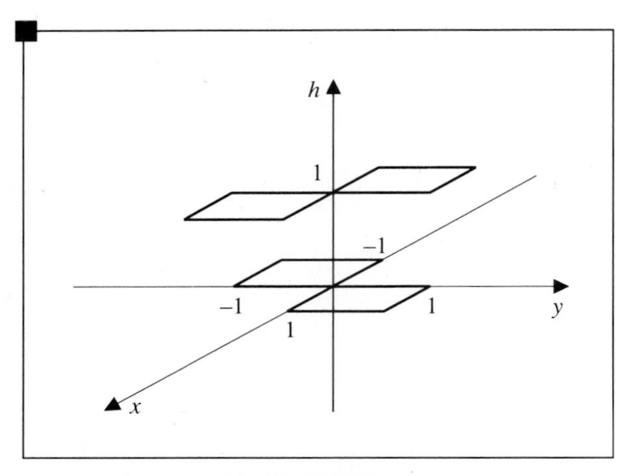

FIGURE 1 **Graph of $h(x, y)$**

In order to find the value of a_{mn}, b_{mn}, and so on (m, $n = 0, 1, 2, \ldots$), we first look at the double-iterated integral from $-a$ to a and $-b$ to b for various combinations of the sine and cosine functions.

THEOREM 1. The integral

$$\int_{-a}^{a}\int_{-b}^{b} \cos^2 0 \, dy \, dx = \int_{-a}^{a}\int_{-b}^{b} dy \, dx = 4ab \qquad \blacksquare$$

In Theorems 2–4, only one case will be proved. The other cases are similar.

THEOREM 2. The integral from $-a$ to a and $-b$ to b of the square of $\sin(m\pi x/a)$, $\sin(m\pi y/b)$, $\cos(m\pi x/a)$, or $\cos(m\pi y/b)$ ($m = 1, 2, \ldots$) equals $2ab$.

Proof. The integral

$$\int_{-a}^{a}\int_{-b}^{b} \sin^2 \frac{m\pi x}{a} \, dy \, dx = 2b \int_{-a}^{a} \sin^2 \frac{m\pi x}{a} \, dx = 2ab \qquad \blacksquare$$

THEOREM 3. The integral from $-a$ to a and $-b$ to b of the square of the product of $\sin(m\pi x/a)$ or $\cos(m\pi x/a)$ with $\sin(n\pi y/b)$ or $\cos(n\pi y/b)$ equals ab ($m, n = 1, 2, \ldots$).

Proof. The integral

$$\int_{-a}^{a}\int_{-b}^{b} \sin^2 \frac{m\pi x}{a} \cos^2 \frac{n\pi y}{b} \, dy \, dx = b \int_{-a}^{a} \sin^2 \frac{m\pi x}{a} \, dx = ab \qquad \blacksquare$$

THEOREM 4. (Orthogonal property.) All other double-iterated integrals from $-a$ to a and $-b$ to b of the product of a pair of functions selected from the set

$$\left\{\sin\frac{m\pi x}{a}\sin\frac{n\pi y}{b},\ \sin\frac{m\pi x}{a}\cos\frac{n\pi y}{b},\ \cos\frac{m\pi x}{a}\sin\frac{n\pi y}{b},\right.$$

$$\left.\cos\frac{m\pi x}{a}\cos\frac{n\pi y}{b}\right\} \qquad m, n = 0, 1, 2, \ldots$$

are zero except the nine cases covered in Theorems 1–3.

Proof. The integral

$$\int_{-a}^{a}\int_{-b}^{b}\sin\frac{m\pi x}{a}\sin\frac{n\pi y}{b}\sin\frac{p\pi x}{a}\sin\frac{q\pi y}{b}\,dy\,dx$$

can be written in the form

$$\int_{-a}^{a}\left[\int_{-b}^{b}\sin\frac{n\pi y}{b}\sin\frac{q\pi y}{b}\,dy\right]\sin\frac{m\pi x}{a}\sin\frac{p\pi x}{a}\,dx$$

Recall from our discussion of the Fourier series of one variable that the integral in the brackets must be equal to zero if $n \neq q$, which is a case not covered in Theorems 1–3. Therefore, the double-iterated integral is zero. ■

We are now at a point where we wish to represent $f(x, y)$ as a periodic function of period $2a$ in x and period $2b$ in y by a double infinite series of sines and cosines; that is,

$$f(x, y) = \sum_{m=0}^{\infty}\sum_{n=0}^{\infty}\left\{a_{mn}\cos\frac{m\pi x}{a}\cos\frac{n\pi y}{b} + b_{mn}\sin\frac{m\pi x}{a}\sin\frac{n\pi y}{b}\right.$$

$$\left. + c_{mn}\sin\frac{m\pi x}{a}\cos\frac{n\pi y}{b} + d_{mn}\cos\frac{m\pi x}{a}\sin\frac{n\pi y}{b}\right\} \qquad \textbf{(8.1)}$$

As in the case of the single Fourier series, we assume that the series converges to $f(x, y)$ and the operations used below are permissible.

To find a_{mn}, multiply both sides of Equation (8.1) by

$$\cos\frac{p\pi x}{a}\cos\frac{q\pi y}{b}$$

and integrate each term from $-a$ to a and $-b$ to b. Using Theorems 3 and 4 we can easily see that all the integrals are zero except the integral associated with the a_{pq} coefficient. We can then write

$$\int_{-a}^{a}\int_{-b}^{b} f(x, y)\cos\frac{p\pi y}{a}\cos\frac{q\pi y}{b}\,dy\,dx = a_{pq}(ab)$$

Reindexing by replacing p with m and q with n, we find

$$a_{mn} = \frac{1}{ab}\int_{-a}^{a}\int_{-b}^{b} f(x, y)\cos\frac{m\pi x}{a}\cos\frac{n\pi y}{b}\,dy\,dx$$

The method for finding other coefficients is similar.

Using the previous argument we state the following definition.

Definition 1. If the series represented by Equation (8.1) converges to $f(x, y)$, a periodic function of $2a$ in x and $2b$ in y, and the coefficients are represented as in Table 1, the series is called a **double Fourier series** and the coefficients are known as **Fourier coefficients.**

TABLE 1 **Fourier Coefficients**

$$\text{Coefficient} = \frac{1}{\kappa ab}\int_{-a}^{a}\int_{-b}^{b} f(x, y)u(x)v(y)\,dy\,dx$$

Coefficient $m, n = 1, 2, \ldots$	κ	$u(x)$	$v(y)$
a_{mn}	1	$\cos\frac{m\pi x}{a}$	$\cos\frac{n\pi y}{b}$
b_{mn}	1	$\sin\frac{m\pi x}{a}$	$\sin\frac{n\pi y}{b}$
c_{mn}	1	$\sin\frac{m\pi x}{a}$	$\cos\frac{n\pi y}{b}$
d_{mn}	1	$\cos\frac{m\pi x}{a}$	$\sin\frac{n\pi y}{b}$
a_{mo}	2	$\cos\frac{m\pi x}{a}$	1
c_{mo}	2	$\sin\frac{m\pi x}{a}$	1
a_{on}	2	1	$\cos\frac{n\pi y}{b}$
d_{on}	2	1	$\sin\frac{n\pi y}{b}$
a_{oo}	4	1	1

All other coefficients are always zero

As you can see, finding the double Fourier series for a function usually involves a great deal of effort. Nine coefficients must be evaluated. If, however, $f(x, y)$ is an even or odd function in x or y, the work of evaluating the coefficients can be reduced considerably.

Rule 1. If $f(x, y)$ is even (odd) in one of its variables, then the Fourier coefficient for a term containing a cosine (sine) function in that variable can be found by halving the interval and doubling the value of the integral.

Rule 2. If $f(x, y)$ is even (odd) in one of its variables, then the Fourier coefficient for a term containing a sine (cosine) function in that variable equals zero.

EXAMPLE 2. Suppose a periodic function $f(x, y)$ is even in x and odd in y. Then using Table 1, the Fourier coefficients

$$d_{mn} = \frac{4}{ab}\int_0^a \int_0^b f(x, y)\cos\frac{m\pi x}{a}\sin\frac{n\pi y}{b}\,dy\,dx$$

$$d_{on} = \frac{2}{ab}\int_0^a \int_0^b f(x, y)\sin\frac{n\pi y}{b}\,dy\,dx$$

All other coefficients are zero. ■

EXAMPLE 3. Find the double Fourier series of the function

$$f(x, y) = \begin{cases} 1 & 0 < x < 3, \quad 0 < y < 3 \\ -1 & 0 < x < 3, \quad -3 < y < 0 \\ -1 & -3 < x < 0, \quad 0 < y < 3 \\ 1 & -3 < x < 0, \quad -3 < y < 0 \end{cases}$$

by extending $f(x, y)$ as a periodic function of period 6 in x and y over the remainder of the xy plane (see Figure 2).

Since $f(x, y)$ is odd in x and y and $a = b = 3$, we see with the help of Table 1 that

$$\begin{aligned} b_{mn} &= \frac{4}{3\cdot 3}\int_0^3 \int_0^3 1\cdot\sin\frac{m\pi x}{3}\sin\frac{n\pi y}{3}\,dy\,dx \\ &= \frac{4}{9}\int_0^3 -\frac{3}{n\pi}\cos\frac{n\pi y}{3}\bigg|_0^3 \sin\frac{m\pi x}{3}\,dx \\ &= \frac{4}{3n\pi}\int_0^3 (1-\cos n\pi)\sin\frac{m\pi x}{3}\,dx \\ &= \frac{4}{3n\pi}(1-\cos n\pi)\frac{3}{m\pi}\left(-\cos\frac{m\pi x}{3}\right)\bigg|_0^3 \\ &= \frac{4}{mn\pi^2}(1-\cos n\pi)(1-\cos m\pi) \\ &= \frac{4}{mn\pi^2}[1-(-1)^n][1-(-1)^m] = \begin{cases} \dfrac{16}{mn\pi^2} & m, n \text{ odd} \\ 0 & m \text{ or } n \text{ even} \end{cases} \end{aligned}$$

The double Fourier series is

$$f(x, y) = \frac{16}{\pi^2}\sum_{m=1,3,5,\ldots}^{\infty}\ \sum_{n=1,3,5,\ldots}^{\infty} \frac{1}{m\cdot n}\sin\frac{m\pi x}{3}\sin\frac{n\pi y}{3}$$ ■

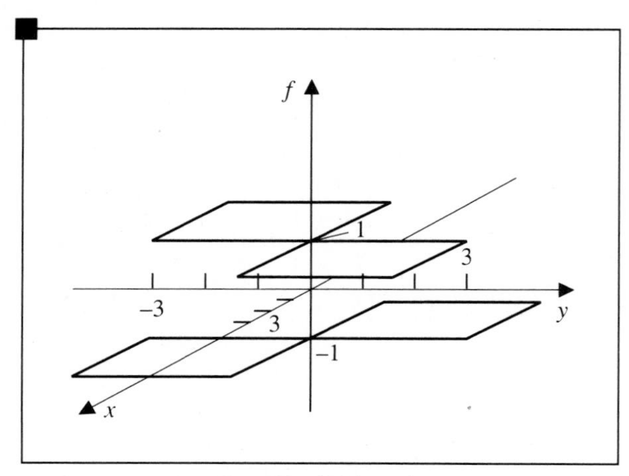

FIGURE 2 **Graph of $f(x, y)$**

SECTION 1 EXERCISES

1. Show that

$$\int_{-a}^{a}\int_{-b}^{b} \cos^2 \frac{m\pi y}{b}\, dy\, dx = 2ab \qquad m = 1, 2, 3, \ldots$$

2. Show that

$$\int_{-a}^{a}\int_{-b}^{b} \cos^2 \frac{m\pi x}{a} \cos^2 \frac{n\pi y}{b}\, dy\, dx = ab \qquad m, n = 1, 2, 3, \ldots$$

3. Show that

$$\int_{-a}^{a}\int_{-b}^{b} \sin \frac{m\pi x}{a} \cos \frac{n\pi y}{b} \cos \frac{p\pi x}{a} \cos \frac{q\pi y}{b}\, dy\, dx = 0 \qquad \begin{matrix} m, p = 0, 1, 2, 3, \ldots \\ n, q = 0, 1, 2, 3, \ldots \end{matrix}$$

4. Find the formula for the following Fourier coefficients:

(a) b_{mn} if $f(x, y)$ is even in x and even in y.

(b) $\left.\begin{matrix} a_{mn} \\ a_{m0} \\ a_{00} \end{matrix}\right\}$ if $f(x, y)$ is odd in x and neither odd nor even in y.

(c) c_{m0} if $f(x, y)$ is odd in x and even in y.

5. Tell whether $f(x, y)$ is even, odd, or neither in x and y for all real x, y.

(a) $f(x, y) = \sin x \cos y$
(b) $g(x, y) = y \sin^2 x$
(c) $h(x, y) = e^x \sin y$
(d) $\varphi(x, y) = e^{x+y}$

(e) $\psi(x, y) = \tan xy$
(f) $F(x, y) = x^2 y^3$
(g) $G(x, y) = x + y - 3$

6. The region $-2 \le x \le 2$, $-3 \le y \le 3$ is pictured below. The value of each function $f(x, y)$, $g(x, y)$, $h(x, y)$, and $\varphi(x, y)$ in the four rectangles is shown. Tell whether the functions are even, odd, or neither in the region.

$-2 < x < 0$, $0 < y < 3$	$0 < x < 2$, $0 < y < 3$
(a) $f(x, y) = x$ (b) $g(x, y) = x - y$ (c) $h(x, y) = 0$ (d) $\varphi(x, y) = xy$	$f(x, y) = x$ $g(x, y) = x + y$ $h(x, y) = 1$ $\varphi(x, y) = xy$
$f(x, y) = -x$ $g(x, y) = -x - y$ $h(x, y) = 0$ $\varphi(x, y) = -xy$	$f(x, y) = -x$ $g(x, y) = -x + y$ $h(x, y) = 1$ $\varphi(x, y) = -xy$

7. Write the double Fourier series for the periodic extension of the following functions. Also sketch one period in x and y.

(a) $f(x, y) = \dfrac{y}{2} \begin{cases} -6 < x < 6 & \text{Period } 12 \\ -4 < y < 4 & \text{Period } 8 \end{cases}$

(b) $g(x, y) = 4\pi - (x + y)$ in $0 < x < 2\pi$, $0 < y < 2\pi$. Extend $g(x, y)$ so that it is even in x and y into the intervals $-2\pi < x < 0$, $-2\pi < y < 0$. Both periods are 4π in x and y.

(c) $$h(x, y) = \begin{cases} 1 & 0 < x < 2, \quad 0 < y < 2 \\ 0 & 0 < x < 2, \quad -2 < y < 0 \\ 0 & -2 < x < 0, \quad -2 < y < 0 \\ -1 & -2 < x < 0, \quad 0 < y < 2 \end{cases}$$ The period in both x and y is 4.

8. Write the double Fourier series for the function $\varphi(x, y) = \sin 2x$. The period in both x and y in the (x, y) plane is π.

9. Find the double Fourier series for the functions defined in the region $0 < x < \pi$, $0 < y < \pi$, if it is extended into the full Fourier region as indicated.

	$f(x, y)$	Extend in x direction as	Extend in y direction as
(a)	xy	Even	Even
(b)	xy	Odd	Even
(c)	x^2y	Odd	Odd
(d)	$1 - xy$	Odd	Odd
(e)	x^2y^2	Even	Odd

SECTION 2

SOLUTIONS OF BOUNDARY VALUE PROBLEMS

Consider the temperature distribution in a rectangular plate a by b that is perfectly insulated on both lateral faces. The temperature initially in the plate is given by the function $10xy$. At $t > 0$, the edges of the plate are set to zero. Find the temperature in the plate at any point for any $t \geq 0$.

This problem can be written formally as follows:

$$\frac{\partial^2 u}{\partial x^2} + \frac{\partial^2 u}{\partial y^2} = \frac{1}{k}\frac{\partial u}{\partial t} \qquad 0 < x < a,\ 0 < y < b,\ t > 0$$

$$u(x, 0, t) = 0 \qquad 0 < x < a,\ t > 0$$

$$u(a, y, t) = 0 \qquad 0 < y < b,\ t > 0$$

$$u(x, b, t) = 0 \qquad 0 < x < a,\ t > 0$$

$$u(0, y, t) = 0 \qquad 0 < y < b,\ t > 0$$

$$u(x, y, 0) = 10xy \qquad 0 < x < a,\ 0 < y < b$$

We choose $u(x, y, t) = X(x)Y(y)T(t)$ as the form our solution will take and substitute into the partial differential equation, where we find

$$X''YT + XY''T = \frac{1}{k}XYT'$$

We divide both sides by XYT, which yields

$$\frac{X''}{X} + \frac{Y''}{Y} = \frac{1}{k}\frac{T'}{T}$$

or

$$\frac{X''}{X} = \frac{1}{k}\frac{T'}{T} - \frac{Y''}{Y}$$

Following the method of separation of variables as shown in Chapter 4, we conclude that

$$\frac{X''}{X} = \frac{1}{k}\frac{T'}{T} - \frac{Y''}{Y} = -\lambda = \text{an arbitrary constant} \tag{8.2}$$

We can rewrite the second and third expressions in Equation (8.2) as

$$\frac{Y''}{Y} = \frac{1}{k}\frac{T'}{T} + \lambda = -\mu = \text{an arbitrary constant}$$

Collecting all these data we arrive at three ordinary differential equations with their associated boundary conditions:

$$X''(x) + \lambda X(x) = 0 \qquad X(0) = X(a) = 0 \tag{8.3}$$

$$Y''(y) + \mu Y(y) = 0 \qquad Y(0) = Y(b) = 0 \tag{8.4}$$

$$T'(t) + k(\lambda + \mu)T(t) = 0 \tag{8.5}$$

Notice in this situation that there are *two* eigenvalue problems. Letting $\lambda = \alpha^2 > 0$, we find the general solution of Equation (8.3) is

$$X(x) = A \cos \alpha x + B \sin \alpha x$$

Using the first boundary condition, we have

$$X(0) = A = 0$$

Using the second boundary condition, it follows that

$$X(a) = B \sin \alpha a = 0$$

Now $B \neq 0$; therefore, $\sin \alpha a = 0$, which tells us that

$$\alpha a = m\pi \qquad \text{or} \qquad \alpha_m = \frac{m\pi}{a} \qquad m = 1, 2, 3, \ldots$$

The eigenvalues are

$$\lambda_m = \alpha_m^2 = \frac{m^2\pi^2}{a^2} \qquad m = 1, 2, 3, \ldots$$

and the eigenfunctions are

$$X_m(x) = \sin \frac{m\pi x}{a}$$

It is easy to show that there are no more eigenvalues for λ.

In a similar way Equation (8.4) yields the eigenvalues

$$\mu_n = \frac{n^2\pi^2}{b^2} \qquad n = 1, 2, 3, \ldots$$

and

$$Y_n(y) = \sin \frac{n\pi y}{b}$$

Next we solve the ordinary differential equation in t, Equation (8.5), which yields

$$T_{mn}(t) = \exp\left[-k(\lambda_m + \mu_n)t\right] = \exp\left[-k\pi^2\left(\frac{m^2}{a^2} + \frac{n^2}{b^2}\right)t\right]$$

Combining all these solutions into a double series, we assume the solution to our problem can be written in the form

$$u(x, y, t) = \sum_{m=1}^{\infty} \sum_{n=1}^{\infty} b_{mn} \sin \frac{m\pi x}{a} \sin \frac{n\pi y}{b} \exp\left[-k(\lambda_m + \mu_n)t\right]$$

We still have to meet the initial condition $u(x, y, 0) = 10xy$. Substituting our assumed solution into this condition, we have

$$u(x, y, 0) = 10xy = \sum_{m=1}^{\infty} \sum_{n=1}^{\infty} b_{mn} \sin \frac{m\pi x}{a} \sin \frac{n\pi y}{b}$$

We notice the series is a double Fourier series whose Fourier coefficients are given by

$$b_{mn} = \frac{4}{ab} \int_0^a \int_0^b 10xy \sin \frac{m\pi x}{a} \sin \frac{n\pi y}{b}\, dy\, dx = (-1)^{m+n} \frac{40ab}{mn\pi^2}$$

Our solution takes the form

$$u(x, y, t) = \frac{40ab}{\pi^2} \sum_{m=1}^{\infty} \sum_{n=1}^{\infty} \frac{(-1)^{m+n}}{mn} \sin \frac{m\pi x}{a} \sin \frac{n\pi y}{b} \exp\left[-k(\lambda_m + \mu_n)t\right]$$

SECTION 2 EXERCISES

10. A rectangular plate a by b is perfectly insulated above and below, and its conditions are described in Table 2. The function $f(x, y)$ is the initial temperature.

(a) Write the two-dimensional heat equation for finding the temperature at any point in the plate at any time $t \geq 0$.
(b) Write the boundary and initial conditions.
(c) Solve the problem for the temperature at any point in the plate in terms of x, y, and t (see Figure 3).

TABLE 2 **Conditions for Exercise 10**

	a	b	①	②	③	④	$f(x, y)$
(a)	2	6	0	0	0	0	100
(b)	10	20	0	0	0	0	$50x$
(c)	2π	π	0	0	0	0	xe^y
(d)	4	8	0	0	ins	ins	$T =$ constant
(e)	4	1	0	ins	0	0	xy
(f)	π	π	ins	0	ins	0	$x + y$
(g)	π	π	0	0	ins	20	0
(h)	π	π	50	ins	0	ins	$10x$
(i)	4	4	ins	ins	ins	ins	50

ins = edge perfectly insulated

FIGURE 3 **Temperature in a Plate**

11. Solve the boundary value problem

$$c^2\left(\frac{\partial^2 u}{\partial x^2} + \frac{\partial^2 u}{\partial y^2}\right) = \frac{\partial^2 u}{\partial t^2}$$
$$u(0, y, t) = 0$$
$$u(\pi, y, t) = 0$$
$$u(x, 0, t) = 0$$
$$u(x, \pi, t) = 0$$
$$u(x, y, 0) = 100$$
$$\frac{\partial u}{\partial t}(x, y, 0) = 0$$

12. Solve the boundary value problem

$$c^2\left(\frac{\partial^2 u}{\partial x^2} + \frac{\partial^2 u}{\partial y^2}\right) = \frac{\partial^2 u}{\partial t^2}$$
$$u(0, y, t) = 0$$
$$u(2, y, t) = 0$$
$$\frac{\partial u}{\partial y}(x, 0, t) = 0$$
$$\frac{\partial u}{\partial y}(x, 3, t) = 0$$
$$u(x, y, 0) = 0$$
$$\frac{\partial u}{\partial t}(x, y, 0) = x$$

13. Find the steady-state temperature $u(x, y, z)$ in a rectangular prism 4 by 4 by 8 if $\nabla^2 u = 0$ and

$$u(0, y, z) = 0$$
$$u(4, y, z) = 0$$
$$u(x, 0, z) = 0$$
$$u(x, 4, z) = 0$$
$$u(x, y, 0) = 10$$
$$u(x\ y, 8) = 50$$

14. A flexible sheet of plastic is stretched across a square window 2 meters by 2 meters. The edges are taped to the frame. At $t = 0$ while the sheet is at rest, a gust of wind imparts a velocity equal to 4 meters per second to the entire surface. What is the displacement in terms of x, y and t?

15. A trampoline 4 meters by 6 meters is fixed along its edges. A clown jumps up and down until he finally flies off the trampoline. At this time ($t = 0$) the displacement is $xy(6 - x)(4 - y)$ and the velocity is zero. Find the displacement for any x, y, t. Let the long axis be in the x direction.

16. A cube 8 centimeters on an edge is insulated on three sides. The top has a temperature of 500° K, and two adjacent sides have a temperature of 0° K. Find the steady-state temperature inside the cube (see Figure 4).

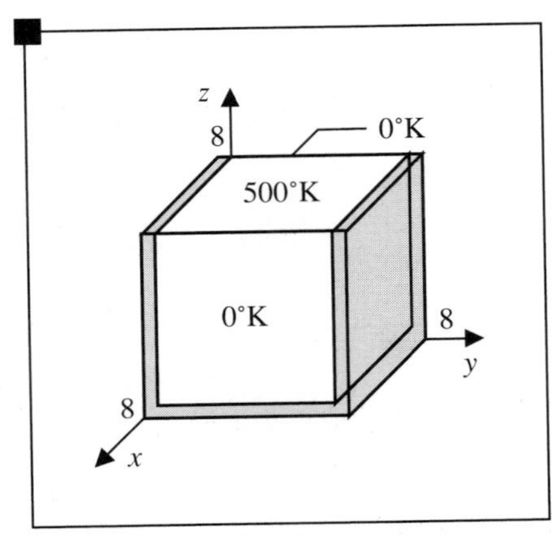

FIGURE 4 **Steady-State Temperature in a Cube**

SECTION 3
VIBRATIONS IN A CIRCULAR MEMBRANE

We now turn to a more difficult but interesting problem involving double Fourier series. In this case, however, the series will consist of factors taken from Bessel functions as well as trigonometric functions. In our example some details will be omitted.

Suppose we have a round drumhead of radius a that is fixed on the circular boundary. At $t = 0$ we will strike the drumhead with a mallet and place the membrane into vibration. What is the transverse displacement of the membrane at any point and time? Since the drumhead is circular, it is more convenient to write the wave equation in terms of polar coordinates for its two spatial dimensions:

$$u_{rr} + \frac{1}{r} u_r + \frac{1}{r^2} u_{\theta\theta} = \frac{1}{c^2} u_{tt}$$

where u is a function of three variables; that is, $u(r, \theta, t)$.

The boundary and initial conditions are given by

$$u(a, \theta, t) = 0 \qquad 0 \le \theta \le 2\pi,\ 0 < t$$

$$u(r, \theta, 0) = 0 \qquad 0 \le r < a,\ 0 \le \theta \le 2\pi$$

$$\frac{\partial u}{\partial t}(r, \theta, 0) = f(r, \theta) \qquad 0 \le r < a,\ 0 \le \theta \le 2\pi$$

Using separation of variables we find by letting $u(r, \theta, t) = R(r)\Theta(\theta)T(t)$ and substituting into the differential equation, that

$$\frac{R''}{R} + \frac{1}{r}\frac{R'}{R} + \frac{1}{r^2}\frac{\Theta''}{\Theta} = \frac{1}{c^2}\frac{T''}{T} = -\lambda$$

Next, equating the left-hand expression to $-\lambda$, we see that

$$r^2 \frac{R''}{R} + r\frac{R'}{R} + r^2\lambda = -\frac{\theta''}{\theta} = \mu$$

Collecting all this information we arrive at three ordinary differential equations along with certain conditions:

$$\theta'' + \mu\theta = 0 \tag{8.6}$$

$$r^2R'' + rR' + (r^2\lambda - \mu)R = 0 \qquad R(a) = 0 \tag{8.7}$$

$$T'' + c^2\lambda T = 0 \qquad T(0) = 0 \tag{8.8}$$

Notice that we have two eigenvalues, λ and μ, to consider. We will start by solving the θ equation. At first glance it would appear that we are in trouble because there are no obvious boundary conditions. But if we realize that θ takes us around the drumhead, we know that when we come back to our starting place both the displacement and rate of change of displacement with respect to θ must be the same. Stating this idea mathematically means u and $\partial u/\partial\theta$ must be continuous at this point. We can use any θ, but it appears that if we let $\theta = \pi$, the work is somewhat simplified. Stating these ideas formally we have

$$u(r, \pi, t) = u(r, -\pi, t)$$

$$\frac{\partial u}{\partial\theta}(r, \pi, t) = \frac{\partial u}{\partial\theta}(r, -\pi, t)$$

which yield the following boundary conditions to go with Equation (8.6):

$$\theta(\pi) = \theta(-\pi) \tag{8.9}$$

$$\theta'(\pi) = \theta'(-\pi) \tag{8.10}$$

If $\mu = \alpha^2 > 0$, the general solution of Equation (8.6) is

$$\Theta(\theta) = A\cos\alpha\theta + B\sin\alpha\theta$$

and

$$\Theta'(\theta) = -\alpha A\sin\alpha\theta + \alpha B\cos\alpha\theta$$

Using Equations (8.9) and (8.10), we find

$$A\cos\alpha\pi + B\sin\alpha\pi = A\cos\alpha(-\pi) + B\sin\alpha(-\pi)$$

$$-\alpha A\sin\alpha\pi + \alpha B\cos\alpha\pi = -\alpha A\sin\alpha(-\pi) + \alpha B\cos\alpha(-\pi)$$

From the first equation we have $B\sin\alpha\pi = 0$, and from the second equation $\alpha A\sin\alpha\pi = 0$.

If $\sin\alpha\pi \neq 0$, then A and B must both equal zero, which leads to the trivial solution. (Remember $\alpha \neq 0$.) Therefore, $\sin\alpha\pi = 0$, from which it follows that

$$\alpha\pi = n\pi \qquad n = 1, 2, \ldots$$

$$\alpha = n$$

and $\mu_n = \alpha_n^2 = n^2$ are eigenvalues and

$$\Theta_n(\theta) = A_n\cos n\theta + B_n\sin n\theta$$

are the eigenfunctions. In studying this problem it will be more useful if we write $\Theta_n(\theta)$ in the form

$$\Theta_n(\theta) = E_n \sin(n\theta + d_n)$$

where E_n and d_n are constants. What about $\mu = 0$ and $\mu < 0$? It can be shown that $\mu = 0$ is an eigenvalue and $\Theta_0 = 1$ its corresponding eigenfunction. However, as the problem proceeds, this eigenfunction will be eliminated. There are no eigenvalues for $\mu < 0$.

We now move on to solve Bessel's equation and the other eigenvalue problem. The general solution to Equation (8.7) is

$$R(r) = CJ_n(\beta r) + DY_n(\beta r)$$

where we have taken $\lambda = \beta^2 > 0$ and $n = 0, 1, 2, 3, \ldots$.

Since $r = 0$ is in the drumhead, we set $D = 0$ since $\lim_{r\to 0} Y_n(r) = -\infty$. Our solution can be written

$$R_n(r) = C_n J_n(\beta r)$$

Using the given boundary condition, we find

$$R_n(a) = C_n J_n(\beta a) = 0$$

Let x_{nj} $(j = 1, 2, \ldots)$ be the jth positive roots of $J_n(x) = 0$; then

$$\beta a = x_{nj}$$

or

$$\beta_{nj} = \frac{x_{nj}}{a}$$

The eigenvalues are

$$\lambda_{nj} = \beta_{nj}^2 = \frac{x_{nj}^2}{a^2}$$

and the eigenfunctions are

$$J_n(\beta_{nj} r)$$

If we check to see if $\lambda = 0$ or $\lambda < 0$ are eigenvalues, we will find that they are not. Therefore, we have taken all the information we can from the equations in Equation (8.7).

Finally we solve Equation (8.8). The equation is relatively easy to handle because λ has already been determined. Equation (8.8) looks like

$$T'' + c^2\beta_{nj}^2 T = 0$$

The general solution of this equation is

$$T(t) = F \cos c\beta_{nj} t + G \sin c\beta_{nj} t$$

Using the homogeneous initial condition associated with this equation, we have

$$T(0) = F = 0$$

Therefore,

$$T_{nj}(t) = \sin c\beta_{nj} t$$

Combining all these solutions of the ordinary differential equations, we see that our assumed solution takes the form

$$u(r, \theta, t) = \sum_{n=0}^{\infty} \sum_{j=1}^{\infty} E_{nj} \sin (n\theta + d_{nj}) J_n(\beta_{nj} r) \sin c\beta_{nj} t$$

where E_{nj} and d_{nj} are constants yet to be determined. Since d_{nj} is the phase shift in the θ direction and it does not matter which radial line we take as our base line, let us set $d_{nj} = 0$. Our starting line will then be the polar axis. If we do this the case $n = 0$ can be dropped.

To find E_{nj} we use the nonhomogeneous initial condition and see that

$$\frac{\partial u}{\partial t}(r, \theta, t) = \sum_{n=1}^{\infty} \sum_{j=1}^{\infty} E_{nj} c\beta_{nj} \sin n\theta J_n(\beta_{nj} r) \cos c\beta_{nj} t$$

and

$$\frac{\partial u}{\partial t}(r, \theta, 0) = f(r, \theta) = \sum_{n=1}^{\infty} \sum_{j=1}^{\infty} E_{nj} c\beta_{nj} \sin n\theta J_n(\beta_{nj} r)$$

The interesting thing about the double series on the right-hand side is that it is a double Fourier series in sine and Bessel functions. We can show that such terms possess an orthogonality property. Therefore, we find

$$E_{nj} c\beta_{nj} = \frac{2}{a^2\pi[J_{n+1}(\beta_{nj} a)]^2} \int_{-\pi}^{\pi} \int_0^a r f(r, \theta) \sin n\theta J_n(\beta_{nj} r)\, dr\, d\theta$$

Our final answer can be written

$$u(r, \theta, t) = \frac{2}{ca^2\pi} \sum_{n=1}^{\infty} \sum_{j=1}^{\infty} \frac{\int_{-\pi}^{\pi} \int_0^a s f(s, \varphi) \sin n\varphi J_n(\beta_{nj} s)\, ds\, d\varphi}{\beta_{nj}[J_{n+1}(\beta_{nj} a)]^2} \times \sin n\theta J_n(\beta_{nj} r) \sin c\beta_{nj} t$$

SECTION 4*
PATTERNS IN A DRUMHEAD

Recall in the problem of the vibrating string (Chapter 4, Section 4) that we spoke of the fundamental frequency and its first, second, third, and so on harmonics. Graphically adding these frequencies together, we arrive at the shape of our wave form.

We would like to give a mathematical reason why a violin sounds more musical than a drum. (The mathematical reason may not be the only reason!) In a vibrating string the harmonics are always an *integer* multiples of the fundamental. But as we will see, this is not the case in a vibrating drumhead.

Let us look at the individual terms of our solution to the drumhead problem derived in Section 8.3. They are

$$u_{nj} = \sin n\theta J_n(\beta_{nj} r) \sin c\beta_{nj} t \tag{8.11}$$

where $n = 1, 2, \ldots$ and $j = 1, 2, \ldots$. Notice that $\sin n\theta J_n(\beta_{nj} r)$ can be thought of as the amplitude of the $\sin c\beta_{nj} t$ factor. In this situation the lowest or fundamental frequency of vibration occurs when $n = j = 1$. Therefore,

$$u_{11} = \sin \theta J_1(\beta_{11} r) \sin c\beta_{11} t$$

The fundamental period is given by

$$T_{11} = \frac{2\pi}{c\beta_{11}}$$

Remember that $\beta_{11} = x_{11}/a$ where x_{11} is the first positive root of $J_1(x) = 0$. From tables listing zeros of Bessel functions,* $x_{11} = 3.83$ and the fundamental period is $2\pi a/(3.83c)$, from which it follows that the fundamental frequency is

$$f_0 = f_{11} = \frac{1}{T_{11}} = \frac{3.83c}{2\pi a}$$

Let us look at the next higher frequency. The frequency of vibration is given by the formula

$$f_{nj} = \frac{x_{nj} c}{2\pi a}$$

and the next higher frequency above the fundmanetal occurs when $n = 2$, $j = 1$. In this case, $\beta_{21} = x_{21}/a$ where x_{21} is the first positive root of $J_2(x) = 0$. (In order to evaluate this frequency, it is necessary to look up x_{nj} in the tables.*) Thus,

$$f_{21} = \frac{5.14c}{2\pi a}$$

The ratio of f_{21} to f_0 is

$$\frac{f_{21}}{f_0} = \frac{5.14}{3.83} = 1.34$$

Unlike a violin, whose ratio of a higher frequency to the frequency of the fundamental is always an integer, in a drum this is not the case. Because

---

* See "Zeros of Bessel Functions" found in such books as *Handbook of Mathematical Functions*, Applied Mathematical Series 55, U.S. Department of Commerce, 1964.

higher frequencies of a drum are not integer multiples of the fundamental, a vibrating drumhead does not sound as musical as a violin.

Another interesting study of a drumhead involves locating the nodes or points on a drumhead where the displacement is zero. The question we ask is this: When the drum is vibrating at a certain frequency, does the amplitude remain zero at some points for all t? In other words, does

$$\sin n\theta J_n(\beta_{nj} r) = 0$$

Let us examine the answer to this question for the fundamental frequency f_0, f_{12}, and f_{21}.

The amplitude of the fundamental frequency is $\sin \theta J_1(3.83r/a)$; therefore, when $\theta = 0$ or π, the amplitude is zero. Now the first and only zero of $J_1(3.83r/a)$, $r \le a$, occurs at $r = a$, which is the edge of the drum. So in this case the node pattern is like the one shown in Figure 5.

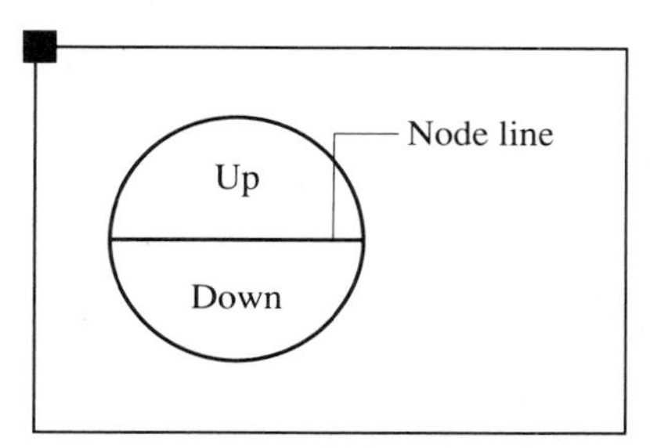

FIGURE 5 **Vibration Pattern of Circular Drumhead**

Let us look at the two other situations. The first will be when $n = 1$, $j = 2$, and the second will be when $n = 2$, $j = 1$ (see Table 3). These vibration patterns are pictured in Figure 6.

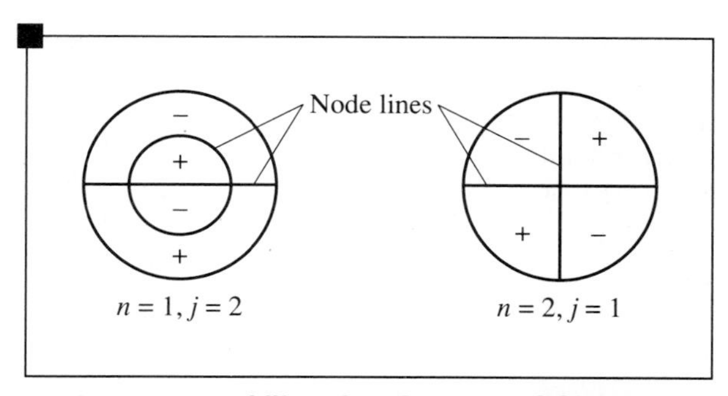

FIGURE 6 **Vibration Pattern of Circular Drumhead**

The analysis of a circular drumhead is fascinating, but a great deal of effort is required to derive the final node lines. The same idea can be applied to a rectangular or even square drumhead (not too common!) with considerably less effort because the answer depends only on sines or cosines.

TABLE 3 **Properties and Patterns of a Drumhead**

	$n = 1, j = 2$	$n = 2, j = 1$
u_{nj} [from Equation (8.11)]	$\sin \theta J_1(\beta_{12} r)$	$\sin 2\theta J_2(\beta_{21} r)$
$\beta_{nj} = \dfrac{x_{nj}}{a}$	$\beta_{12} = \dfrac{x_{12}}{a} = \dfrac{7.02}{a}$	$\beta_{21} = \dfrac{x_{21}}{a} = \dfrac{5.14}{a}$
Period	$\dfrac{2\pi a}{7.02c}$	$\dfrac{2\pi a}{5.14c}$
Frequency	$\dfrac{7.02c}{2\pi a}$	$\dfrac{5.14c}{2\pi a}$
Frequency in terms of f_0	$1.83f_0$	$1.34f_0$
Nodes	$\left.\begin{array}{l}\theta = 0, \pi \\ r = 0.55a, a\end{array}\right\}^*$	$\left.\begin{array}{l}\theta = 0, \frac{\pi}{2}, \pi, \frac{3\pi}{2} \\ r = a\end{array}\right\}^{**}$
	* $\sin \theta = 0$ implies $\theta = 0, \pi$	** $\sin 2\theta = 0$ implies $\theta = 0, \frac{\pi}{2}, \pi, \frac{3\pi}{2}$
	If $J_1\left(\dfrac{7.02}{a} r\right) = 0$	If $J_2\left(\dfrac{5.14}{a} r\right) = 0$
	$\dfrac{7.02}{a} r = 3.83$ or 7.02	$\dfrac{5.14}{a} r = 5.14$
	$r = 0.55a, a$	$r = a$

SECTION 4 EXERCISES

In Exercises 17–19, a circular drumhead of radius a fixed on its boundary is put into motion by striking it with a mallet. If the individual terms of the solution are given by

$$\sin n\theta J_n(\beta_{nj} r) \sin c\beta_{nj} t \qquad n, j = 1, 2, 3, \ldots$$

(a) find the fundamental frequency f_0.
(b) find the higher frequencies f_{mn} indicated.
(c) find the ratios of these higher frequencies to the fundamental.
(d) find the nodes for the frequencies and sketch them.

	a in meters	Velocity c in m/s	Frequencies
17.	1	30	f_{21}, f_{32}, f_{44}
18.	0.1	15	f_{12}, f_{21}, f_{51}
19.	0.5	25	f_{23}, f_{27}, f_{14}

In Exercises 20–24, a rectangular drumhead $a \times b$ fixed on its boundary is put into motion by striking. If the individual terms of the solutions are

$$\sin \frac{m\pi x}{a} \sin \frac{n\pi y}{b} \sin c\pi t \sqrt{\left(\frac{m}{a}\right)^2 + \left(\frac{n}{b}\right)^2} \qquad m, n = 1, 2, 3, \ldots$$

(a) find the fundamental period f_0.
(b) find the higher frequencies f_{mn} indicated.
(c) find the ratios of these higher frequencies to the fundamental.
(d) find the nodes for the frequencies and sketch them.

	a in meters	b in meters	Velocity c in m/s	Frequencies
20.	2	1	20	f_{12}, f_{21}, f_{31}
21.	2	1	20	f_{22}, f_{14}, f_{53}
22.	5	4	30	f_{23}, f_{32}, f_{22}
23.	3	3	25	f_{33}, f_{45}, f_{41}
24.	8	1	12	f_{12}, f_{22}, f_{23}

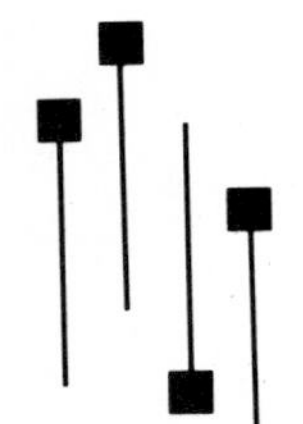

CHAPTER NINE

Green's Functions

SECTION 1
INTRODUCTION

In the preceding chapters we discussed the solution to boundary value problems in a series form. However, a series form is neither the only possible approach to these problems nor always the best. Our purpose in this chapter is to introduce a technique that represents the solution to these problems in terms of an integral.

EXAMPLE 1. For the Poisson equation

$$\nabla^2 u(\mathbf{x}) = f(\mathbf{x}) \tag{9.1}$$

on a domain D of R^2 with the boundary conditions

$$u\bigg|_{\partial D} = g(\mathbf{x}) \tag{9.2}$$

(where ∂D is the boundary of D), we want to find a function $G = G(\mathbf{x}, \boldsymbol{\eta})$, $\boldsymbol{\eta} \in R^2$, which is called the **Green's function**, for the problem (and depends only on the differential operator and the domain D) so that the solution of Equations (9.1) and (9.2) can be represented as

$$u(\mathbf{x}) = \int_D G(\mathbf{x}, \boldsymbol{\eta}) f(\boldsymbol{\eta})\, d\boldsymbol{\eta} + \int_{\partial D} g(\boldsymbol{\eta}) \frac{\partial G}{\partial \boldsymbol{\eta}}\, ds \tag{9.3}$$

where

$$\frac{\partial G}{\partial \boldsymbol{\eta}} = \text{grad } G \cdot n = \left[\frac{\partial G}{\partial \eta_1}, \frac{\partial G}{\partial \eta_2}\right] \cdot \mathbf{n}$$

and $\mathbf{n}$ is the outward normal to the boundary ∂D. ■

To see some of the advantages for using such an integral representation for the solution, we note that Equation (9.3) holds for all possible $f(\mathbf{x})$ and $g(\mathbf{x})$, thus providing us with the solution of Equations (9.1) and (9.2) even when the method of separation of variables is not applicable. Furthermore, the dependence of the solution on $f(\mathbf{x})$ and $g(\mathbf{x})$ appears in Equation (9.3) explicitly and in closed form and hence enables us to investigate the dependence of the solution on these elements of the problem directly.

From an applied point of view, Green's functions have a natural physical interpretation as the influence of a unit source located at $\boldsymbol{\eta}$ on the point $\mathbf{x}$. Accordingly, we can refer to $G(\mathbf{x}, \boldsymbol{\eta})$ as the **influence function**. We emphasize this interpretation of the Green's functions in the rest of the chapter (especially Sections 3 and 4) and use it to infer some of their properties.

The major undertaking in this method is computing the Green's function, which appears as a kernel in the integral representation of the solution. In order to understand how the Green's function can be computed however, we have to digress and briefly discuss the Dirac delta-function and related topics.

SECTION 2
THE DIRAC δ-FUNCTION

In the following discussion we introduce the Dirac δ-function as a "limit" of some sequences of functions and thus provide an intuitive though not rigorous definition of this function and its properties. A more abstract and elegant presentation of this topic can be made from a functional analysis point of view, but we do not pursue this here.

To begin with, consider the sequence

$$f_n(x) = \begin{cases} \dfrac{n}{2} & \text{for } x \in \left[-\dfrac{1}{n}, \dfrac{1}{n}\right] \\ 0 & \text{otherwise} \end{cases}$$

Then it is easy to verify that the total area under the graph of f_n is 1 for all n; that is,

$$\int_{-\infty}^{\infty} f_n(x)\, dx = 1$$

Moreover, by the mean value theorem for integrals we infer that for any continuous function g,

$$\int_{-\infty}^{\infty} f_n(x)g(x)\, dx = \frac{n}{2}\int_{-1/n}^{1/n} g(x)\, dx = g(\xi_n)\,\frac{n}{2}\cdot\frac{2}{n} = g(\xi_n) \tag{9.4}$$

where

$$\xi_n \in \left[-\frac{1}{n}, \frac{1}{n}\right]$$

□ ***Remark.*** The mean value theorem for integrals states that under proper assumptions on $g(x)$ there exists a point ξ on the interval $[a, b]$ so that

$$\int_a^b g(x)\,dx = g(\xi)(b-a) \quad \Box$$

From Equation (9.4) we deduce that $\lim_{n\to\infty} \xi_n = 0$ and hence

$$\lim_{n\to\infty} \int_{-\infty}^{\infty} f_n(x)g(x)\,dx = g(0) \tag{9.5}$$

Thus although the sequence f_n has no classical limit [since $f_n(0) \to \infty$], the limit of the integrals that appear in Equation (9.5) is well defined. We are tempted, therefore, to introduce a "limit" function $\delta(x)$, which would have existed if the interchange of the limit and the integral operations in Equation (9.5) was permissible, so that

$$\int_{-\infty}^{\infty} \delta(x)g(x)\,dx = g(0)$$

The sequence $\{f_n\}$ is not the only one that "converges" to the Dirac δ-function. In fact, although the functions $f_n(x)$ are not differentiable everywhere, there exist other sequences of smooth functions that lead to the same result.

EXAMPLE 2. Consider the sequence

$$F_n(x) = \frac{n}{\pi(1 + n^2x^2)} \qquad n = 1, 2, \ldots$$

Then for all $x \neq 0$, $\lim_{n\to\infty} F_n(x) = 0$, and yet

$$\int_{-\infty}^{\infty} F_n(x)\,dx = 1 \qquad n = 1, 2, \ldots$$

for all n. Hence by the same procedure that we applied to the sequence $\{f_n\}$, it follows that for any (continuous) $g(x)$,

$$\lim_{n\to\infty} \int_{-\infty}^{\infty} F_n(x)g(x)\,dx = g(0) \qquad \blacksquare$$

Thus we may infer that an equivalent, sequence-independent, "definition" of the δ-function is given by

1. $\delta(x) = 0 \qquad$ for $x \neq 0$.
2. $\delta(x)$ has a singularity at $x = 0$.
3. $\displaystyle\int_{-\infty}^{\infty} \delta(x)\,dx = 1$.

Although we refer to $\delta(x)$ as a "function," a more appropriate name is a "generalized function" or a distribution.

We now discuss some of the properties of the δ-function and introduce its generalized derivatives.

THEOREM 1.

$$\int_{-\infty}^{\infty} f(x)\,\delta(x-x_0)\,dx = f(x_0) \tag{9.6}$$

$$\int_{-\infty}^{\infty} f(x)\,\delta(\lambda x)\,dx = \frac{1}{\lambda} f(0) \qquad 0<\lambda$$

$$\int_{-\infty}^{\infty} f(x)\,\delta(-x)\,dx = f(0)$$

Proof. To prove the first property, we define $\bar{x} = x - x_0$ and substitute in Equation (9.6):

$$\int_{-\infty}^{\infty} f(x)\,\delta(x-x_0)\,dx = \int_{-\infty}^{\infty} f(x+x_0)\,\delta(\bar{x})\,d\bar{x} = f(x_0)$$

Similarly, the second and third properties can be easily proved by introducing $\bar{x} = \lambda x$ and $\bar{x} = -x$, respectively. ■

Although $\delta(x)$ is not continuous and, therefore, not differentiable in the classical sense, we can define the generalized (or weak) derivatives of $\delta(x)$ to any order by repeatedly applying the formula for integration by parts. Thus, since $\delta(x) = 0$ for $x \neq 0$, we have (formally)

$$\int_{-\infty}^{\infty} f(x)\,\delta'(x)\,dx = f(x)\,\delta(x)\Big|_{-\infty}^{\infty} - \int_{-\infty}^{\infty} f'(x)\,\delta(x)\,dx = -f'(0)$$

Similarly, for $\delta^{(n)}(x)$, the nth derivative of $\delta(x)$, we have

$$\int_{-\infty}^{\infty} f(x)\,\delta^{(n)}(x)\,dx = (-1)^n f^{(n)}(0) \tag{9.7}$$

We can, therefore, define $\delta^{(n)}(x)$, $n = 1, 2, \ldots$, as generalized functions that are zero for $x \neq 0$, singular at $x = 0$, and satisfy Equation (9.7).

THEOREM 2. $\delta(x)$ is equal to the generalized derivative of the Heaviside function

$$H(x) = \begin{cases} 1 & x \geq 0 \\ 0 & x < 0 \end{cases}$$

Proof. To prove our statement we apply integration by parts to

$$\int_{-\infty}^{\infty} f(x)H'(x)\,dx$$

We obtain

$$\int_{-\infty}^{\infty} f(x)H'(x)\,dx = f(x)H(x)\Big|_{-\infty}^{\infty} - \int_{-\infty}^{\infty} f'(x)H(x)\,dx$$

But $\lim_{x\to\infty} f(x) = 0$ since $f(x)$ is integrable on $[-\infty, \infty]$, and, therefore

$$\int_{-\infty}^{\infty} f(x)H'(x)\,dx = -\int_{-\infty}^{\infty} f'(x)H(x)\,dx$$
$$= -\int_{0}^{\infty} f'(x)\,dx = -f(x)\Big|_{0}^{\infty} = f(0)$$

which proves the theorem. ■

Finally, we introduce the n-dimensional δ-function.

Definition 1. We define $\delta(\mathbf{x} - \mathbf{a})$ where $\mathbf{x} = (x_1, \ldots, x_n)$, $\mathbf{a} = (a_1, \ldots, a_n)$, as

$$\delta(\mathbf{x} - \mathbf{a}) = \delta(x_1 - a_1) \cdot \ldots \cdot \delta(x_n - a_n)$$

It is obvious from this definition that

$$\int_{R^n} f(\mathbf{x})\,\delta(\mathbf{x} - \mathbf{a})\,d\mathbf{x} = f(\mathbf{a})$$

SECTION 2 EXERCISES

1. Carry out the proof of Equation (9.7).

2. Let

$$f(x) = \begin{cases} 0 & x < 0 \\ 1 & 0 \le x < 2 \\ 2 & 2 \le x < \infty \end{cases}$$

Express the generalized derivative of f in terms of Dirac δ-functions.

3. Generalize the results of Exercise 2 to a function with n jump discontinuities.

4. Show that the sequence

$$f_n(x) = \frac{n}{\sqrt{2\pi}} \exp\left[-\frac{(nx)^2}{2}\right]$$

converges to $\delta(x)$.

□ ***Hint.*** Note that each f_n is a normal distribution with mean zero and standard deviation $1/n$. □

5. Find a continuous function whose generalized derivative is the Heaviside function $H(x)$.

6. Let $\mathbf{x} = \mathbf{x}(u_1, \ldots, u_n)$ be a coordinate transformation in R^n that does not change the origin of the coordinate system (i.e., the origin in the x system corresponds to the origin in the u system). Show that

$$\delta(x) = \frac{1}{|J|}\,\delta(\mathbf{u})$$

where J is the Jacobian of the coordinate transformation.

7. Specialize the results of Exercise 6 to the following coordinate transformations:
   (a) Cartesian to polar coordinates in R^2
   (b) Cartesian to cylindrical in R^3
   (c) Cartesian to spherical in R^3

8. Generalize Exercise 6 to the case where the origin in the x-coordinates is transformed to the point $\mathbf{a}$ in the u system.

SECTION 3
GREEN'S FUNCTION FOR THE LAPLACE OPERATOR

Definition 2. The solution of

$$\nabla^2 G(\mathbf{x}, \boldsymbol{\eta}) = \frac{\partial^2 G}{\partial x_1^2} + \cdots + \frac{\partial^2 G}{\partial x_n^2} = \delta(\mathbf{x} - \boldsymbol{\eta}) \qquad \mathbf{x}, \boldsymbol{\eta} \in D \tag{9.8}$$

on a domain D of R^n satisfying the boundary condition

$$G(\mathbf{x}, \boldsymbol{\eta})\bigg|_{\partial D} = 0 \qquad \mathbf{x} \in \partial D, \boldsymbol{\eta} \in D \tag{9.9}$$

is called the **Green's function for the Laplace operator in *D*.**

□ ***Remark.*** For the special case $D = R^n$, the domain has no boundaries and the corresponding solution of Equation (9.8), which does not take into account Equation (9.9), is called the **infinite space Green's function.** □

In the following discussion we cover explicitly the properties and computation of G in R^2 only. However, similar (but not the same) results hold in R^n.

THEOREM 3. G is symmetric in $\mathbf{x}$ and $\boldsymbol{\eta}$; that is

$$G(\mathbf{x}, \boldsymbol{\eta}) = G(\boldsymbol{\eta}, \mathbf{x})$$

Proof. To prove this theorem we use Green's lemma, which states that under proper restrictions on φ, ψ and D, the following formula holds:

$$\int_D (\psi \nabla^2 \varphi - \varphi \nabla^2 \psi)\, dA = \int_{\partial D} \left(\psi \frac{\partial \varphi}{\partial \mathbf{n}} - \varphi \frac{\partial \psi}{\partial \mathbf{n}} \right) ds \tag{9.10}$$

Substituting $\psi = G(\mathbf{x}, \boldsymbol{\eta})$ and $\varphi = G(\mathbf{x}, \boldsymbol{\eta}^*)$ (where $\boldsymbol{\eta}$, $\boldsymbol{\eta}^*$ are arbitrary but fixed points in D) in Equation (9.10), we obtain

$$\int_D [G(\mathbf{x}, \boldsymbol{\eta}) \nabla^2 G(\mathbf{x}, \boldsymbol{\eta}^*) - G(\mathbf{x}, \boldsymbol{\eta}^*) \nabla^2 G(\mathbf{x}, \boldsymbol{\eta})]\, dA$$

$$= \int_{\partial D} \left[G(\mathbf{x}, \boldsymbol{\eta}) \frac{\partial G}{\partial \mathbf{n}} (\mathbf{x}, \boldsymbol{\eta}^*) - G(\mathbf{x}, \boldsymbol{\eta}^*) \frac{\partial G}{\partial \mathbf{n}} (\mathbf{x}, \boldsymbol{\eta}) \right] ds$$

Using Equations (9.8) and (9.9), this yields

$$\int_D [G(\mathbf{x}, \boldsymbol{\eta})\, \delta(\mathbf{x} - \boldsymbol{\eta}^*) - G(\mathbf{x}, \boldsymbol{\eta}^*)\, \delta(\mathbf{x} - \boldsymbol{\eta})]\, dA = 0$$

and hence

$$G(\boldsymbol{\eta}^*, \boldsymbol{\eta}) = G(\boldsymbol{\eta}, \boldsymbol{\eta}^*)$$

Since $\boldsymbol{\eta}$, $\boldsymbol{\eta}^*$ are arbitrary points in D, this proves the theorem. ■

THEOREM 4. $\partial G/\partial \mathbf{n}$ has a discontinuity at $\mathbf{x} = \boldsymbol{\eta}$. Moreover,

$$\lim_{\varepsilon \to 0} \int_{C_\varepsilon} \frac{\partial G}{\partial \mathbf{n}}\, ds = 1 \tag{9.11}$$

where C_ε is the circle of radius ε around $\mathrm{x} = (x_1, x_2)$; that is,

$$(x_1 - \eta_1)^2 + (x_2 - \eta_2)^2 = \varepsilon^2$$

Proof. Let D_ε be the domain bounded by the circle C_ε. Then from Equation (9.8) we infer that

$$\int_{D_\varepsilon} \nabla^2 G\, dA = \int_{D_\varepsilon} \delta(\mathbf{x} - \boldsymbol{\eta})\, dA = 1$$

[since $\delta(\mathbf{x} - \boldsymbol{\eta})$ is zero except at $\mathbf{x} = \boldsymbol{\eta}$]. Hence, using the divergence theorem we infer that

$$1 = \Big|_{D_\varepsilon} \nabla^2 G\, dA = \int_{C_\varepsilon} \nabla G \cdot \mathbf{n}\, ds = \int_{C_\varepsilon} \frac{\partial G}{\partial \mathbf{n}}\, ds$$

Thus,

$$\lim_{\varepsilon \to 0} \int_{C_\varepsilon} \frac{\partial G}{\partial \mathbf{n}}\, ds = 1 \tag{9.12}$$

This shows that $\partial G/\partial \mathbf{n}$ has a discontinuity at $\mathbf{x} = \boldsymbol{\eta}$ since otherwise (i.e., if $\partial G/\partial \mathbf{n}$ was continuous) the integral in Equation (9.12) would have had to equal 0. ■

THEOREM 5. The solution of the Dirichlet problem

$$\nabla^2 u = f(\mathbf{x}) \qquad \text{on } D \subset R^2 \tag{9.13}$$

subject to the boundary conditions

$$u\Big|_{\partial D} = g(x) \tag{9.14}$$

is given by

$$u(\mathbf{x}) = \int_D G(\mathbf{x}, \boldsymbol{\eta}) f(\boldsymbol{\eta})\, d\boldsymbol{\eta} + \int_{\partial D} g(\boldsymbol{\eta}) \frac{\partial G}{\partial \mathbf{n}}\, ds \tag{9.15}$$

where $G(\mathbf{x}, \boldsymbol{\eta})$ is the Green's function for the Laplace operator on D.

Proof. To prove this formula we use Green's lemma with

$$\psi(\boldsymbol{\eta}) = G(\boldsymbol{\eta}, \mathbf{x}) \qquad \varphi(\boldsymbol{\eta}) = u(\boldsymbol{\eta})$$

We obtain

$$\int_D [G(\boldsymbol{\eta}, \mathbf{x})\nabla^2 u(\boldsymbol{\eta}) - u(\boldsymbol{\eta})\nabla^2 G(\boldsymbol{\eta}, \mathbf{x})]\, d\boldsymbol{\eta} = \int_{\partial D} \left[G(\boldsymbol{\eta}, \mathbf{x}) \frac{\partial u(\boldsymbol{\eta})}{\partial \mathbf{n}} - u(\boldsymbol{\eta}) \frac{\partial G(\boldsymbol{\eta}, \mathbf{x})}{\partial \mathbf{n}} \right] ds$$

Using Equations (9.8), (9.9), (9.13), and (9.14), this reduces to

$$\int_D [G(\boldsymbol{\eta}, \mathbf{x}) f(\boldsymbol{\eta}) - u(\boldsymbol{\eta})\, \delta(\boldsymbol{\eta} - \mathbf{x})]\, d\boldsymbol{\eta} = -\int_{\partial D} g(\boldsymbol{\eta}) \frac{\partial G}{\partial \mathbf{n}}\, ds \tag{9.16}$$

The desired result now follows from Equation (9.16) by virtue of the symmetry of $G(x, \boldsymbol{\eta})$ (Theorem 3) and the fact that

$$u(x) = \int_D u(\boldsymbol{\eta})\, \delta(\boldsymbol{\eta} - \mathbf{x})\, d\boldsymbol{\eta}$$

■

At this juncture it is natural to inquire into the intuitive meaning of G and Equation (9.15). To discuss this subject we note that the solution u of the Poisson equation

$$\nabla^2 u = f(x)$$

represents the gravitational potential due to a source distribution $f(x)$. It follows then that the Green's function that satisfies Equation (9.8) represents the gravitational potential at $\mathbf{x}$ due to a unit source at $\boldsymbol{\eta}$. Equation (9.15) can be interpreted simply as a restatement of the superposition principle; that is, the total field due to a volume distribution $f(\mathbf{x})$ and surface distribution g is equal to the sum (which in this case is represented by an integral) of the pointwise contributions of these sources.

Finally, we point out that in the preceding discussion, as well as in the rest of this chapter, we consider only the Green's functions related to the Dirichlet problem. The modifications required to represent the solutions of the Neumann problem for the Laplace operator are given in the exercises throughout the chapter.

SECTION 3 EXERCISES

9. Prove that the solution of the Dirichlet problem

$$\nabla^2 u = f \qquad \text{on } D \subset R^3$$

$$u\Big|_{\partial D} = g$$

is unique.

10. Let u be a solution of $\nabla^2 u = 0$ on $D \subset R^2$. Show that

$$\iint_D |\nabla u|^2 \, dx \, dy = \int_{\partial D} u \frac{\partial u}{\partial \mathbf{n}} \, ds$$

□ ***Hint.*** Use the following Green's identity:

$$\iint_D f\nabla^2 g + \nabla f \cdot \nabla g) \, dx \, dy = \int_{\partial G} f \frac{\partial g}{\partial \mathbf{n}} \, ds \; \square$$

11. Show that the Neumann problem for the Poisson equation

$$\nabla^2 u = f \qquad \text{on } D \subset R^3$$

$$\left.\frac{\partial u}{\partial \mathbf{n}}\right|_{\partial D} = g$$

has a solution only if

$$\int_D f \, dv = \int_{\partial D} g \, dS$$

□ ***Hint.*** Use the divergence theorem in the form

$$\int_D \operatorname{div}(\operatorname{grad} u) \, dV = \int_{\partial D} \operatorname{grad} u \cdot \mathbf{n} \, dS \; \square$$

12. Let $G(\mathbf{x}, \boldsymbol{\eta})$ be the solution of

$$\nabla^2 G(\mathbf{x}, \boldsymbol{\eta}) = \delta(\mathbf{x} - \boldsymbol{\eta}) \qquad \text{on } D \subset R^3$$

$$\left.\frac{\partial G}{\partial \mathbf{n}}\right|_{\partial D} = S^{-1}$$

where $S = \int_{\partial D} dS$. Show that the solution of the Neumann problem

$$\nabla^2 u = f \qquad \text{on } D \subset R^3$$

$$\int_D f \, dV = 0$$

$$\left.\frac{\partial u}{\partial \mathbf{n}}\right|_{\partial D} = 0$$

satisfies

$$u(\mathbf{x}) = \int_D G(\mathbf{x}, \boldsymbol{\eta}) f(\boldsymbol{\eta}) \, dV + \frac{1}{S} \int_{\partial D} u(\boldsymbol{\eta}) \, dS$$

(Explain why the second condition is required.)

□ ***Note.*** The solution of this problem contains an arbitrary constant of integration. Hence, the last term in this equation can be absorbed by this constant. □

SECTION 4

SOME OTHER GREEN'S FUNCTIONS

In this section we define and discuss Green's functions related to several differential operators that appear in applications, namely the Helmholtz and heat and wave equations operators.

Helmholtz Operator

Definition 3. The Green's function for the Helmholtz operator on a domain D of R^2 is the solution $G(\mathbf{x}, \boldsymbol{\eta})$ of

$$(\nabla^2 + k^2)G(\mathbf{x}, \boldsymbol{\eta}) = \delta(\mathbf{x} - \boldsymbol{\eta}) \qquad \text{on } D \tag{9.17}$$

and

$$G(\mathbf{x}, \boldsymbol{\eta})\Big|_{\partial D} = 0 \tag{9.18}$$

The Green's function for the Laplace operator is a particular case of this definition. Moreover, it can be easily proved that the Green's function for the Helmholtz operator has the same properties as those for the Green's function for the Laplace operator; that is, (in R^2)

$$G(\mathbf{x}, \boldsymbol{\eta}) = G(\boldsymbol{\eta}, x) \tag{9.19}$$

$$\lim_{\varepsilon \to 0} \int_{C_\varepsilon} \frac{\partial G}{\partial \mathbf{n}}\, ds = 1$$

where C_ε is the circle $(x_1 - \eta_1)^2 + (x_2 - \eta_2)^2 = \varepsilon^2$.

THEOREM 6. The solution of

$$(\nabla^2 + k^2)u(\mathbf{x}) = f(\mathbf{x}) \qquad \text{on } D \subset R^2 \tag{9.20}$$

$$u(\mathbf{x})\Big|_{\partial D} = g(\mathbf{x}) \tag{9.21}$$

is given by

$$u(\mathbf{x}) = \int_D G(\mathbf{x}, \boldsymbol{\eta}) f(\boldsymbol{\eta})\, d\boldsymbol{\eta} + \int_{\partial D} g(\boldsymbol{\eta}) \frac{\partial G}{\partial \mathbf{n}}\, ds$$

Proof. If we apply Green's lemma to $\psi(\boldsymbol{\eta}) = G(\boldsymbol{\eta}, \mathbf{x})$ and $\varphi(\boldsymbol{\eta}) = u(\boldsymbol{\eta})$, we obtain

$$\int_D [G(\boldsymbol{\eta}, \mathbf{x})\nabla^2 u(\boldsymbol{\eta}) - u(\boldsymbol{\eta})\nabla^2 G(\boldsymbol{\eta}, \mathbf{x})]\, d\boldsymbol{\eta} = \int_{\partial D} \left[G(\boldsymbol{\eta}, \mathbf{x}) \frac{\partial u}{\partial \mathbf{n}} - u(\boldsymbol{\eta}) \frac{\partial G}{\partial \mathbf{n}} \right] ds \tag{9.22}$$

Adding and subtracting $k^2 Gu$ in the left-hand side of Equation (9.22) and using Equations (9.18) and (9.21) to simplify the right-hand side, we obtain

$$\int_D [G(\boldsymbol{\eta}, \mathbf{x})(\nabla^2 + k^2)u(\boldsymbol{\eta}) - u(\boldsymbol{\eta})(\nabla^2 + k^2)G(\boldsymbol{\eta}, \mathbf{x})]\, d\boldsymbol{\eta} = -\int_{\partial D} g(\boldsymbol{\eta}) \frac{\partial G}{\partial \mathbf{n}}\, ds \tag{9.23}$$

Using Equations (9.17) and (9.20) we infer from Equation (9.23) that

$$\int_D [G(\boldsymbol{\eta}, \mathbf{x}) f(\boldsymbol{\eta}) - u(\boldsymbol{\eta})\, \delta(\mathbf{x} - \boldsymbol{\eta})]\, d\boldsymbol{\eta} = -\int_{\partial D} g(\boldsymbol{\eta}) \frac{\partial G}{\partial \mathbf{n}}\, ds$$

which yields the desired result using Equation (9.19). ■

Green's Function for the Heat Equation Operator

Definition 4. Green's function for the heat equation operator on $D \subset R^n$ and $t > t_0$ is the solution $G(\mathbf{x}, t, \boldsymbol{\eta}, \tau)$ of

$$\frac{\partial G}{\partial t} - k\nabla^2 G = \delta(\mathbf{x} - \boldsymbol{\eta})\, \delta(t - \tau) \tag{9.24}$$

on D subject to the boundary condition

$$G(\mathbf{x}, t, \boldsymbol{\eta}, \tau)\Big|_{\partial D} = 0 \tag{9.25}$$

We recall that the heat equation with a source is

$$\frac{\partial u}{\partial t} - k\nabla^2 u = r(\mathbf{x}, t)$$

Hence, the term $\delta(\mathbf{x} - \boldsymbol{\eta})\, \delta(t - \tau)$ represents a localized heat source at $\mathbf{x} = \eta$ that acts at time $t = \tau$, and G is the system response to this impulse of heat. We infer therefore by the **causality principle** (which states that the reaction cannot precede the action) that

$$G(\mathbf{x}, t, \boldsymbol{\eta}, \tau) = 0 \qquad \text{for } t < \tau$$

Another important property of G is stated in Theorem 7.

THEOREM 7. $G(\mathbf{x}, t, \boldsymbol{\eta}, \tau) = G(\mathbf{x}, t - \tau, \boldsymbol{\eta}, 0)$.

Proof. If we substitute $T = t - \tau$ in Equation (9.24) we obtain

$$\frac{\partial G}{\partial T} - k\nabla^2 G = \delta(\mathbf{x} - \boldsymbol{\eta})\, \delta(T)$$

Hence $G(\mathbf{x}, t, \boldsymbol{\eta}, \tau)$ and $G(\mathbf{x}, t - \tau, \boldsymbol{\eta}, 0)$ satisfy the same equation (and boundary conditions) and therefore are identical. ■

Finally, we state (without proof) the relationship between the Green's function in Equations (9.24) and (9.25) and the solutions of the heat equation in one space dimension. Similar but more complicated expressions can be derived in higher space dimensions.

THEOREM 8. The solution of

$$\frac{\partial u}{\partial t} - k\frac{\partial^2 u}{\partial x^2} = r(x, t) \qquad t > 0$$

subject to the boundary conditions

$$u(a, t) = f_1(t), \qquad u(b, t) = f_2(t)$$
$$u(x, 0) = g(x)$$

is

$$u(x, t) = \int_0^t \int_a^b G(x, t, \eta, \tau) r(\eta, \tau)\, d\eta\, d\tau + \int_a^b G(x, t, \eta, 0) g(\eta)\, d\eta$$
$$+ k\int_0^t \left[f_2(\tau)\frac{\partial G}{\partial \eta}(x, t, b, \tau) - f_1(\tau)\frac{\partial G}{\partial \eta}(x, t, a, \tau)\right] d\tau \qquad \blacksquare$$

□ ***Remark.*** $G(\mathbf{x}, t, \boldsymbol{\eta}, \tau)$ is symmetric in the space variables; that is,

$G(\boldsymbol{x}, t, \boldsymbol{\eta}, \tau) = G(\boldsymbol{\eta}, t, \mathbf{x}, \tau)$ □

Green's Function for the Wave Equation

Definition 5. The Green's function for the wave equation is the solution of

$$\frac{\partial^2 u}{\partial t^2} - c^2 \nabla^2 u = \delta(\mathbf{x} - \boldsymbol{\eta})\, \delta(t - \tau) \qquad \text{on } D \subset R^n$$

subject to the boundary conditions

$$G(\mathbf{x}, t, \boldsymbol{\eta}, \tau)\Big|_{\partial D} = 0$$

The three basic properties of the Green's function for the heat equation are also satisfied by the Green's function of the wave equation; that is,

1. (Causality) $G(\mathbf{x}, t, \boldsymbol{\eta}, \tau) = 0 \quad$ for $t < \tau$
2. $G(\boldsymbol{x}, t, \boldsymbol{\eta}, \tau) = G(\mathbf{x}, t - \tau, \boldsymbol{\eta}, 0)$
3. (Symmetry) $G(\mathbf{x}, t, \boldsymbol{\eta}, \tau) = G(\boldsymbol{\eta}, t, \mathbf{x}, \tau)$

In one space dimension the relationship between the solution of the wave equation and the Green's function is given by Theorem 9.

THEOREM 9. The solution of

$$\frac{\partial^2 u}{\partial t^2} - c^2\frac{\partial^2 u}{\partial x^2} = f(x, t)$$

subject to the boundary conditions

$$u(a, t) = u(b, t) = 0$$

$$u(x, 0) = g_1(x), \qquad \frac{\partial u}{\partial t}(x, 0) = g_2(x)$$

is

$$u(x, t) = \int_0^t \int_a^b G(x, t, \eta, \tau) f(\eta, \tau)\, d\eta\, d\tau + \int_a^b \left[g_2(\eta) G(x, t, \eta, 0) - g_1(\eta) \frac{\partial G}{\partial \tau}(x, t, \eta, 0) \right] dx$$ ■

Similar but more complicated expressions for $u(\mathbf{x}, t)$ exist in higher-dimension or other types of boundary conditions.

SECTION 4 EXERCISES

13. Find the appropriate definition of the Green's function for the Dirichlet problem

$$\nabla^2 u - k^2 u = f \qquad \text{on } D \subset R^3$$

$$u\Big|_{\partial D} = h$$

and obtain the explicit expression of u in terms of G, f, and h.

14. Let a_i, $i = 1, \ldots, 6$, be constants and let

$$Lu = \left(a_1 \frac{\partial^2}{\partial x^2} + 2a_2 \frac{\partial^2}{\partial x \partial y} + a_3 \frac{\partial^2}{\partial y^2} + a_4 \frac{\partial}{\partial x} + a_5 \frac{\partial}{\partial y} + a_6 \right) u$$

be elliptic (i.e., $a_2^2 - a_1 a_3 < 0$) on R^2. Show that

$$\int_D (uLv - vL^*u) = \int_{\partial D} (U\mathbf{i} + V\mathbf{j}) \cdot \mathbf{n}\, ds$$

where

$$L^*u = \left(a_1 \frac{\partial^2}{\partial x^2} + 2a_2 \frac{\partial^2}{\partial x \partial y} + a_3 \frac{\partial^2}{\partial y^2} - a_4 \frac{\partial}{\partial x} - a_5 \frac{\partial}{\partial y} + a_6 \right) u$$

$$U = a_1(uv_x - vu_x) - 2a_2 vu_y + a_4 uv$$

$$V = a_3(uv_y - vu_y) + 2a_2 uv_x + a_5 uv$$

L^* is called the adjoint of L.

□ ***Hint.*** Use the identity

$$uv_{xy} - vu_{xy} = (uv_x)_y - (vu_y)_x$$

and Green's theorem

$$\iint_D \left(\frac{\partial f}{\partial x} + \frac{\partial g}{\partial y} \right) dx\, dy = \int_{\partial D} (f\, dy - g\, dx)$$ □

15. Find the proper definition of the Green's function for the operator L in Exercise 14 (consider the Dirichlet problem only).

SECTION 5
DIRECT COMPUTATION OF GREEN'S FUNCTION

In this section we will directly compute (i.e., by solving the appropriate differential equation) the infinite space Green's function for each of the operators discussed in Sections 3 and 4. We will also compute the Green's function for the Laplace operator on a disk of radius 1.

EXAMPLE 3. Compute the infinite space Green's function for the Laplace operator in two dimensions.

Solution. Since we are considering the Green's function for the whole space R^2, the boundary conditions on G can be ignored (the space has no boundary). To solve

$$\nabla^2 G = \delta(\mathbf{x} - \boldsymbol{\eta})$$

we attempt to find G in the form

$$G(x, y, \eta_1, \eta_2) = G(r) \tag{9.26}$$

where

$$r = [(x - \eta_1)^2 + (y - \eta_2)^2]^{1/2}$$

Since $\delta(\mathbf{x}, - \boldsymbol{\eta})$ is zero for $x \neq \eta$, we infer that G must satisfy

$$\nabla^2 G = 0 \qquad \text{for } \mathbf{x} \neq \boldsymbol{\eta}$$

Or using Equation (9.26)

$$\frac{1}{r}\frac{\partial}{\partial r}\left(r\frac{\partial G}{\partial r}\right) = 0 \qquad \text{for } r \neq 0$$

The solution of this equation is

$$G = C_1 + C_2 \ln r$$

To determine C_2 we now use Equation (9.11). We obtain

$$1 = \lim_{\varepsilon \to 0} \int_{C_\varepsilon} \frac{\partial G}{\partial \mathbf{n}}\, ds = \lim_{\varepsilon \to 0} \int_0^{2\pi} \frac{C_2}{r}\, r\, d\theta = 2\pi C_2$$

Hence,

$$G = C_1 + \frac{1}{2\pi} \ln r \tag{9.27}$$

Thus although C_1 remains arbitrary, for simplicity (and convenience) we set it to zero. ■

EXAMPLE 4. Evaluate the infinite space Green's function for the Helmholtz operator in two dimensions.

Solution. As in Example 3 we attempt to find G, which depends on r only. The differential equation for G is then

$$\frac{1}{r}\frac{\partial}{\partial r}\left(r\,\frac{\partial G}{\partial r}\right) + k^2 G = 0 \qquad r \neq 0$$

This is a zero-order Bessel equation whose solution is given by

$$G(r) = C_1 J_0(kr) + C_2\, Y_0(kr)$$

where J_0 and Y_0 are Bessel functions of the first and second kind. However, since J_0 is regular at $r = 0$ whereas G must have a singularity there, we set $C_1 = 0$. To evaluate C_2 we use again the condition in Equation (9.11). Thus, we must have

$$\lim_{r\to 0} \int_0^{2\pi} C_2 \cdot \frac{\partial Y_0(kr)}{\partial r}\, r\, d\theta = 1 \tag{9.28}$$

But for $r \ll 1$, we can approximate Y_0 as

$$Y_0(kr) \approx \frac{2}{\pi} \ln r$$

Hence,

$$C_2 = \frac{1}{4}$$

and

$$G = \frac{1}{4}\, Y_0(kr)$$

■

EXAMPLE 5. Compute the infinite space Green's function for the heat equation in one space dimension.

Solution. To evaluate the Green's function in this case we must solve

$$\frac{\partial G}{\partial t} - k\,\frac{\partial^2 G}{\partial x^2} = \delta(x-\eta)\,\delta(t-\tau) \qquad -\infty < x < \infty \tag{9.29}$$

subject to the conditions

$$G(x, t, \eta, \tau) = 0 \qquad \text{for } t < \tau \tag{9.30}$$

and

$$G(x, t, \eta, \tau) = G(x, t-\tau, \eta, 0) \tag{9.31}$$

We solve this problem by applying the Fourier transform in x to Equation (9.29). We obtain (since $\int_{-\infty}^{\infty} e^{+i\omega x}\,\delta(x-\eta)\,dx = e^{i\omega\eta}$)

$$\frac{\partial \bar{G}}{\partial t} + k\omega^2 \bar{G} = \frac{e^{i\omega\eta}}{2\pi}\,\delta(t-\tau) \tag{9.32}$$

where $\bar{G}$ (the Fourier transform of G) is

$$\bar{G}(\omega, t, \eta, \tau) = \frac{1}{2\pi}\int_{-\infty}^{\infty} G(x, t, \eta, \tau)e^{i\omega x}\,dx$$

Introducing $T = t - \tau$, Equation (9.32) for $T \neq 0$ reduces to

$$\frac{\partial \bar{G}}{\partial T} + k\omega^2 \bar{G} = 0$$

whose general solution is

$$\bar{G} = C(\omega)e^{-k\omega 2T}$$

But from Equation (9.30) we infer that $\bar{G} = 0$ when $t < \tau$; hence,

$$\bar{G}(\omega, t, \eta, \tau) = C(\omega)e^{-k\omega^2(t-\tau)}H(t-\tau) \tag{9.33}$$

where H is the Heaviside function.

To determine $C(\omega)$ we integrate Equation (9.32) (with respect to T) on the interval $-\varepsilon < T < \varepsilon$ where $0 < \varepsilon \ll 1$. This yields

$$\bar{G}(T = \varepsilon) - \bar{G}(T = -\varepsilon) = \frac{e^{i\omega\eta}}{2\pi} - k\omega^2 \int_{-\varepsilon}^{\varepsilon} \bar{G}\,dT$$

But $\bar{G}(T = -\varepsilon) = 0$ [from Equation (9.33)] and $\bar{G}$ is assumed to be bounded; therefore, as $\varepsilon \to 0$ we obtain

$$\bar{G}(T = 0+) = \frac{e^{i\omega\eta}}{2\pi}$$

that is,

$$C(\omega) = \frac{e^{i\omega\eta}}{2\pi}$$

Finally, we apply the inverse Fourier transform to $\bar{G}$ to obtain G:

$$\begin{aligned} G(x, t, \eta, \tau) &= \frac{H(t-\tau)}{2\pi}\int_{-\infty}^{\infty} e^{-i\omega(x-\eta)}e^{-k\omega^2(t-\tau)}\,d\omega \\ &= \frac{H(t-\tau)}{2\sqrt{\pi k(t-\tau)}}\exp\left\{-\left[\frac{(x-\eta)^2}{4k(t-\tau)}\right]\right\} \end{aligned}$$

■

COROLLARY. The solution of

$$\frac{\partial u}{\partial t} - k\frac{\partial^2 u}{\partial x^2} = r(x, t) \qquad t > 0$$

with the initial conditions

$$u(x, 0) = f(x) \qquad -\infty < x < \infty$$

is

$$u(x, t) = \int_0^t \int_{-\infty}^{\infty} \frac{\exp\left(-\dfrac{(x-\eta)^2}{4k(t-\tau)}\right) r(\eta, \tau)}{2\sqrt{\pi k(t-\tau)}}\, d\eta\, d\tau + \int_{-\infty}^{\infty} \frac{1}{2\sqrt{\pi k t}} \exp\left(-\frac{(x-\eta)^2}{4kt}\right) f(\eta)\, d\eta$$

(where $t > 0$). ■

EXAMPLE 6. Compute the infinite space Green's function for the wave equation in one space dimension.

Solution. To solve

$$\frac{\partial^2 G}{\partial t^2} - c^2 \frac{\partial^2 G}{\partial x^2} = \delta(x-\eta)\, \delta(t-\tau) \tag{9.34}$$

subject to the conditions in Equations (9.30) and (9.31), we apply, as in Example 5, the Fourier transform to Equation (9.34). We obtain

$$\frac{\partial^2 \bar{G}}{\partial t^2} + c^2\omega^2 \bar{G} = \frac{e^{i\omega\eta}}{2\pi}\, \delta(t-\tau) \tag{9.35}$$

whose solution for $T = t - \tau \neq 0$ [taking into account the conditions in Equation (9.30) is

$$\bar{G} = [C_1 \cos c\omega(t-\tau) + C_2 \sin c\omega(t-\tau)]H(t-\tau) \tag{9.36}$$

To determine C_1 and C_2 we first use the fact that $G(t = \tau) = 0$ to infer that $C_1 = 0$. As for C_2, we integrate Equation (9.35) (once) over $-\varepsilon < T < \varepsilon$, $0 < \varepsilon \ll 1$, and obtain

$$\lim_{\varepsilon \to 0} \frac{\partial G}{\partial T}(T = \varepsilon) = \frac{e^{i\omega\eta}}{2\pi}$$

Hence, by using Equation (9.36) it follows that

$$C_2 = \frac{e^{i\omega\eta}}{2\pi c\omega}$$

Finally, to obtain G explicitly we apply the inverse Fourier transform to $\bar{G}$:

$$G(x, t, \eta, \tau) = \frac{H(t-\tau)}{2\pi} \int_{-\infty}^{\infty} \frac{e^{-i\omega(x-\eta)} \sin c\omega(t-\tau)}{c\omega}\, d\omega = \begin{cases} \dfrac{1}{2c} & |x-\eta| < c(t-\tau) \\ 0 & |x-\eta| > c(t-\tau) \end{cases} \qquad t > \tau$$

(and $G = 0$ for $t < \tau$). This can be reexpressed in terms of the Heaviside functions as follows:

$$G(x, t, \eta, \tau) = \frac{1}{2c}\{H[(x - \eta) + c(t - \tau)] - [H(x - \eta) - c(t - \tau)]\} \qquad t > \tau$$

■

EXAMPLE 7. Compute the Green's function for the Laplace operators on the unit disk (in R^2).

Solution. The usual procedure to evaluate the Green's function on domains with boundary is to rewrite it as a sum

$$G = G_1 + v$$

where G_1 is the infinite space Green's function. Thus, for the Laplace operator we obtain from Equation (9.8) that

$$\nabla^2 G = \nabla^2 G_1 + \nabla^2 v = \delta(\mathbf{x} - \boldsymbol{\eta})$$

But $\nabla^2 G_1 = \delta(\mathbf{x} - \boldsymbol{\eta})$; hence, v must satisfy

$$\nabla^2 v = 0$$

Moreover, from Equation (9.9) we now infer that on the boundary

$$\mathbf{v}\Big|_{\partial D} = -G_1\Big|_{\partial D}$$

Thus, v is a regular solution of the Laplace equation whose values on the boundary of D are equal to those of $-G_1$ on the boundary.

Applying this technique to our present problem, it follows that we must solve the equations

$$\frac{\partial^2 v}{\partial x^2} + \frac{\partial^2 v}{\partial y^2} = 0 \tag{9.37}$$

with the boundary conditions

$$\mathbf{v}\Big|_{x^2+y^2=1} = -\frac{1}{4\pi}\ln r\Big|_{x^2+y^2=1}$$

where $r = \sqrt{(x - \eta_1)^2 + (y - \eta_2)^2}$. (Note that $\boldsymbol{\eta}$ is considered as a fixed but arbitrary point on the disk.)

The polar coordinates for x, y and η_1, η_2,

$$x = \rho\cos\theta \qquad y = \rho\sin\theta$$

$$\eta_1 = \sigma\cos\varphi \qquad \eta_2 = \sigma\sin\varphi$$

lead to the general solution of Equation (9.37) in the form

$$v = \frac{a_0}{2} + \sum_{n=1}^{\infty}\rho^n(a_n\cos n\theta + b_n\sin n\theta) \tag{9.38}$$

As to the boundary conditions, we obtain

$$\mathbf{v}\Big|_{\rho=1} = -\frac{1}{4\pi}\ln\left[1+\sigma^2-2\sigma\cos(\theta-\varphi)\right] = -\sum_{n=1}^{\infty}\frac{\sigma^n\cos n(\theta-\varphi)}{n}$$

We infer then that

$$a_0 = 0, \qquad a_n = \frac{\sigma^n\cos n\varphi}{2\pi n}, \qquad b_n = \frac{\sigma^n \sin n\varphi}{2\pi n} \qquad n = 1, 2, \ldots \tag{9.39}$$

Combining Equations (9.38) and (9.39) it follows that

$$v(\rho, \theta, \sigma, \varphi) = \frac{1}{2\pi}\sum_{n=1}^{\infty}\frac{(\rho\sigma)^n}{n}\cos n(\theta-\varphi) = -\frac{1}{4\pi}\ln\left[1+(\rho\sigma)^2-2(\rho\sigma)\cos(\theta-\varphi)\right]$$

Hence

$$G = \frac{1}{4\pi}\ln\left[\rho^2+\sigma^2-2\rho\sigma\cos(\theta-\varphi)\right] - \frac{1}{4\pi}\ln\left[1+(\rho\sigma)^2-2\rho\sigma\cos(\theta-\varphi)\right]$$

■

COROLLARY. The solution on the unit disk of

$$\nabla^2 u = 0$$

subject to the boundary condition

$$u\Big|_{x^2+y^2=1} = g$$

is

$$u(x, y) = \int_0^{2\pi} P(\rho, \theta-\varphi)g(\varphi)\,d\varphi$$

where P is the Poisson kernel

$$P(\rho, \theta-\varphi) = \frac{1-\rho^2}{1+\rho^2-2\rho\cos(\theta-\varphi)}$$

Proof. From Equation (9.15) we know that the solution can be expressed as

$$u(x, y) = \int_{\partial D} g(\boldsymbol{\eta})\frac{\partial G}{\partial \boldsymbol{\eta}}\,ds \tag{9.40}$$

In this case, however,

$$\frac{\partial G}{\partial \mathbf{n}}\Big|_{\partial D} = \left(\frac{\partial G}{\partial \sigma}\right)\Big|_{\sigma=1} = \frac{1-\rho^2}{1+\rho^2-2\rho\cos(\theta-\varphi)}$$

which yields the desired result. [Note that in Equation (9.40) x, y are fixed and η_1, η_2 are the variables. Hence we must compute $\partial G/\partial \mathbf{n}$ with respect to these variables.]

■

SECTION 5 EXERCISES

16. Show that the infinite space Green's function for the Laplace operator in R^3 is

$$G(\mathbf{x}, \boldsymbol{\eta}) = \frac{-1}{4\pi r}$$

where $r = [(x - \eta_1)^2 + (y - \eta_2)^2 + (z - \eta_3)^3]^{1/2}$.

17. Show that the infinite space Green's function for the operator $(\nabla^2 - k^2)$ in R^3 (see Exercise 13) is of the form Ce^{-kr}/r (where r is defined as in Exercise 16). Compute the value of C.

18. Compute the Green's function for the Laplace operator in R^2 on a disk of radius a.

19. Repeat Exercise 18 when $D = \{x, y \mid x^2 + y^2 > 1\}$.

20. Repeat Exercise 18 when D is the semicircle $0 < r < 1$ and $0 \leq \theta \leq \pi$.

21. Compute the infinite space Green's function for the heat and wave operator in R^n.

SECTION 6
THE EIGENFUNCTION METHOD

In this section we describe a method to compute the Green's function for a differential operator on a domain D by expressing it in the form of an infinite series that consists of the eigenfunctions of a related differential operator on D.

As a prototype example of this procedure, consider the Green's function for the Laplace operator on a domain D of R^n, which requires the solution of

$$\nabla^2 G = \delta(\mathbf{x} - \boldsymbol{\eta}) \qquad \text{on } D \tag{9.41}$$

subject to the boundary conditions

$$G\Big|_{\partial D} = 0$$

To provide a solution to this problem using the eigenfunction method, we assume that the eigenfunctions and eigenvalues of

$$(\nabla^2 + \lambda)\psi(\mathbf{x}) = 0 \qquad \text{on } D \tag{9.42}$$

$$\psi\Big|_{\partial D} = 0 \tag{9.43}$$

are known as $\psi_{\mathbf{m}}(\mathbf{x})$ and $\lambda_{\mathbf{m}}$, respectively. (In general it is natural and convenient to label the eigenfunction by using several indices.) Furthermore, let us assume that these eigenfunctions have the following two properties:

1. $\psi_{\mathbf{m}}$ are orthogonal to each other on D; that is,

$$\int_D \psi_{\mathbf{m}}(\mathbf{x})\psi_{\mathbf{k}}(\mathbf{x})\, d\mathbf{x} = 0 \qquad \text{if } \mathbf{m} \neq \mathbf{k} \tag{9.44}$$

2. $\{\psi_{\mathbf{m}}\}$ form a complete set of functions on D; that is, for (any) given function $f(x)$ on D there exist $a_{\mathbf{m}}$ (constants) so that

$$f(\mathbf{x}) = \sum_{\mathbf{m}} a_{\mathbf{m}} \psi_{\mathbf{m}}(\mathbf{x}) \tag{9.45}$$

To compute the coefficients $a_{\mathbf{m}}$ in this series, we multiply both sides of Equation (9.26) by $\psi_{\mathbf{k}}$ (where $\mathbf{k}$ is arbitrary but fixed) and integrate over D. Using Equation (9.44) we obtain

$$\int_D f(\mathbf{x})\psi_{\mathbf{k}}(\mathbf{x})\, d\mathbf{x} = a_{\mathbf{k}} \int_D \psi_{\mathbf{k}}^2(\mathbf{x})\, d\mathbf{x}$$

that is,

$$a_{\mathbf{k}} = \frac{\int_D f(\mathbf{x})\psi_{\mathbf{k}}(\mathbf{x})\, d\mathbf{x}}{\|\psi_{\mathbf{x}}\|^2}$$

where

$$\|\psi_{\mathbf{k}}\| = \left(\int_D \psi_{\mathbf{k}}^2\, d\mathbf{x}\right)^{1/2}$$

is called the (L_2) norm of $\psi_{\mathbf{k}}$.

Returning to our original problem of computing the Green's function on D, we expand G using the eigenfunctions $\psi_{\mathbf{m}}$:

$$G(\mathbf{x}, \boldsymbol{\eta}) = \sum_{\mathbf{m}} a_{\mathbf{m}} \psi_{\mathbf{m}}(\mathbf{x}) \tag{9.46}$$

[In this equation we assume, for the moment, that $\boldsymbol{\eta}$ is fixed but arbitrary; that is $a_{\mathbf{m}} = a_{\mathbf{m}}(\boldsymbol{\eta})$]

Substituting Equation (9.46) in Equation (9.41) and using Equation (9.42), we obtain

$$-\sum_{\mathbf{m}} \lambda_{\mathbf{m}} a_{\mathbf{m}} \psi_{\mathbf{m}} = \delta(\mathbf{x} - \boldsymbol{\eta})$$

Multiplying Equation (9.28) by $\psi_{\mathbf{k}}$ (where $\mathbf{k}$ is arbitrary but fixed) and integrating over D, we obtain

$$-\sum_{\mathbf{m}} \lambda_{\mathbf{m}} a_{\mathbf{m}} \int_D \psi_{\mathbf{m}} \psi_{\mathbf{k}}\, d\mathbf{x} = \int_D \delta(\mathbf{x} - \boldsymbol{\eta})\psi_{\mathbf{k}}(\mathbf{x})\, dx$$

Using Equation (9.44) and the basic properties of the δ-function yields

$$a_{\mathbf{k}} = -\frac{\psi_{\mathbf{k}}(\boldsymbol{\eta})}{\lambda_k \|\psi_{\mathbf{k}}\|^2}$$

Hence

$$G(\mathbf{x}, \boldsymbol{\eta}) = -\sum_{\mathbf{m}} \frac{\psi_{\mathbf{m}}(\mathbf{x})\psi_{\mathbf{m}}(\boldsymbol{\eta})}{\lambda_{\mathbf{m}} \|\psi_{\mathbf{m}}\|^2} \tag{9.47}$$

EXAMPLE 8. Compute the Green's function for the Laplace operator on the rectangular domain shown in Figure 1.

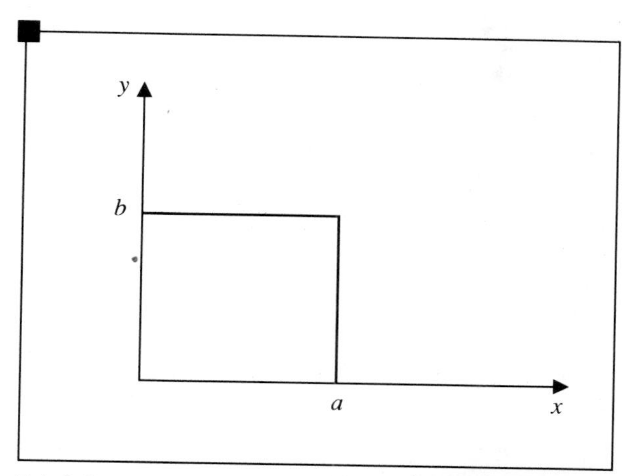

FIGURE 1 **Domain for the Laplace Operator in Example 8**

Solution. As a first step we must evaluate the eigenfunctions and eigenvalues of Equations (9.42) and (9.43) on the domain D. Using the method of separation of variables, these can be readily found to be

$$\psi_{n,m}(x, y) = \sin\frac{n\pi x}{a} \sin\frac{m\pi y}{b}$$

$$\lambda_{n,m} = \pi^2\left(\frac{n^2}{a^2} + \frac{m^2}{b^2}\right)$$

Moreover,

$$\|\psi_{n,m}\|^2 = \int_0^b \int_0^a \sin^2\frac{n\pi x}{a} \sin^2\frac{m\pi y}{b}\, dx\, dy = \frac{ab}{4}$$

Hence, using Equation (9.47) we obtain

$$G = \frac{-4ab}{\pi^2} \sum_{n=1}^{\infty} \sum_{m=1}^{\infty} \frac{\psi_{n,m}(x, y)\psi_{n,m}(\eta_1, \eta_2)}{n^2b^2 + m^2a^2}$$ ■

EXAMPLE 9. Compute the general expression of the Green's function for the Helmholtz operator on a domain D by the eigenfunction method.

Solution. The Green's function for the Helmholtz operator on D must satisfy

$$\nabla^2 G + k^2 G = \delta(\mathbf{x} - \boldsymbol{\eta}) \qquad \text{on } D \tag{9.48}$$

and

$$G\Big|_{\partial D} = 0$$

In Equation (9.48), k is a fixed number. Therefore, for clarity we refer to the corresponding Green's function as G_k.

To obtain the desired expression, we expand $G_k(\mathbf{x}, \boldsymbol{\eta})$ using the eigenfunctions $\psi_{\mathbf{m}}$ of Equations (9.42) and (9.43):

$$G_k(\mathbf{x}, \boldsymbol{\eta}) = \sum_{\mathbf{m}} b_{\mathbf{m}} \psi_{\mathbf{m}}(\mathbf{x}) \tag{9.49}$$

Substituting Equation (9.49) in Equation (9.48) and using Equation (9.42), we obtain

$$\sum_{\mathbf{m}} b_{\mathbf{m}}(k^2 - \lambda_{\mathbf{m}})\psi_{\mathbf{m}}(\mathbf{x}) = \delta(\mathbf{x} - \boldsymbol{\eta})$$

Multiplying this equation by $\psi_{\mathbf{n}}(\mathbf{x})$ and integrating over D yields

$$b_{\mathbf{n}} = \frac{\psi_{\mathbf{n}}(\boldsymbol{\eta})}{(k^2 - \lambda_{\mathbf{n}})\|\psi_{\mathbf{n}}\|^2}$$

Hence,

$$G_k(\mathbf{x}, \boldsymbol{\eta}) = \sum_{\mathbf{m}} \frac{\psi_{\mathbf{m}}(\mathbf{x})\psi_{\mathbf{m}}(\boldsymbol{\eta})}{(k^2 - \lambda_{\mathbf{m}})\|\psi_{\mathbf{m}}\|^2}$$

Obviously this result is valid only if k^2 is not an eigenvalue of Equations (9.42) and (9.43). ■

SECTION 6 EXERCISES

22. Compute the Green's functions for the Laplace and Helmholtz operators in R^2 when

(a) $D = \{x, y),\quad 0 < x < a, -\infty < y < \infty\}$ (Stripe)

(b) $D = \{(x, y),\quad 0 < x, 0 < y\}$ (Quarter plan)

(c) $D = \{(r, \theta),\quad 0 < r, 0 < \theta < \theta_0\}$ (Wedge)

23. Compute the Green's function for $(\nabla^2 - k^2)$ on the rectangle

$$0 \le x \le a, 0 \le y \le b.$$

SECTION 7
THE METHOD OF IMAGES

The method of images is usually applied to the computation of the Green's function for the Laplace operator on domains with simple boundaries. This use will be the only case considered in this section.

The basic idea of this technique is to interpret the infinite space Green's function $G^I(\mathbf{x}, \boldsymbol{\eta})$ to be the influence of a "source" located at $\mathbf{x}$ on the point $\boldsymbol{\eta}$. Hence, in order to evaluate the Green's function for a domain D, we attempt to find one or several additional sources located at points y_i outside D, to which correspond infinite space Green's functions $G^I(\mathbf{y}_i, \boldsymbol{\eta})$, so that the total influence of

$$G(\mathbf{x}, \boldsymbol{\eta}) = G^I(\mathbf{x}, \boldsymbol{\eta}) + \sum_{i=1}^{k} (\pm) G^I(\mathbf{y}_i(\mathbf{x}), \boldsymbol{\eta}) \tag{9.50}$$

on the boundary of D is zero. The points $\mathbf{y}_i$ are referred to as the **images** of x and they are positive or negative according to the sign of $G^I(y_i, \eta)$ in Equation (9.50).

If this construction is successful, then $G(\mathbf{x}, \boldsymbol{\eta})$ is the required Green's function (for the Laplace operator) on D. In fact, by construction G satisfies the boundary condition

$$G\Big|_{\partial D} = 0$$

Moreover,

$$\begin{aligned}\nabla^2 G &= \nabla^2 G^I(\mathbf{x}, \boldsymbol{\eta}) + \sum_{i=1}^{k} \nabla^2 G^I(\mathbf{y}_i, \boldsymbol{\eta}) \\ &= \delta(\mathbf{x} - \boldsymbol{\eta}) + \sum_{i=1}^{k} \delta(\mathbf{y}_i - \boldsymbol{\eta})\end{aligned}$$

But $\mathbf{y}_i \notin D$, whereas $\boldsymbol{\eta} \in D$; hence, $\delta(\mathbf{y}_i - \boldsymbol{\eta}) = 0$, $i = 1, \ldots, k$; that is,

$$\nabla^2 G = \delta(\mathbf{x} - \boldsymbol{\eta})$$

EXAMPLE 10. Compute the Green's function for the half space $x > 0$ in R^2.

Solution. To compensate for the influence of $G^I(\mathbf{x}, \boldsymbol{\eta})$ on the boundary $x = 0$, it is obvious that we must consider a negative source located at $(-x, y)$ (see Figure 2).

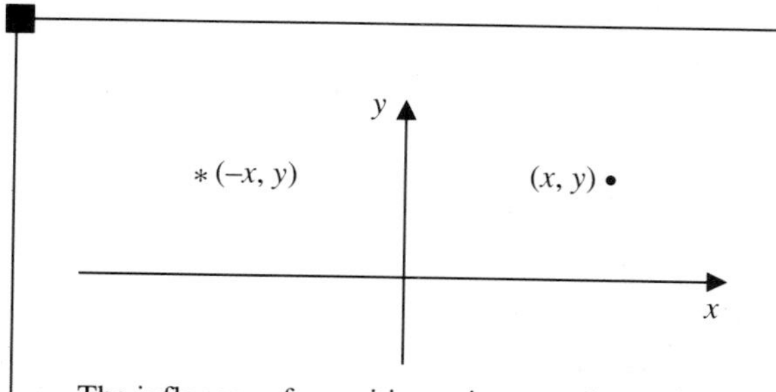

The influence of a positive unit source located at (x, y) on the boundary is being compensated by a negative unit source at the image point $(-x, y)$.

FIGURE 2

Hence

$$G = G^I(x, y, \eta_1, \eta_2) - G^I(-x, y, \eta_1, \eta_2)$$

But $G^I(\mathbf{x}, \boldsymbol{\eta}) = (1/2\pi) \ln r$ [see Equation (9.27)]; hence, the explicit expression for G is

$$G(x, y, \eta_1, \eta_2) = \frac{1}{4\pi} \ln \frac{[(x - \eta_1)^2 + (y - \eta_2)^2]}{[(x + \eta_1)^2 + (y - \eta_2)^2]}$$ ■

EXAMPLE 11. Compute the Green's function for the quarter space $x > 0$, $y > 0$, in R^2.

Solution. To compensate for the influence of a source at (x, y) on the boundary, we need three images located at $(x, -y)$, $(-x, -y)$, and $(-x, y)$ (see Figure 3). Those at $(x, -y)$ and $(-x, y)$ must be negative sources, whereas the last at $(-x, -y)$ is positive. It follows then that the required Green's function for D is

$$G = G^I(x, y, \eta_1, \eta_2) + G^I(-x, -y, \eta_1, \eta_2) - G^I(-x, y, \eta_1, \eta_2) - G^I(x, -y, \eta_1, \eta_2)$$

or, explicitly,

$$G(x, y, \eta_1, \eta_2) = \frac{1}{4\pi} \ln \frac{[(x - \eta_1)^2 + (y - \eta_2)^2][(x + \eta_1)^2 + (y + \eta_2)^2]}{[(x + \eta_1)^2 + (y - \eta_2)^2][(x - \eta_1)^2 + (y + \eta_2)^2]}$$

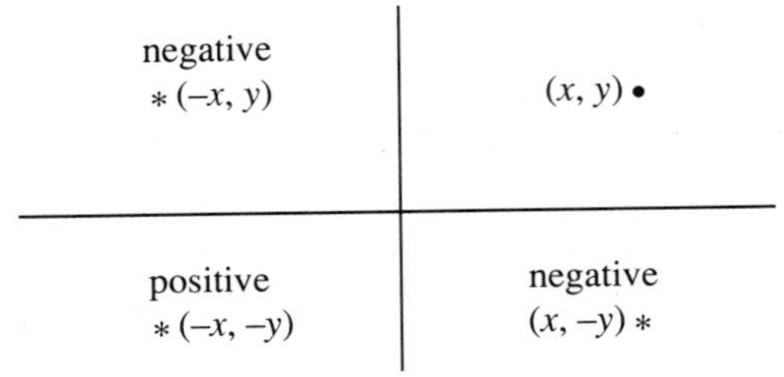

To compensate for the unit source at (x, y) on the boundary, three additional sources are needed.

FIGURE 3 ■

SECTION 7 EXERCISES

All Green's functions referred to in this set of exercises are for the Laplace operator.

24. Find the Green's function on the domain $D \subset R^2$, which is enclosed by $y = 0$ and $y = 2x$, $x > 0$, $y > 0$.

□ ***Hint.*** Construct five additional image points. □

25. Repeat Exercise 24 when D is the wedge $0 < r$, $0 < \theta < (2\pi/n)$ where n is an integer, $n > 1$.

26. Compute the Green's function in R^3 when

$$D = \{(x, y, z) \qquad 0 \leq x, 0 < y, 0 < z\}$$

27. Compute the Green's function for the unit sphere in R^3.

CHAPTER TEN

The Laplace Transform

In this brief chapter we use several examples to illustrate the application of the Laplace transform to the solution of some boundary value problems. Although we assume that the reader is already familiar with the definition, properties, and inversion of this transform, we do provide a short review of these topics in Section 1. The rest of the chapter is devoted to specific elementary applications of this transform that do not require the theory of functions with complex variables. The Fourier transform that is related to the Laplace transform was introduced in Chapter 7.

SECTION 1
THE LAPLACE TRANSFORM

Definition 1. The Laplace transform $\mathscr{L}(f)$ of the function $f(t)$ is defined as

$$\mathscr{L}(f)(s) = \int_0^\infty e^{-st} f(t)\, dt$$

Obviously, the Laplace transform is a linear operator; that is,

$$\mathscr{L}(af + bg) = a\mathscr{L}(f) + b\mathscr{L}(g)$$

where a and b are constants.

EXAMPLE 1. Compute the Laplace transform of $f(x) = \delta(t - a)$, $a > 0$.

Solution. By definition,

$$\mathscr{L}(f) = \int_0^\infty e^{-st}\, \delta(t - a)\, dt = e^{-sa}$$ ■

EXAMPLE 2. Compute the Laplace transform of t^k, $k > -1$.

Solution

$$\mathscr{L}(t^k) = \int_0^\infty t^k e^{-st}\,dt = \frac{1}{s^{k+1}} \int_0^\infty r^k e^{-r}\,dr = \frac{\Gamma(k+1)}{s^{k+1}}$$

where $r = st$. ■

EXAMPLE 3. Find the Laplace transform of the translated Heaviside function

$$H(t-a) = \begin{cases} 1 & t \geq a \\ 0 & t < a \end{cases}$$

Solution

$$\mathscr{L}[H(t-a)] = \int_0^\infty H(t-a)e^{-st}\,dt = \int_a^\infty e^{-st}\,dt = \frac{e^{-as}}{s}$$ ■

The most important property of the Laplace transform is the relation between $\mathscr{L}(f)$ and $\mathscr{L}(f')$, $\mathscr{L}(f'')$, and so on.

THEOREM 1

$$\mathscr{L}(f') = s\mathscr{L}(f) - f(0) \tag{10.1}$$

$$\mathscr{L}(f'') = s^2\mathscr{L}(f) - sf(0) - f'(0) \tag{10.2}$$

and so on.

Proof. By repeated application of integration by parts,

$$\mathscr{L}(f') = \int_0^\infty e^{-st}f'(t)\,dt = f(t)e^{-st}\Big|_0^\infty + s\int_0^\infty e^{-st}f(t)\,dt$$

$$= s\mathscr{L}(f) - f(0)$$ ■

As you might recall from a course in ordinary differential equations, the final step in the application of the Laplace transform to differential equations requires the inversion of the transform. This is usually done with the aid of a table of Laplace transforms and some "factor theorems."

THEOREM 2. If $\mathscr{L}(f)(s) = g(s)$, then

$$\mathscr{L}[e^{at}f(t)] = g(s-a) \tag{10.3}$$

$$\mathscr{L}[H(t-a)f(t-a)] = e^{-sa}g(s) \tag{10.4}$$

$$\mathscr{L}[tf(t)] = -\frac{dg}{ds}$$ ■

EXAMPLE 4. Find

$$\mathscr{L}^{-1}\left[\frac{s+a}{(s+a)^2+k^2}\right]$$

Solution. Since

$$\mathscr{L}(\cos kt) = \frac{s}{s^2+k^2} = g(s)$$

we see that

$$\frac{s+a}{(s+a)^2+k^2} = g(s+a)$$

Hence from Theorem 2 we infer that

$$\mathscr{L}^{-1}\left[\frac{s+a}{(s+a)^2+k^2}\right] = e^{-at}\cos kt$$

■

Another important property of the Laplace transform is related to the convolution of two functions.

Definition 2. The convolution of two functions f, g is defined as

$$f * g = \int_0^t f(\tau)g(t-\tau)\,d\tau$$

THEOREM 3. Let $F(s)$, $G(s)$ be the Laplace transform of f and g, respectively. Then

$$\mathscr{L}^{-1}[F(s)G(s)] = f * g$$

■

EXAMPLE 5. Use the Laplace transform to solve the equation

$$\ddot{x} + 4\dot{x} + 3x = 0, \qquad x(0) = 2, \qquad \dot{x}(0) = -4 \tag{10.5}$$

Solution. Applying the Laplace transform to Equation (10.5) using Equations (10.1) and (10.2), we obtain

$$g(s) = \frac{2s+4}{s^2+4s+3} = \frac{1}{s+1} + \frac{1}{s+3} \tag{10.6}$$

where $g(s) = \mathscr{L}(x)(s)$. But

$$\mathscr{L}(e^{at})(s) = \frac{1}{s-a}$$

Hence, applying the inverse Laplace transform to Equation (10.6) yields

$$x(t) = \mathcal{L}^{-1}[g(s)] = \mathcal{L}^{-1}\left(\frac{1}{s+1}\right) + \mathcal{L}^{-1}\left(\frac{1}{s+3}\right) = e^{-t} + e^{-3t}$$ ■

SECTION 1 EXERCISES

1. Show that $f * g = g * f$.

2. Show that if $f(0) = g(0) = 0$, then

$$\frac{d}{dt}(f * g) = \frac{df}{dt} * g = f * \frac{dg}{dt}$$

3. Let D be a linear differential operator of order n with a constant coefficient. Show that if $f^{(i)}(0) = g^{(i)}(0)$, $i = 0, \ldots, (n-1)$, then

$$D(f * g) = (Df) * g = f * (Dg)$$

4. Use the Laplace transform to solve the following system of differential equations:

$$\dot{x} = x + y$$
$$\dot{y} = -4x - 3y \qquad x(0) = 1,\ y(0) = -1$$

SECTION 2
APPLICATIONS TO THE HEAT EQUATION

In this section we solve two boundary value problems related to the heat equation in one space dimension using the Laplace transform.

EXAMPLE 6. This example concerns a cooling rod. In Exercise 5 in Chapter 1, we showed that the heat conduction equation for a thin homogeneous rod that is not insulated laterally is

$$\rho c \frac{\partial u}{\partial t} = \kappa \frac{\partial^2 u}{\partial x^2} - a(u - T_0)$$

Redefining u as $u - T_0$ (i.e., defining $\bar{u} = u - T_0$ and then dropping the bar over $\bar{u}$) and dividing by ρc, this equation can be rewritten as

$$\frac{\partial u}{\partial t} = k \frac{\partial^2 u}{\partial x^2} - bu \tag{10.7}$$

We now want to solve this equation for a semi-infinite rod, $0 < x < \infty$, subject to the following initial and boundary conditions:

$$u(x, 0) = 0 \tag{10.8}$$
$$u(0, t) = C = \text{constant} \tag{10.9}$$

$u(x, t)$ is bounded as $x \to \infty$ for all t; that is, there exists $M > 0$ so that

$$\lim_{x \to \infty} |u(x, t)| < M \tag{10.10}$$

Solution. Let $g(x, s)$ be the Laplace transform of u with respect to t:

$$g(x, s) = \mathscr{L}[u(x, t)] = \int_0^\infty e^{-st} u(x, t)\, dt$$

Then

$$\mathscr{L}\left[\frac{\partial^2 u}{\partial x^2}\right] = \frac{\partial^2}{\partial x^2} \mathscr{L}(u) = \frac{\partial^2 g}{\partial x^2}$$

From Equation (10.1) and the boundary condition in Equation (10.8), it follows that

$$\mathscr{L}\left[\frac{\partial u}{\partial t}\right] = sg(x, s) - u(x, 0) = sg(x, s)$$

Hence, if we apply the Laplace transform to Equation (10.7), we obtain

$$kg_{xx}(x, s) - (s + b)g(x, s) = 0 \tag{10.11}$$

Moreover, the boundary conditions in Equations (10.9) and (10.10) yield

$$g(0, s) = \frac{C}{s} \tag{10.12}$$

$$\lim_{x \to \infty} g(x, s) = 0 \tag{10.13}$$

Equation (10.11) is an ordinary differential equation with constant coefficients in x (s is considered as a "constant") whose solution subject to the boundary conditions in Equations (10.12) and (10.13) is

$$g(x, s) = \frac{C}{s} e^{-\alpha x}$$

where

$$\alpha = \left(\frac{s + b}{k}\right)^{1/2}$$

Thus

$$\frac{\partial u}{\partial t} = \mathscr{L}^{-1}[sg(x, s)] = \mathscr{L}^{-1}(Ce^{-\alpha x})$$

But by using a table of Laplace transforms and the property in Equation (10.3), we infer that

$$\mathscr{L}^{-1}(e^{-\alpha x}) = \frac{x}{2\sqrt{\pi k t^3}} \exp\left[-bt - \frac{x^2}{4kt}\right]$$

Hence, the required solution is

$$u(x, t) = \frac{Cx}{2\sqrt{\pi k}} \int_0^t \frac{\exp\left(-b\tau - \dfrac{x^2}{4k\tau}\right)}{\tau^{-3/2}}\, d\tau$$

■

EXAMPLE 7. The heat equation for a thin homogeneous and insulated rod in which heat is generated at the rate $r(t)$ per unit volume is

$$\frac{\partial u}{\partial t} = k\frac{\partial u^2}{\partial x^2} + r(t) \tag{10.14}$$

(see Exercise 1 in Chapter 1). We solve this equation for a semi-infinite rod, $x > 0$, subject to the following boundary conditions:

$$u(0, t) = 0$$
$$u(x, 0) = 0$$

$u(x, t)$ is bounded for all x and t

(These boundary conditions imply that heat is flowing out at $x = 0$.)

Solution. Applying the Laplace transform to Equation (10.14) and using the notations of Example 6, we obtain

$$kg_{xx}(x, s) - sg(x, s) = R(s) \tag{10.15}$$

where

$$R(s) = \mathscr{L}[r(t)] \tag{10.16}$$

The boundary conditions on g are

$$g(0, s) = 0$$
$$\lim_{x\to\infty} g(x, s) = 0$$

The solution of Equation (10.15) subject to the boundary conditions is

$$g(x, s) = \frac{R(s)}{s}(1 - e^{-x\sqrt{s/k}})$$

Hence,

$$\frac{\partial u}{\partial t} = \mathscr{L}^{-1}[sg(x, s)] = \mathscr{L}^{-1}[R(s)] - \mathscr{L}^{-1}\left\{R(s)\exp\left[-x\sqrt{\frac{s}{k}}\right]\right\}$$

Using the convolution theorem and Equation (10.16) we obtain

$$\frac{\partial u}{\partial t} = r(t) - r(t) * \left[\frac{x}{2\sqrt{\pi k t^3}}\exp\left(-\frac{x^2}{4kt}\right)\right]$$

Hence, $u(x, t)$ is given explicitly as

$$u(x, t) = \int_0^t \left\{r(\tau) - r(\tau) * \left[\frac{x}{2\sqrt{\pi k \tau^3}}\exp\left(-\frac{x^2}{4k\tau}\right)\right]\right\} d\tau \tag{10.17}$$

■

SECTION 3
APPLICATIONS TO THE WAVE EQUATION

As in Section 2 we consider two simple examples that demonstrate the general principles for the application of the Laplace transform to the wave equation.

EXAMPLE 8. This example concerns forced vibrations. The vibrations of a string under the influence of an external force (e.g., gravity) are given by

$$\frac{\partial^2 u}{\partial t^2} = c^2 \frac{\partial^2 u}{\partial x^2} + f(x, t) \qquad \textbf{(10.18)}$$

(see Exercise 3 in Chapter 1). In this example we solve this equation for a semi-infinite string $0 < x$ when $f = f(t)$ and subject to the following initial and boundary conditions:

$$u(0, t) = 0$$

$$u(x, 0) = \frac{\partial u}{\partial t}(x, 0) = 0$$

$u(x, t)$ is always bounded

Solution. As in Example 6 we let $g(x, s)$ be the Laplace transform of $u(x, t)$ with respect to t. Using Equation (10.6) and the second boundary condition, we obtain

$$\mathscr{L}\left(\frac{\partial^2 u}{\partial t^2}\right) = s^2\mathscr{L}(u) - su(x, 0) - \frac{\partial u}{\partial t}(x, 0) = s^2 g(x, s)$$

Applying the Laplace transform to Equation (10.18) yields

$$c^2 g_{xx}(x, s) - s^2 g(x, s) + F(s) = 0 \qquad \textbf{(10.19)}$$

where

$$F(s) = \mathscr{L}[f(t)]$$

The boundary conditions on $g(x, s)$ are

$$g(0, s) = 0$$

$$\lim_{x \to \infty} g(x, s) = 0$$

The solution of Equation (10.19) subject to the boundary conditions is

$$g(x, s) = F(s)\,\frac{1 - e^{-(xs/c)}}{s^2}$$

By the convolution theorem we infer that

$$u(x, t) = \mathscr{L}^{-1}[g(x, s)] = f(t) * \mathscr{L}^{-1}\left(\frac{1}{s^2} - \frac{e^{-xs/c}}{s^2}\right)$$
$$= f(t) * t - f(t) * \left[\left(t - \frac{x}{c}\right)H\left(t - \frac{x}{c}\right)\right]$$

where H is the Heaviside function. We can write this result more explicitly as

$$u(x, t) = \int_0^t f(\tau)(t - \tau)\, d\tau - \int_0^t f(\tau)\left(t - \tau - \frac{x}{c}\right)H\left(t - \tau - \frac{x}{c}\right) d\tau$$

In the particular case when $f(t) = -g$ (string falling under gravity), this yields

$$u(x, t) = \begin{cases} -\dfrac{gt^2}{2} & x > ct \\ \dfrac{g}{2c^2}(x^2 - 2cxt) & x < ct \end{cases}$$

■

EXAMPLE 9. Solve the boundary value problem for the semi-infinite string

$$\frac{\partial^2 u}{\partial t^2} = c^2 \frac{\partial^2 u}{\partial x^2} \qquad x > 0,\ t > 0 \tag{10.20}$$

and the conditions

$$u(0, t) = 0$$

$$u(x, 0) = \cos x, \qquad \frac{\partial u}{\partial t}(x, 0) = 0$$

u is bounded

Solution. Applying the Laplace transform to Equation (10.20) and using the boundary conditions, we obtain (using the notation of Example 8)

$$c^2 g_{xx} - s^2 g + s \cdot \cos x = 0 \tag{10.21}$$
$$g(0, s) = 0, \qquad \lim_{x \to \infty} g(x, s) = 0 \tag{10.22}$$

The solution of Equations (10.21) and (10.22) is

$$g(x, s) = \frac{s}{s^2 + c^2}(\cos x - e^{-sx/c})$$

Hence

$$u(x, t) = \mathscr{L}^{-1}[g(x, s)] = \cos(ct)\cos x - \cos(x - ct)H\left(t - \frac{x}{c}\right)$$

■

EXAMPLE 10. This example concerns a semi-infinite transmission line. Model equations for a transmission line were derived in Section 5 in Chapter 1. We showed that in the high-frequency limit both potential and current satisfy the wave equation

$$\frac{1}{c^2} e_{tt} = e_{xx} \qquad c = \sqrt{LC} \tag{10.23}$$

In this example we discuss the solution to this problem for the semi-infinite line $x > 0$ when the potential at $x = 0$ is prescribed; that is,

$$e(0, t) = V(t) \tag{10.24}$$

In addition we assume the boundary conditions

$$e(x, 0) = \frac{\partial e}{\partial t}(x, 0) = 0 \tag{10.25}$$

$e(x, t)$ is always bounded

Applying the Laplace transform to Equation (10.23), we obtain, using the initial conditions in Equation (10.25),

$$g_{xx} - \frac{s^2}{c^2} g = 0$$

where $g = \mathscr{L}(e)(s)$. Hence,

$$g(x, s) = c_1(s)e^{-sx/c} + c_2(s)e^{sx/c}$$

However, because the solution has to be bounded for all x, it follows that $c_2(s) = 0$. Also from Equation (10.24) we infer that

$$c_1(s) = \mathscr{L}(V)(s)$$

Hence,

$$g(x, s) = \mathscr{L}(V)e^{-sx/c}$$

Using Equation (10.4) we can invert this relation to obtain

$$e(x, t) = f\left(t - \frac{x}{c}\right)H\left(t - \frac{x}{c}\right)$$

Thus if $f(t) = \delta(t)$, then the pulse will propagate unchanged with speed c. ■

SECTIONS 2 AND 3 EXERCISES

5. Show that the solution of the heat equation on $-\infty < x < \infty$ with the initial condition $u(x, 0) = f(x)$ is

$$u(x, t) = F(x) * \frac{1}{2\sqrt{\pi t}} \exp\left(-\frac{x^2}{4t}\right)$$

(Assume that $k = 1$.)

6. Show that the solution of $\nabla^2 u = f(\mathbf{x})$ in R^3 can be written as

$$u(x, y, z) = f * \left(-\frac{1}{4\pi r}\right)$$

7. Solve

$$\frac{\partial^2 u}{\partial x^2} + 2a\frac{\partial^2 u}{\partial x \partial t} + \frac{\partial^2 u}{\partial t^2} = 0 \qquad a = \text{constant}$$

$0 < x < L$, $t > 0$ are subject to the boundary conditions

$$u(x, 0) = \frac{\partial u}{\partial t}(x, 0) = u(0, t) = 0$$

and $u(L, t) = f(t)$.

8. Simplify Equation (10.17) when $r =$ constant.

9. Solve the heat equation

$$\frac{1}{u}\frac{\partial u}{\partial t} = \frac{\partial^2 u}{\partial x^2} \qquad 0 < x < a, \alpha t$$

subject to the conditions

$$u(0, t) = u(a, t) = 0$$
$$u(x, 0) = c$$

where c is a constant.

10. Solve the equation

$$\frac{\partial^2 u}{\partial t^2} = \frac{\partial^2 u}{\partial x^2} + \cos t \qquad 0 < x < \pi, 0 < t$$

with the boundary conditions

$$u(0, t) = u(\pi, t) = u(x, 0) = \frac{\partial u}{\partial t}(x, 0) = 0$$

*11. Discuss the finite transmission line $(0 < x < a)$ subject to the same boundary conditions as in Example 10.

12. Solve

$$\frac{\partial u}{\partial x} + ax\frac{\partial u}{\partial t} = ax$$
$$u(x, 0) = u(0, t) = c$$

where a, c are constants.

Numerical Methods

With the advent of fast digital computers, numerical solutions to various difficult mathematical problems became feasible and popular. In this chapter we describe how numerical methods can be used to derive approximate solutions to boundary value problems. The technique we expound is called the **finite difference method**. In recent years other powerful methods christened "the finite element method" and "boundary elements" have been used to solve many complicated boundary value problems in engineering applications. However, we do not discuss these other techniques in this chapter.

The finite difference method reduces a linear boundary value problem (i.e., one with linear partial differential equations) to an algebraic system of linear equations that have to be solved numerically. In Section 1 we discuss numerical algorithms that accomplish this task.

SECTION 1
SYSTEMS OF LINEAR EQUATIONS

Numerical algorithms to solve the system

$$A\mathbf{x} = \mathbf{b} \tag{11.1}$$

where $A = (a_{ij})$ is an $n \times n$ matrix, and $\mathbf{x}$, $\mathbf{b} \varepsilon R^n$, can be divided broadly into two categories: direct and iterative. In this section we describe algorithms from each of these categories.

Direct Algorithms

One of the oldest and most well-known algorithms to solve the system in Equation (11.1) when the determinant at A is nonzero, is Cramer's method.

However, numerically this method is inefficient compared to other algorithms, such as Gauss elimination, since it requires much more computation.

The **Gauss elimination algorithm**, which is often used when A is a general matrix, can be described as follows:

1. Arrange, if necessary, the equations so that $a_{11} \neq 0$, and then multiply the first equation by

$$-\frac{a_{21}}{a_{11}}, \ldots, -\frac{a_{n1}}{a_{11}}$$

and add to the second through the nth equation, respectively (thus eliminating x_1 from these equations). The resulting system, which is equivalent to the original, can be written as

$$\begin{aligned} a_{1,1}x_1 + a_{1,2}x_2 + \cdots + a_{1,n}x_n &= b_1 \\ c_{2,2}x_2 + \cdots + c_{2,n}x_n &= d_2 \\ c_{n,2}x_2 + \cdots + c_{n,n}x_n &= d_n \end{aligned} \tag{11.2}$$

2. Consider now the subsystem consisting of the second to the nth equations in Equation (11.2) and use the same procedure as in the first step to eliminate x_2 from these equations except the first (in the subsystem).
3. Continuing in this fashion we finally bring the original system to one of the following:

   a. **Triangular form**—when A is nonsingular ($\equiv$ determinant $A \neq 0$). The final system will then be

$$\begin{aligned} \alpha_{1,1}x_1 + \alpha_{1,2}x_2 + \cdots + \alpha_{1,n}x_n &= \beta_1 \\ \alpha_{2,2}x_2 + \cdots + \alpha_{2,n}x_n &= \beta_2 \\ \cdots \alpha_{n,n}x_n &= \beta_n \end{aligned}$$

   The solution of this system can be obtained easily from the bottom up; that is,

$$x_n = \frac{\beta_n}{\alpha_{n,n}}, \qquad x_{n-1} = \frac{1}{\alpha_{n-1,n-1}}(\beta_{n-1} - \alpha_{n-1,m}x_n)$$

   and so on.

   b. **Trapezoidal form**—when A is singular ($\equiv$ determinant $A = 0$). In this case the equations of the system are not independent of each other, and the final form of the system will be

$$\begin{aligned} \alpha_{1,1}x_1 + \cdots + \alpha_{1,n}x_n &= \beta_1 \\ \alpha_{2,2}x_2 + \cdots + \alpha_{2,n}x_n &= \beta_2 \\ \alpha_{k,k}x_k + \cdots + \alpha_{k,n}x_n &= \beta_k \end{aligned} \tag{11.3}$$

   where $k < n$. The general solution of the system can be expressed in terms of $(n - k + 1)$ arbitrary parameters, and it can be obtained by solving Equation (11.3) from the bottom up:

$$x_k = \frac{1}{\alpha_{k,k}}[\beta_k - \alpha_{k,k+1}x_{k+1} \cdots - \alpha_{k,n}x_n]$$

   and so on.

Although this description of the Gauss elimination algorithm is complete from a mathematical point of view, it has to be complemented numerically by two steps: namely, pivoting and scaling, the purpose of which is to reduce the numerical errors in the calculations. (These steps also reduce the numerical errors for other algorithms described in this section.)

1. **Pivoting** requires the rearrangement of the equations in the system so that the elements in the diagonal of A are as large as possible (in absolute value). Especially $a_{k,k}$ should be (whenever possible) the largest coefficient in the kth equation.
2. **Scaling** consists of dividing each equation by a_{kk} (after pivoting) so that "1's" stand along the diagonal of the system.

EXAMPLE 1. Use Gauss elimination to solve

$$\begin{aligned} 2x_1 + x_2 - x_3 &= 0 \\ x_1 - x_2 + 2x_3 &= 4 \\ x_1 + 4x_2 + x_3 &= -2 \end{aligned}$$

using three-decimal-floating arithmetic.

Solution. We pivot the system by interchanging the second and third equations, and then scale it. We obtain

$$\begin{aligned} x_1 + 0.5x_2 - 0.5x_3 &= 0 \\ 0.25x_1 + x_2 + 0.25x_3 &= -0.5 \\ 0.5x_1 - 0.5x_2 + x_3 &= 2 \end{aligned}$$

Eliminating x_1 from the second and third equations yields

$$\begin{aligned} x_1 + 0.5x_2 - 0.5x_3 &= 0 \\ 0.875x_2 + 0.375x_3 &= -0.5 \\ -0.75x_2 + 1.25x_3 &= 2 \end{aligned}$$

Scaling the least two equations and eliminating x_2 from the third equation leads to

$$\begin{aligned} x_1 + 0.5x_2 - 0.5x_3 &= 0 \\ x_2 + 0.428x_3 &= -0.571 \\ 1.25x_3 &= 1.26 \end{aligned}$$

Hence,

$$x_3 = 1, \qquad x_2 = -1, \qquad x_1 = 1$$

which is the exact solution. ■

Although the Gauss elimination algorithm is optimal when A is a general matrix, in many applications A is a sparse matrix (i.e., a matrix in which most of the off-diagonal elements are zero). Accordingly, several algorithms were devised to take advantage of this structure in order to decrease the amount of numerical computations required to solve the system.

A case of particular interest in obtaining numerical solutions to boundary value problems arises when A is tridiagonal; that is, all the entries in A are zero except those of the main diagonal and the two adjacent to it (one from above and the other from below).

A direct compact algorithm to solve such a system requires only $5n$ multiplications and divisions, compared to the Gauss elimination method for a general matrix, which requires $n^3/2$ such operations.

To describe this algorithm consider the triple diagonal system

$$\begin{aligned}
a_{1,1}x_1 + a_{1,2}x_2 &= b_1\\
a_{2,1}x_1 + a_{2,2}x_2 + a_{2,3}x_3 &= b_2\\
\cdots\\
a_{k,k-1}x_{k-1} + a_{k,k}x_k + a_{k,k+1}x_{k+1} &= b_k\\
\vdots\\
a_{n,n-1}x_{n-1} + a_{n,n}x_n &= b_n
\end{aligned} \tag{11.4}$$

We can use the first equation to express x_1 in terms of x_2:

$$x_1 = \frac{b_1}{a_{11}} - \frac{a_{12}}{a_{22}} x_2 \tag{11.5}$$

Substituting Equation (11.5) in the second equation of Equation (11.4), we can express x_2 in terms of x_3, and so on. We see then that in general there is a linear relationship between x_k and x_{k+1}, which is given by

$$x_k = \alpha_k x_{k+1} + \beta_k \qquad k = 1, \ldots, n \tag{11.6}$$

To compute the coefficients α_k and β_k, we derive a recursion formula for them. Thus, substituting Equation (11.6) in the $(k + 1)$th equation in Equation (11.4), we obtain

$$x_{k+1} = \frac{b_{k+1} - a_{k+1,k}\beta_k}{a_{k+1,k+1} + a_{k+1,k}\alpha_k} - \frac{a_{k+1,k+2}}{a_{k+1,k+1} + a_{k+1,k}\alpha_k} x_{k+2}$$

Hence,

$$\alpha_{k+1} = -\frac{a_{k+1,k+2}}{a_{k+1,k+1} + a_{k+1,k}\alpha_k} \tag{11.7}$$

$$\beta_{k+1} = \frac{b_{k+1} - a_{k+1,k}\beta_k}{a_{k+1,k+1} + a_{k+1,k}\alpha_k} \tag{11.8}$$

Since α_1 and β_1 are known from Equation (11.5), the computation of α_k, β_k, $k = 2, \ldots, n$, using Equations (11.7) and (11.8) is straightforward. Moreover, since $\alpha_n = 0$ it follows that $x_n = \beta_n$, and the rest of the unknowns can be obtained by back substitution.

Iterative Algorithms

To solve a system of linear equations by iterations, we first pivot and scale the equations and then rewrite them as follows:

$$
\begin{aligned}
x_1 &= b_1 - a_{1,2}x_2 - \cdots - a_{1,n}x_n \\
x_2 &= b_2 - a_{2,1}x_1 - a_{2,3}x_3 - \cdots - a_{2,n}x_n \\
&\vdots \\
x_n &= b_n - a_{n,1}x_1 - \cdots - a_{n,n-1}x_{n-1}
\end{aligned}
\tag{11.9}
$$

Starting now with an arbitrary guess for the solution

$$\mathbf{x}^{(0)} = [x_1^{(0)}, \ldots, x_n^{(0)}]$$

we substitute these values in the right-hand side of the system in Equation (11.9) to obtain a new vector,

$$\mathbf{x}^{(1)} = [x_1^{(1)}, \ldots, x_n^{(1)}]$$

Using $\mathbf{x}^{(1)}$ as a new guess for the solution and repeating the process described above, we obtain $\mathbf{x}^{(2)}$, and so on. We terminate these iterations and say that $\mathbf{x}^{(m)}$ is the (approximate) solution of the system in Equation (11.9) if for some prescribed $\varepsilon > 0$,

$$\|\mathbf{x}^{(m)} - \mathbf{x}^{(m-1)}\| = \sum_{i=1}^{n} |x_i^{(m)} - x_i^{(m-1)}| < \varepsilon \tag{11.10}$$

We can try to accelerate the convergence of this algorithm, that is, decrease the number of iterations needed until the solution is obtained, as follows. We observe that when we have to calculate $x_i^{(k+1)}$, $i > 1$, we have already computed $x_1^{(k+1)}, \ldots, x_{i-1}^{(k+1)}$; therefore, these improved values for the solution can be used for this purpose (rather than $x_1^{(k)}, \ldots, x_{i-1}^{(k)}$). The iterative formulas for the new algorithm are given by

$$
\begin{aligned}
x_1^{(k+1)} &= b_1 - a_{1,2}x_2^{(k)} - \cdots - a_{1,n}x_n^{(k)} \\
x_i^{(k+1)} &= b_i - [a_{i,1}x_1^{(k+1)} + \cdots + a_{i,i-1}x_{i-1}^{(k+1)}] \\
&\quad - (a_{i,i+1}x_{i+1}^{(k)} + \cdots + a_{i,n}x_n^{(k)}) \qquad 1 < i < n \\
x_n^{(k+1)} &= b_n - a_{n,1}x_1^{(k+1)} - \cdots - a_{n,n-1}x_{n-1}^{(k+1)}
\end{aligned}
$$

□ ***Remarks***

1. The two algorithms described above are called the **Jacobi algorithm** and the **Gauss–Seidel algorithm** respectively.
2. It might seem as though the Gauss–Seidel algorithm should always converge faster than the Jacobi algorithm. However, it can be shown that the two algorithms are not comparable; that is, in some instances the Jacobi algorithm is better than the Gauss–Seidel whereas in other cases the reverse is true.
3. Although the iterations can start with an arbitrary $\mathbf{x}^0$, you should use the best available approximation to the solution as the initial guess in order to reduce the computational effort. □

Finally, we observe that these two algorithms will diverge in some instances; that is, for some systems, $\|\mathbf{x}^{(m)} - \mathbf{x}^{(m-1)}\|$ will never satisfy the condition in Equation (11.10) for ε small enough. A sufficient condition for the convergence of these two algorithms is

$$\sum_{\substack{k=1\\k\neq i}}^{n} |a_{ik}| < |a_{ii}| \qquad i = 1, \ldots, n$$

However, this is not a necessary condition for convergence.

SECTION 1 EXERCISES

Write computer programs to solve the following systems of linear equations by the algorithms described in this section. For the Jacobi and Gauss–Seidel algorithms, use $\varepsilon = 10^{-3}$ to terminate the iterations.

1. $$\begin{bmatrix} 5 & -2 & 3 & 1 \\ 1 & 6 & 2 & -5 \\ 2 & -3 & -4 & 7 \\ 3 & 2 & 6 & -4 \end{bmatrix} \cdot \begin{bmatrix} x \\ y \\ z \\ w \end{bmatrix} = \begin{bmatrix} 1 \\ 0 \\ -1 \\ 0 \end{bmatrix}$$

2. $$\begin{bmatrix} 1 & -2 & 3 \\ -2 & 1 & -1 \\ 3 & -1 & 1 \end{bmatrix} \cdot \begin{bmatrix} x \\ y \\ z \end{bmatrix} = \begin{bmatrix} 1 \\ 10 \\ 100 \end{bmatrix}$$

3. $$\begin{bmatrix} 1 & 0 & 3 & 0 \\ 0 & 4 & 0 & 2 \\ 1 & 0 & 0 & 4 \\ 2 & 5 & 0 & 0 \end{bmatrix} \cdot \begin{bmatrix} x \\ y \\ z \\ w \end{bmatrix} = \begin{bmatrix} 1 \\ -1 \\ 0 \\ 0 \end{bmatrix}$$

4. $$\begin{bmatrix} 1 & 2 & 0 & 0 \\ 1 & 3 & -1 & 0 \\ 0 & 1 & 4 & 2 \\ 0 & 0 & 1 & 3 \end{bmatrix} \cdot \begin{bmatrix} x \\ y \\ z \\ w \end{bmatrix} = \begin{bmatrix} 1 \\ 2 \\ -2 \\ 4 \end{bmatrix}$$

SECTION 2
FINITE DIFFERENCE SCHEMES

From a numerical point of view, we consider a function u as known on a domain D if its values on a grid of points in this domain are known and if by using these values we can calculate an acceptable approximation to the value of u at any other point in D (by interpolation). It follows, then, that in order to solve a differential equation on D we have to introduce a grid of points on this domain and construct an algorithm to calculate the values of the unknown function u at these points. Thus, if $\mathbf{x}_i$, $i = 1, \ldots, m$, are the grid points, then the fundamental unknowns that have to be computed are

$$u_i = u(\mathbf{x}_i) \qquad i = 1, \ldots, m \tag{11.11}$$

Since the differential equation and the boundary conditions at hand are the only means to compute these quantities, we must develop an approximation scheme for the derivatives of u in terms of these unknowns. To see how this can be done, consider the one-dimensional case where, to begin with, we assume that the x_i's are equispaced:

$$x_i - x_{i-1} = h$$

$u_{i-2}\ u_{i-1}\ u_i\ u_{i+1}\ u_{i+2}$ Values of function

$x_{i-2}\ x_{i-1}\ x_i\ x_{i+1}\ x_{i+2}$ Grid points

To derive an approximation for the first-order derivative of u at x_i, we note that (using Taylor's expansion)

$$u_{i+1} = u(x_i + h) = u(x_i) + hu'(x_i) + \frac{h^2}{2} u''(x_i) + O(h^3) \tag{11.12}$$

and

$$u_{i-1} = u(x_i - h) = u(x_i) - hu'(x_i) + \frac{h^2}{2} u''(x_i) + O(h^3) \tag{11.13}$$

□ ***Remark.*** The notation $O(h^n)$ means that the terms in the remainder of the series are of order h^n or higher. Equivalently, this means that the error committed by neglecting these terms is of order h^n. □

Subtracting Equation (11.13) from Equation (11.11), we obtain

$$2hu'(x_i) = u_{i+1} - u_{i-1} + O(h^3)$$

Hence,

$$u_i' = u'(x_i) = \frac{u_{i+1} - u_{i-1}}{2h} + O(h^2) \tag{11.14}$$

This formula is called the **central difference approximation for u_i'** since x_i is in the center of the interval $[x_{i-1}, x_{i+1}]$.

Obviously, we can derive other approximations for u_i'. Thus, if we subtract u_i from Equation (11.12), we obtain

$$hu_i' = u_{i+1} - u_i + O(h^2)$$

Therefore,

$$u_i' = \frac{u_{i+1} - u_i}{h} + O(h) \tag{11.15}$$

This approximation is called the **forward difference formula for u_i**.

The error term in Equation (11.14) is of order h^2, whereas in Equation (11.15) it is of order h. Thus, if $h < 1$, the approximation scheme given by Equation (11.14) is superior to the one given by Equation (11.15). However,

Equation (11.15) is still useful whenever Equation (11.14) is not applicable (e.g., at the boundary where u_{-1} is outside the domain of interest).

We can obtain similar approximation schemes for the second-order derivative by adding Equations (11.12) and (11.13) and subtracting $2u_i$. This yields

$$h^2 u''(x_i) = u_{i+1} + u_{i-1} - 2u_i + O(h^4)$$

[Note that the remainder is of order h^4 since the third-order terms in Equations (11.12) and (11.13) cancel out]. Thus,

$$u_i'' = u''(x_i) = \frac{u_{i-1} + u_{i+1} - 2u_i}{h^2} + O(h^2) \tag{11.16}$$

This is the **central difference approximation for u_i''**. A forward approximation scheme for u_i'' can be derived by noting that

$$u_{i+2} = u(x_i + 2h) = u(x_i) + 2hu'(x_i) + 2h^2 u''(x_i) + O(h^3)$$

Therefore, by using Equation (11.12) we obtain

$$u_i'' = \frac{u_{i+2} - 2u_{i+1} + u_i}{h^2} + O(h)$$

When the grid points are not equispaced, the general technique described above can be easily adapted to derive appropriate approximations. Thus, if

$$x_i - x_{i-1} = h_i$$

then

$$u_{i+1} = u(x_i + h_{i+1}) = u(x_i) + h_{i+1}u'(x_i) + O(h^2)$$
$$u_{i-1} = u(x_i - h_i) = u(x_i) - h_i u'(x_i) + O(h^2)$$

which yields

$$u_i' = \frac{u_{i+1} - u_{i-1}}{h_i + h_{i+1}} + O(h) \tag{11.17}$$

Similarly, for the second-order derivative,

$$u_i'' = 2\,\frac{h_i u_{i+1} + h_{i+1}u_{i-1} - (h_i + h_{i+1})u_i}{(h_i h_{i+1})(h_i + h_{i+1})} + O(h) \tag{11.18}$$

From Equations (11.17) and (11.18) we infer that in general the accuracy of our difference approximations is better with an equispaced grid. Equations (11.17) and (11.18) or their generalizations to higher dimensions are, therefore, useful near irregular boundaries where their use is mandatory.

Finally, the formulas derived above have a natural extension to higher dimensions. For example, for an equispaced grid in two dimensions with step size h we have (see Figure 1)

$$u_{i-1,j} = u(x_i - h, y_j) = u(x_i, y_j) - h\left[\frac{\partial u}{\partial x}\right]_{ij} + \frac{h^2}{2}\left[\frac{\partial^2 u}{\partial x^2}\right]_{ij} + O(h^3)$$

$$u_{i+1,j} = u(x_i + h, y_j) = u(x_i, y_j) + h\left[\frac{\partial u}{\partial x}\right]_{ij} + \frac{h^2}{2}\left[\frac{\partial^2 u}{\partial x^2}\right]_{ij} + O(h^3)$$

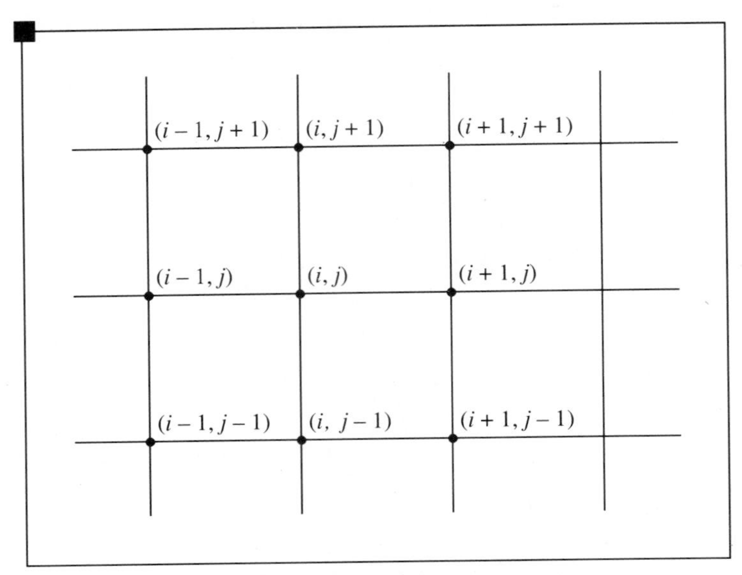

FIGURE 1 **A Grid Around the Point (i, j)**

Therefore,

$$\left[\frac{\partial u}{\partial x}\right]_{ij} = \frac{u_{i+1,j} - u_{i-1,j}}{2h} + O(h^2)$$

$$\left[\frac{\partial^2 u}{\partial x^2}\right]_{ij} = \frac{u_{i+1,j} + u_{i-1,j} - 2u_{ij}}{h^2} + O(h^2) \tag{11.19}$$

The corresponding formulas for $[\partial u/\partial y]_{ij}$ and $[\partial^2 u/\partial y^2]_{ij}$ should be obvious.

We now demonstrate how these formulas are used to solve numerically a boundary value problem in one dimension.

EXAMPLE 2. Solve

$$u'' + (1 + x^2)u' + u = x \qquad u(0) = 0,\ u(1) = 1 \tag{11.20}$$

on [0, 1) with step size $h = 0.25$.

Solution. Since $h = 0.25$, the numerical solution of this problem is equivalent to the computation of the three unknowns

$$u_1 = u(0.25), \qquad u_2 = u(0.5), \qquad u_3 = u(0.75)$$

(Note that $u_0 = u(0) = 0$ and $u_4 = u(1) = 1$.)

u_0	u_1	u_2	u_3	u_4
0	0.25	0.5	0.75	1

Using Equation (11.20) and the finite difference formulas in Equations (11.14) and (11.16), we obtain the following equations at $x = 0.25$, 0.5 and 0.75, respectively:

$$\frac{u_0 + u_2 - 2u_1}{h^2} + [1 + (0.25)^2]\frac{u_2 - u_0}{h} + u_1 = 0.25$$

$$\frac{u_1 + u_3 - 2u_2}{h^2} + [1 + (0.5)^2]\frac{u_3 - u_1}{h} + u_2 = 0.5$$

$$\frac{u_4 + u_2 - 2u_3}{h^2} + [1 + (0.75)^2]\frac{u_4 - u_2}{h} + u_3 = 0.75$$

Using the boundary conditions and rearranging these equations leads to

$$-31u_1 + 20.25u_2 = 0.25$$
$$11u_1 - 31u_2 + 21u_3 = 0.5$$
$$9.75u_2 - 31u_3 = 21.50$$

This system can be solved by the techniques described in Section 1. ■

SECTION 2 EXERCISES

5. Solve the boundary value problem given by Equation (11.20) with $h = 10^{-1}, 10^{-2}$. What can be said about the coefficient matrix for the resulting system of equations?

6. Prove the following finite difference approximations:

$$u_i' = \frac{-u_{i+2} + 8u_{i+1} - 8u_{i-1} + u_{i-2}}{12h} + O(h^4)$$

$$u_i''' = \frac{u_{i+3} - 3u_{i+2} + 3u_{i+2} - u_i}{h^3} + O(h)$$

$$u_i^{(iv)} = \frac{u_{i+2} - 4u_{i+1} + 6u_i - 4u_{i-1} + u_{i-2}}{h^4} + O(h^2)$$

7. Compare the exact and numerical solution of

$$u'' + 2u' + u = x \qquad u(0) = 1,\ u(1) = 3$$

when $h = 10^{-1}, 10^{-2}$.

SECTION 3 NUMERICAL SOLUTIONS FOR THE POISSON EQUATION

To solve numerically the Dirichlet problem for the Poisson or Laplace equations on a rectangular $D \subset R^2$, that is,

$$\nabla^2 u = f(\mathbf{x}) \qquad \text{on } D$$

and

$$u\Big|_{\partial D} = g(\mathbf{x})$$

we introduce on D an equispaced grid with step size h. Using the difference formula in Equation (11.19) for $\partial^2 u/\partial x^2$ and its equivalent for $\partial^2 y/\partial y^2$, we approximate the differential equation at each interior grid point (x_i, y_j) by

$$\frac{u_{i+1,j} + u_{i-1,j} - 2u_{ij}}{h^2} + \frac{u_{i,j+1} + u_{i,j-1} - 2u_{ij}}{h^2} = f(x_i, y_j)$$

that is,

$$u_{i+1,j} + u_{i-1,j} - 4u_{ij} + u_{i,j+1} + u_{i,j-1} = h^2 f(x_i, y_j) \tag{11.21}$$

Thus, for each interior point we obtain a linear equation in the unknowns u_{ij}. Furthermore, since the values of u on the boundary are known, the total number of equations equals the number of the unknowns; that is, the boundary value problem has been reduced to a system of linear equations.

EXAMPLE 3. Derive a system of linear equations for the solution of

$$\nabla^2 u = 0$$

on

$$D = \{(x, y) \quad 0 \le x \le 1, 0 \le y \le 1\}$$

with the boundary conditions

$$u(x, 0) = 1, \qquad u(1, y) = 1, \qquad u(0, y) = u(x, 1) = 0$$

and $h = \frac{1}{3}$ (see Figure 2).

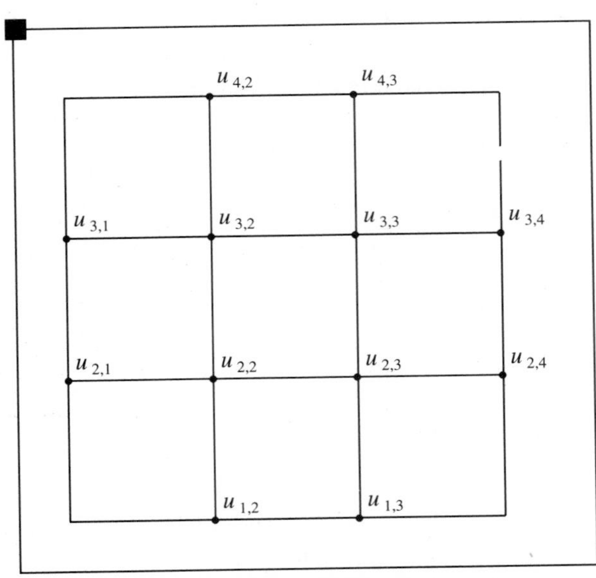

FIGURE 2 **Unknowns for Example 3**

Solution. From the boundary conditions we have

$$u_{1,2} = u_{1,3} = u_{2,4} = u_{3,4} = 1 \qquad \text{and} \qquad u_{2,1} = u_{3,1} = u_{4,2} = u_{4,3} = 0$$

Hence, to solve this problem we have to compute only u_{22}, u_{23}, u_{32}, and u_{33}. From Equation (11.21) we obtain the following equations for the four inner grid points:

$$\begin{aligned}
u_{2,3} + u_{2,1} - 4u_{2,2} + u_{3,2} + u_{1,2} &= 0\\
u_{2,2} + u_{2,4} - 4u_{2,3} + u_{3,3} + u_{1,3} &= 0\\
u_{3,1} + u_{3,3} - 4u_{3,2} + u_{4,2} + u_{2,2} &= 0\\
u_{3,2} + u_{3,4} - 4u_{3,3} + u_{4,3} + u_{2,3} &= 0
\end{aligned} \tag{11.22}$$

Using the boundary conditions this yields

$$\begin{aligned}
u_{2,3} - 4u_{2,2} + u_{3,2} &= -1\\
u_{2,2} - 4u_{2,3} + u_{3,3} &= -2\\
u_{3,3} - 4u_{3,2} + u_{2,2} &= 0\\
u_{3,2} - 4u_{3,3} + u_{2,3} &= -1
\end{aligned}$$

which is the required system of equations. ■

Other Boundary Conditions

When the boundary value problem is of Neumann or mixed type, the number of equations derived using Equation (11.21) at the inner grid points will not be equal to the number of the unknowns since the values of u on the boundary are not known. To overcome this difficulty we add additional grid points outside the region whenever the boundary conditions are given in terms of the derivatives and then use Equation (11.16) to equalize the number of unknowns to the number of equations. The details of this "trick" are demonstrated in Example 4.

EXAMPLE 4. Solve

$$\nabla^2 u = 0$$

on

$$D = \{(x, y) \qquad 0 \le x \le 1, 0 \le y \le 1\}$$

with the boundary conditions

$$\frac{\partial u}{\partial y}(x, 0) = 1, \qquad u(1, y) = 1, \qquad u(0, y) = u(x, 1) = 0$$

and $h = \frac{1}{3}$.

Solution. Since the boundary conditions along $y = 0$ are given in terms of $\partial u/\partial y$, we add the grid points $u_{0,2}$ and $u_{0,3}$ and use Equation (11.21) at the four inner grid points and the two grid points along $x = 0$ [i.e., $(\frac{1}{3}, 0)$ and

$(\frac{3}{2}, 0)$]. We obtain six equations in eight unknowns. These equations consist of the four given by the system in Equation (11.22) and

$$u_{1,1} + u_{1,3} - 4u_{1,2} + u_{2,2} + u_{0,2} = 0$$
$$u_{1,2} + u_{1,4} - 4u_{1,3} + u_{2,3} + u_{0,3} = 0$$

(See Figure 3.)

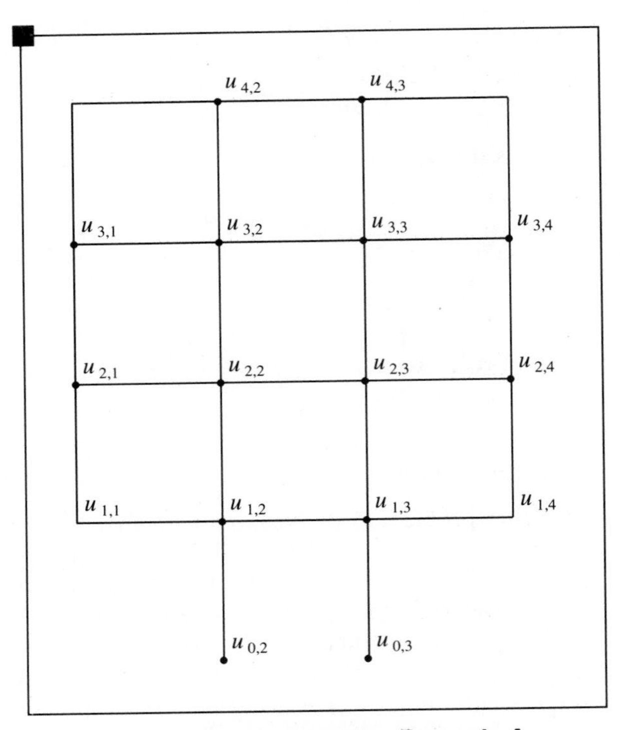

FIGURE 3 **Unknowns for Example 4**

But

$$u_{3,1} = u_{2,1} = u_{1,1} = u_{4,2} = u_{4,3} = 0$$

and

$$u_{1,4} = u_{2,4} = u_{3,4} = 1$$

Hence,

$$u_{2,3} - 4u_{2,2} + u_{3,2} + u_{1,2} = 0$$
$$u_{2,2} - 4u_{2,3} + u_{3,3} + u_{1,3} = -1$$
$$u_{3,3} - 4u_{3,2} + u_{2,2} = 0$$
$$u_{3,2} - 4u_{3,3} + u_{2,3} = -1$$
$$u_{1,3} - 4u_{1,2} + u_{2,2} + u_{0,2} = 0$$
$$u_{1,2} - 4u_{1,3} + u_{2,3} + u_{0,3} = -1$$

To solve this system we need two additional equations. These equations are obtained from the boundary condition at $(\frac{1}{3}, 0)$ and $(\frac{2}{3}, 0)$ using the approximation

$$\left[\frac{\partial u}{\partial y}\right]_{ij} = \frac{u_{i,j+1} - u_{i,j-1}}{2h}$$

This yields

$$u_{2,2} - u_{0,2} = \tfrac{2}{3}, \qquad u_{2,3} - u_{0,3} = \tfrac{2}{3}$$

The system now consists of eight equations in eight unknowns. ■

Irregular Regions

When the domain D has an irregular shape, Equation (11.21) will not be applicable to grid points whose distance from the boundary is less than h. Under these circumstances we must use approximation formulas for the derivatives, which are similar to Equations (11.17) and (11.18). Thus, using the notation shown in Figure 4, we obtain the following approximations at the grid point **a**:

$$\left[\frac{\partial u}{\partial x}\right]_{\mathbf{a}} = \frac{u_2 - u_1}{h(1 + \alpha_2)}, \qquad \left[\frac{\partial u}{\partial y}\right]_{\mathbf{a}} = \frac{u_4 - u_3}{h(\alpha_3 + \alpha_4)}$$

$$\left[\frac{\partial^2 u}{\partial x^2}\right]_{\mathbf{a}} = \frac{2}{h^2}\,\frac{u_2 + \alpha_2 u_1 - (1 + \alpha_2)u_a}{\alpha_2(1 + \alpha_2)}$$

$$\left[\frac{\partial^2 u}{\partial y^2}\right]_{\mathbf{a}} = \frac{2}{h^2}\,\frac{\alpha_3 u_4 + \alpha_4 u_3 - (\alpha_3 + \alpha_4)u_a}{\alpha_3 \alpha_4(\alpha_3 + \alpha_4)}$$

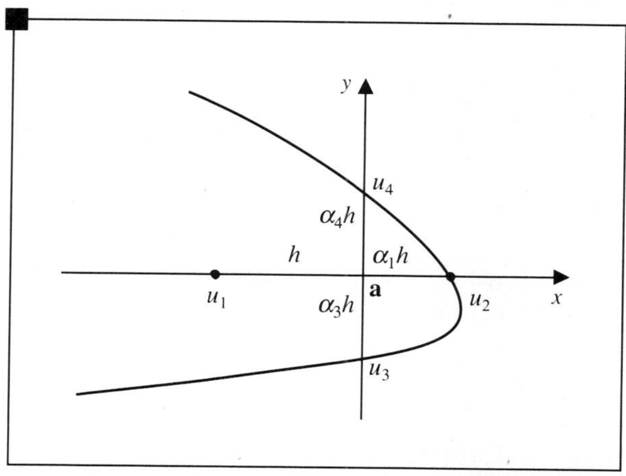

FIGURE 4 **Finite Difference Scheme Near an "Irregular" Boundary**

where $u_a = u(\mathbf{a})$. The finite difference approximation scheme to $\nabla^2 u = f$ at $\mathbf{a}$ is given by

$$\frac{u_2 + \alpha_2 u_1 - (1 + \alpha_2)u_a}{\alpha_2(1 + \alpha_2)} + \frac{\alpha_3 u_4 + \alpha_4 u_3 - (\alpha_3 + \alpha_4)u_a}{\alpha_3 \alpha_4(\alpha_3 + \alpha_4)} = \frac{h^2}{2} f(a)$$

Symmetry Considerations

In practical applications of the finite difference method, the number of equations to be solved can be very large; for example, a two-dimensional grid with 50 divisions along the x- and y-axes will yield 2500 equations. The same grid in three dimensions will lead to 625,000 equations! This number can be reduced, however, when the region D, the boundary conditions, and the differential equation are invariant under certain symmetry operations.

EXAMPLE 5. Since the region and the boundary conditions as shown in Figure 5 are invariant under reflections with respect to the x- and y-axes, it follows that the solution of $\nabla^2 u = 0$, which is also invariant with respect to these operations, will satisfy

$$u_1 = u_2 = u_3 = u_4$$

and so on.

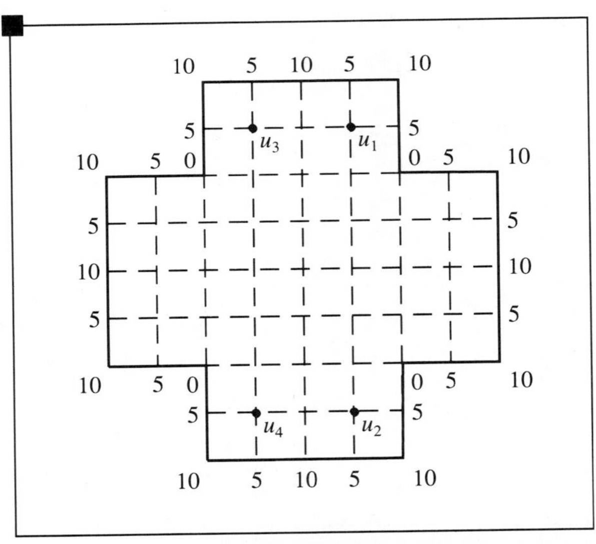

FIGURE 5 **Simplifications Due to Symmetry**

SECTION 3 EXERCISES

8. Derive a finite difference approximation scheme for the Poisson equation in three dimensions.

9. For a boundary value problem with circular symmetry in R^2, it is natural to use polar rather than Cartesian coordinates. Derive a finite difference approximation to $\nabla^2 u = f$ in these coordinates.

□ ***Hint.*** Use a grid with constant $\Delta\theta$ and Δr. Note also that

$$\nabla^2 u = \frac{\partial^2 u}{\partial r^2} + \frac{1}{r}\frac{\partial u}{\partial r} + \frac{1}{r^2}\frac{\partial^2 u}{\partial \theta^2} \quad \square$$

10. Solve

$$\nabla^2 u = r \cos^2 \theta \qquad u(1, \theta) = 0$$

on the unit disk using $\Delta\theta = \pi/6$ and $\Delta r = 0.2$. Compare with the exact solution.

11. Solve

$$\nabla^2 u = x^2 - y^2$$

on

$$D = \{(x, y) \quad 0 \le x \le 1, 0 \le y \le 1\}$$

with the boundary conditions

$$\frac{\partial u}{\partial x}(0, y) = 1, \qquad u(x, 0) = 1, \qquad u(1, y) = u(x, 1) = 0$$

Use different step sizes and compare the solutions obtained.

12. Complete the solutions of Examples 3, 4, and 5. Use several different h.

13. Derive a finite difference scheme to solve

$$\nabla^2 u + (x^2 + y^2)\frac{\partial u}{\partial x} = f(\mathbf{x})$$

14. Solve numerically the differential equations that appear in Exercise 13 if

$$f(\mathbf{x}) = x^2 + y^2$$

and the boundary conditions are the same as in Example 3.

SECTION 4
NUMERICAL SOLUTIONS FOR THE HEAT AND WAVE EQUATIONS

To obtain numerical solutions for the heat equation in one dimension,

$$\frac{\partial^2 u}{\partial x^2} = \frac{1}{k}\frac{\partial u}{\partial t} \tag{11.23}$$

with the boundary conditions

$$u(x, 0) = f(x), \qquad u(0, t) = g_1(t), \qquad u(a, t) = g_2(t)$$

on $0 \leq x \leq a$ and $0 \leq t \leq T$. We introduce a grid with step size Δx, Δt in x and t, respectively, and use Equations (11.19) and (11.15) to approximate Equation (11.23). Denoting $u(x_i, t_j)$ by u_{ij} leads to the following finite difference equation at the grid point (ij):

$$\frac{u_{i+1,j} - 2u_{ij} + u_{i-1,j}}{(\Delta x)^2} = \frac{1}{k}\frac{u_{i,j+1} - u_{i,j}}{\Delta t} \tag{11.24}$$

or

$$u_{i,j+1} = \frac{k\,\Delta t}{(\Delta x)^2}(u_{i+1,j} + u_{i-1,j}) + \left(1 - \frac{2k\,\Delta t}{(\Delta x)^2}\right)u_{i,j} \tag{11.25}$$

We observe that in Equation (11.24) the forward difference formula is used to approximate $\partial u/\partial t$ since otherwise, that is, for the central difference formula, the values of u at $t \pm \Delta t$ are needed to evaluate u at $t + \Delta t$, and these data are not available.

The formula in Equation (11.24) expresses u at t_{j+1} in terms of its values at time t_j; hence, since u at time $t = 0$ is given, we can compute u on $[0, T]$ consecutively, that is, at Δt, $2\Delta t$, and so on (there is no need to solve systems of equations). Furthermore, Equation (11.25) can be simplified by choosing $r = k\,\Delta t/(\Delta x)^2$ to be equal to $\frac{1}{2}$, which leads to

$$u_{i,j+1} = \tfrac{1}{2}(u_{i+1,j} + u_{i-1,j}) \tag{11.26}$$

The meaning of this formula is that on such a grid the value of $u(x_i, t_{j+1})$ is the average of $u(x_{i-1}, t_j)$ and $u(x_{i+1}, t_j)$.

The algorithm defined by Equation (11.25) or Equation (11.26) is called the **explicit method** for the numerical solution of the heat equation. Its drawback is that in order to ensure its stability, that is, to prevent the accumulated numerical error from becoming too large and therefore render the numerical solution meaningless, we must choose $r \leq \frac{1}{2}$. This condition places a restriction on Δt that must satisfy $\Delta t \leq (\Delta x^2)/2k$ and, therefore, increases the computational effort needed to obtain the required solution.

To overcome this restriction, Crank and Nicolson observed that the forward difference formula

$$\left[\frac{\partial u}{\partial t}\right]_{ij} = \frac{u_{i,j+1} - u_{i,j}}{\Delta t}$$

can be interpreted as a central difference formula at time $t_j + \frac{1}{2}\,\Delta t$ and, hence, a consistent approximation scheme must evaluate $\partial^2 u/\partial x^2$ at this time. To accomplish this we average the approximations for $\partial^2 u/\partial x^2$ at t_j and t_{j+1} and, thus, obtain the following finite difference formula for Equation (11.23) at (ij):

$$\frac{1}{2}\left[\frac{u_{i+1,k} - 2u_{i,j} + u_{i-1,j}}{(\Delta x)^2} + \frac{u_{i+1,j+1} - 2u_{i,j+1} + u_{i-1,j+1}}{(\Delta x)^2}\right] = \frac{1}{k}\frac{u_{i,j+1} - u_{i,j}}{\Delta k}$$

that is,

$$-ru_{i+1,j+1} + (2+2r)u_{i,j+1} - ru_{i-1,j+1} = ru_{i+1,j} + (2-2r)u_{i,j} + ru_{i-1,j} \tag{11.27}$$

The algorithm represented by Equation (11.27) is called the **Crank–Nicolson algorithm**. It requires the solution of a system of linear equations at each time step, but it is stable for all values of r.

As for the wave equation

$$\frac{\partial^2 u}{\partial x^2} = \frac{1}{c^2}\frac{\partial^2 u}{\partial t^2} \tag{11.28}$$

with the boundary conditions

$$u(x, 0) = f(x), \qquad \frac{\partial u}{\partial t}(x, 0) = g(x)$$

$$u(0, t) = u(a, t) = 0$$

we use Equation (11.19) to derive a finite difference approximation for Equation (11.28). We obtain

$$\frac{u_{i+1,j} - 2u_{i,j} + u_{i-1,j}}{(\Delta x)^2} = \frac{1}{c^2}\frac{u_{i,j+1} - 2u_{i,j} + u_{i,j-1}}{(\Delta t)^2}$$

Hence,

$$u_{i,j+1} = w^2(u_{i+1,j} + u_{i-1,j}) + 2(1-w^2)u_{i,j} - u_{i,j-1} \tag{11.29}$$

where $w = (c\,\Delta t)/\Delta x$. Furthermore, if we set $w = 1$, the algorithm formulas will be simplified and we obtain

$$u_{i,j+1} = u_{i+1,j} + u_{i-1,j} - u_{i,j-1} \tag{11.30}$$

We infer from Equation (11.29) or Equation (11.30) that we can calculate the values of u at t_{j+1} if these values are known at t_j and t_{j-1}. It follows then that in order to begin the computation of u at Δt, its values at $-\Delta t$ must be known. We can easily overcome this difficulty, however, if we observe that $(\partial u/\partial t)(x, 0) = g(x)$ implies

$$\frac{u_{i,1} - u_{i,-1}}{2\,\Delta t} = g(x_i) \tag{11.31}$$

Thus, using Equations (11.30) and (11.31) we obtain for $u_{i,1}$,

$$\begin{aligned} u_{i,1} &= \tfrac{1}{2}(u_{i+1,0} + u_{i-1,0}) + g(x_i)\,\Delta t \\ &= \tfrac{1}{2}[f(x_{i+1}) + f(x_{i-1})] + g(x_i)\,\Delta t \end{aligned}$$

The algorithm given by Equation (11.29) is stable for $w \le 1$. However, unexpectedly the numerical results obtained by setting $w = 1$ are better than those for $w < 1$; that is, decreasing the time step will not bring the numerical solution closer to the exact solution.

SECTION 4 EXERCISES

15. Solve the heat equation in one dimension with the following boundary conditions using the explicit and Crank–Nicolson algorithms. Compare the accuracy of the solutions and the computational effort.

(a) $u(x, 0) = x, \quad u(0, t) = T, \quad u(a, t) = 0$
(b) $u(x, 0) = \sin x, \quad u(0, t) = T, \quad u(a, t) = 0$
(c) $u(x, 0) = \cos x, \quad \dfrac{\partial u}{\partial x}(0, t) = c, \quad u(a, t) = 0$

16. Solve the wave equation in one dimension with $w = 1$ and $w = \frac{1}{2}$ using the following boundary conditions:

(a) $u(x, 0) = \sin x, \quad \dfrac{\partial u}{\partial t}(x, 0) = 0$

(b) $u(x, 0) = 0, \quad \dfrac{\partial u}{\partial t} = \sin x$

(c) $u(x, 0) = \cos 2x, \quad \dfrac{\partial u}{\partial t}(x, 0) = \sin x$

Always assume that $u(0, t) = u(a, t) = 0$. Compare the numerical and exact solutions.

17. Derive a finite difference algorithm to solve

$$\frac{\partial^2 u}{\partial x^2} + b(x)\frac{\partial u}{\partial x} = \frac{1}{k}\frac{\partial u}{\partial t}$$

subject to the appropriate boundary conditions.

18. Solve numerically the differential equation that appears in Exercise 17 if the boundary conditions on u are $u(x, 0) = \sin 2x$, $u(0, t) = T$, $u(a, t) = 0$, and $b(x) = x$.

19. Derive a finite difference scheme for

$$\frac{\partial^2 u}{\partial x^2} + b(x)\frac{\partial^2 u}{\partial x \partial t} = \frac{1}{c^2}\frac{\partial^2 u}{\partial t^2}$$

if $u(x, 0) = \sin 5x$, $(\partial u/\partial t)(x, 0) = 0$, and $u(0, t) = u(a, t) = 0$.

APPENDIX ONE
Tables

TABLE 1 **Values of Gamma Function $\Gamma(\nu)$**

$$\Gamma(\nu) = \int_0^\infty e^{-t} t^{\nu-1}\, dt$$

ν	$\Gamma(\nu)$
1.00	1.000
1.05	0.974
1.10	0.951
1.15	0.933
1.20	0.918
1.25	0.906
1.30	0.897
1.35	0.891
1.40	0.887
1.45	0.886
1.50	0.886
1.55	0.889
1.60	0.894
1.65	0.900
1.70	0.909
1.75	0.919
1.80	0.931
1.85	0.946
1.90	0.962
1.95	0.980
2.00	1.000

TABLE 2 Properties of Bessel Functions

$$\frac{d}{dx}[x^{\nu}J_{\nu}(x)] = x^{\nu}J_{\nu-1}(x)$$

$$\frac{d}{dx}[x^{-\nu}J_{\nu}(x)] = -x^{-\nu}J_{\nu+1}(x)$$

$$\frac{d}{dx}J_{\nu}(x) = J_{\nu-1}(x) - \frac{\nu}{x}J_{\nu}(x)$$

$$xJ_{\nu+1}(x) = 2\nu J_{\nu}(x) - xJ_{\nu-1}(x)$$

$$2J'_{\nu}(x) = J_{\nu-1}(x) - J_{\nu+1}(x)$$

$$\int x^{\nu}J_{\nu-1}(x)\,dx = x^{\nu}J_{\nu}(x)$$

$$\int x^{-\nu}J_{\nu+1}(x)\,dx = -x^{-\nu}J_{\nu}(x)$$

$$\int xJ_0(x)\,dx = xJ_1(x)$$

$$\int x^2J_1(x)\,dx = x^2J_2(x)$$

$$\int x^3J_2(x)\,dx = x^3J_3(x)$$

$$\int J_1(x)\,dx = -J_0(x)$$

$$\int x^{-1}J_2(x)\,dx = -x^{-1}J_1(x)$$

$$\int x^{-2}J_3(x)\,dx = -x^{-2}J_2(x)$$

$$\int x^2J_0(x)\,dx = x^2J_1(x) + xJ_0(x) + \int J_0(x)\,dx$$

$$\int x^nJ_0(x)\,dx = x^nJ_1(x) + (n-1)x^{n-1}J_0(x) - (n-1)^2\int x^{n-2}J_0(x)\,dx$$

$$\int x^nJ_1(x)\,dx = -x^nJ_0(x) + n\int x^{n-1}J_0(x)\,dx$$

TABLE 3 **Table of Bessel Functions**

x	$J_0(x)$	$J_1(x)$	$Y_0(x)$	$Y_1(x)$	$\int_0^x J_0(s)\,ds$
0	1	0	$-\infty$	$-\infty$	0
0.5	0.94	0.24	−0.44	−1.47	0.49
1.0	0.77	0.44	0.09	−0.78	0.92
1.5	0.51	0.56	0.38	−0.41	1.24
2.0	0.22	0.58	0.51	−0.11	1.43
2.5	−0.05	0.50	0.50	0.15	1.47
3.0	−0.26	0.34	0.38	0.32	1.39
3.5	−0.38	0.14	0.19	0.41	1.22
4.0	−0.40	−0.07	−0.02	0.40	1.02
4.5	−0.32	−0.23	−0.19	0.30	0.84
5.0	−0.18	−0.33	−0.31	0.15	0.72
5.5	−0.01	−0.34	−0.34	−0.02	0.67
6.0	0.15	−0.28	−0.29	−0.18	0.71

TABLE 4 **Table of Zeros of Bessel Functions**

	1	2	3	4	5	6
$J_0(x)$	2.40	5.52	8.65	11.79	14.93	18.07
$J_1(x)$	3.83	7.02	10.17	13.32	16.47	19.62
$Y_0(x)$	0.89	3.96	7.09	10.22	13.36	16.50
$Y_1(x)$	2.20	5.43	8.60	11.75	14.90	18.04

TABLE 5 **Properties of Legendre Polynomials**

$$P_n(1) = 1$$

$$P_n(-1) = (-1)^n$$

$$P_{n+1}(x) = \frac{2n+1}{n+1}\, xP_n(x) - \frac{n}{n+1}\, P_{n-1}(x)$$

$$(1 - x^2)P_n'(x) = nP_{n-1}(x) - nxP_n(x)$$

$$P_{n+1}'(x) - P_{n-1}'(x) = (2n+1)P_n(x)$$

$$\int_x^1 P_n(x)\,dx = \frac{1}{2n+1}\,[P_{n-1}(x) - P_{n+1}(x)]$$

Rodrigues' Formula: $\dfrac{(-1)^n}{2^n n!}\dfrac{d^n}{dx^n}(1 - x^2)^n$

Generating Function: $\dfrac{1}{\sqrt{1 - 2zx + x^2}} = \sum_{n=0}^{\infty} P_n(x) z^n$

TABLE 6 Some Fourier Transforms

	$f(t)$	$F(\omega) = \mathscr{F}[f(t)]$	
1.	$g(t) \pm h(t)$	$G(\omega) \pm H(\omega)$	
2.	$kf(t)$	$kF(\omega)$	
3.	$f(at)$	$\frac{1}{a} F\left(\frac{\omega}{a}\right)$	
4.	$e^{-bt}f(t)$	$F(\omega - ib)$	
5.	$f'(t)\left[\lim_{\lvert t\rvert \to +\infty} f'(t) = 0\right]$	$i\omega F(\omega)$	
6.	$f''(t)$	$-\omega^2 F(\omega)$	
7.	$f^{(n)}(t)$	$(i\omega)^n F(\omega)$	
8.	$\int_{-\infty}^{t} f(s)\, ds$	$\frac{F(\omega)}{i\omega}$	
9.	$u(t)$	$\frac{1}{i2\pi\omega}$	
10.	$u(t)e^{-bt}$	$\frac{1}{2\pi(b + i\omega)}$	$b > 0$
11.	$u(t)t$	$-\frac{1}{2\pi\omega^2}$	
12.	$u(t) \sin at$	$\frac{1}{2\pi} \frac{a}{a^2 - \omega^2}$	
13.	$u(t) \cos at$	$\frac{1}{2\pi} \frac{i\omega}{a^2 - \omega^2}$	
14.	$e^{-bt}u(t)t$	$\frac{1}{2\pi} \frac{1}{(b + i\omega)^2}$	$b > 0$
15.	$e^{-bt}u(t) \sin at$	$\frac{1}{2\pi} \frac{a}{(b + i\omega)^2 + a^2}$	$b > 0$
16.	$e^{-bt}u(t) \cos at$	$\frac{1}{2\pi} \frac{b + i\omega}{(b + i\omega)^2 + a^2}$	$b > 0$
17.	$u(t - \tau)$	$\frac{1}{2\pi} \frac{e^{-i\omega\tau}}{i\omega}$	

APPENDIX TWO

Answers to Selected Exercises

The reader should note that, in addition to the solutions provided here, many exercises contain the final answer as part of the problem.

CHAPTER 1

SECTION 2

1. $\dfrac{1}{k}\dfrac{\partial u}{\partial t} = \dfrac{\partial^2 u}{\partial x^2} + r(x, t)$

4. $\dfrac{1}{k}\dfrac{\partial u}{\partial t} = \dfrac{\partial}{\partial x}\left(A(x)\dfrac{\partial u}{\partial x}\right)$

7. $\dfrac{1}{k}\dfrac{\partial u}{\partial t} = \nabla^2 u\,\dfrac{\partial^2 u}{\partial x^2} + \dfrac{\partial^2 u}{\partial y^2} + \dfrac{\partial^2 u}{\partial z^2}$

SECTION 3

12. $\dfrac{1}{c^2}\dfrac{\partial^2 u}{\partial t^2} = \dfrac{\partial^2 u}{\partial x^2} + g$

13. $\dfrac{\partial^2 u}{\partial t^2} = \rho\,\dfrac{\partial^2 u}{\partial x^2} - ku - b\,\dfrac{\partial u}{\partial t}$

16. $\dfrac{1}{c^2}\dfrac{\partial^2 u}{\partial t^2} = \nabla^2 u$

18. Let $w = x - ct,\ z = x + ct$.

SECTION 4

24. $h_{tt} = g(ax + b)h_{xx} + agh_x$

SECTION 5

26. Mass $\approx LC$, spring constant $\approx RG$, coefficient of friction $\approx LG + RC$

SECTION 6

29. Zero inside the cavity; M/r outside the cavity.

32. $\dfrac{1}{r}\dfrac{\partial}{\partial r}\left(r\,\dfrac{\partial u}{\partial r}\right) + \dfrac{1}{r^2}\dfrac{\partial^2 u}{\partial r^2} = 0$

SECTION 7

36. $u \approx \dfrac{1}{\rho}$

38. $\dfrac{\partial u}{\partial t} + \dfrac{\partial(\rho u)}{\partial x} = -\,a$

40. $$\frac{\partial \rho_1}{\partial t} + \frac{\partial(\rho_1 u_1)}{\partial x} = -a_1(\rho_1) + a_2(\rho_2)$$

$$\frac{\partial \rho_2}{\partial t} + \frac{\partial(\rho_2 u_2)}{\partial x} = -a_2(\rho_2) + a_1(\rho_1)$$

CHAPTER 2

SECTION 1

1. (a) linear; (b) nonlinear; (c) nonlinear

3. $u = \dfrac{x^2}{2}$

5. Substitute $w + v$ in the heat equation.

7. Burger's equation is nonlinear.

SECTION 2

11. Differentiate the first equation with respect to t and the second with respect to x.

13. Equation (2.18) is second order in x and t.

15. Differentiate $u + c$.

SECTION 3

16. $\dfrac{\partial u}{\partial r}(a, \theta) = f(\theta) \qquad u$ is bounded

$$u(r, \pi) = u(r, -\pi) \qquad 0 < r < a$$

$$\frac{\partial u}{\partial \theta}(r, \pi) = \frac{\partial u}{\partial \theta}(r, -\pi)$$

21. The implicit conditions follow:

1. u and its first-order derivative with respect to θ are continuous at $\pm\pi$.
2. $u(0, \theta)$ is bounded.

SECTION 4

26. $\nabla^2 u = 0$

$$u(0, y) = 100, \qquad u(x, 0) = 0$$

$$\frac{\partial u}{\partial x}(a, y) = 0, \qquad u(x, b) = 0$$

$$0 \le x \le a, \qquad 0 \le y \le b$$

29. $\nabla^2 u = 0 \qquad 0 \le x \le 8, 0 \le y \le 6$

$$u(0, y) = 100 \qquad \frac{\partial u}{\partial y}(x, 0) = 0$$

$$u(x, -\tfrac{3}{4}x + 6) = 0$$

35. $\nabla^2 u = 0 \qquad 0 \le r \le a,\ -\pi < \theta \le \pi$

$u(a, \theta) = 100 \qquad 0 \le \theta < \pi$

$u(a, -\theta) = 0 \qquad -\pi \le \theta < 0$

$u(r, \pi) = u(r, -\pi) \qquad \dfrac{\partial u}{\partial \theta}(r, \pi) = \dfrac{\partial u}{\partial \theta}(r, -\pi)$

u is bounded.

36. (a) $u(0, t) = u(L, t) = 0$

$u(x, 0) = 0, \qquad \dfrac{\partial u}{\partial t}(x, 0) = g(x) \qquad 0 \le x \le L,\ 0 \le t$

(b) $u(0, t) = u(L, t) = 0, \qquad u(x, 0) = \sin(x - L),$

$\dfrac{\partial u}{\partial t}(x, 0) = x^2 \qquad 0 \le x \le L,\ 0 \le t$

CHAPTER 3

SECTION 3

	Period	*Primitive Period*
1. (a)	$2n\pi$	2π
(b)	$n\pi/3$	$\pi/3$
(c)	Not periodic	
(d)	n	1
(g)	Any real number $\neq 0$	None
(h)	Not periodic	

2. (a) Yes
(b) No
(d) Yes
(f) Yes

3. $b = 0$

7. (b) Orthogonal
(c) Orthogonal

8. (b) $\sqrt{15/28}(x^2 + 1)$

SECTION 4

11. (a) $-2 \sum_{n=1}^{\infty} \dfrac{\sin nx}{n}$

(b) $-\dfrac{2}{\pi} \sum_{n=1}^{\infty} \dfrac{\sin(n\pi x/2)}{n}$

(c) $\dfrac{1}{2} - \dfrac{4}{\pi^2} \sum_{n=1}^{\infty} \dfrac{\cos(2n-1)\pi x}{(2n-1)^2}$

(d) $\dfrac{\pi^2}{3} + 4 \sum_{n=1}^{\infty} \dfrac{(-1)^n}{n^2} \cos nx$

(e) $\frac{1}{2} + \frac{2}{\pi} \sum_{n=1,3,5,\ldots}^{\infty} \frac{\sin nx}{n}$

12. (a) $\frac{5}{2} - \frac{3}{\pi} \sum_{n=1}^{\infty} \frac{\sin(2n\pi x/3)}{n}$

(c) $\frac{1}{2} + \frac{4}{\pi^2} \sum_{n=1,3,5,\ldots}^{\infty} \frac{\cos n\pi x}{n^2}$

(d) $\sin 3t$ Fourier series same as given function

(j) $-\frac{1}{2} + \frac{1}{\pi} \sum_{n=1}^{\infty} \frac{\sin 2n\pi x}{n}$

(l) $\frac{\pi}{2} - \frac{4}{\pi} \sum_{n=1}^{\infty} \left\{ \frac{\cos(2n-1)r}{(2n-1)^2} - \frac{\pi}{2}(-1)^{n+1} \frac{\sin nr}{n} \right\}$

13. (a) $1 - \frac{2}{\pi} \sum_{n=1}^{\infty} \frac{\sin(n\pi x/2)}{n}$

SECTION 5

14. (a) Odd
(b) Odd
(d) Even
(e) Neither
(g) Even

16. (c) (1) $\frac{1}{3} + \frac{4}{\pi^2} \sum_{n=1}^{\infty} \frac{(-1)^n}{n^2} \cos n\pi x$

(2) $\frac{2}{\pi^3} \sum_{n=1}^{\infty} \frac{1}{n^3} [(-1)^{n+1} n^2\pi^2 + (-1)^n 2 - 2] \sin n\pi x$

(e) (1) $\frac{3}{4} + \frac{4}{\pi^2} \sum_{n=1}^{\infty} \frac{1}{n^2} \left(\cos \frac{n\pi}{2} - 1 \right) \cos \frac{n\pi x}{2}$

(2) $\frac{2}{\pi} \sum_{n=1}^{\infty} \left[\frac{2}{n^2\pi^2} \sin \frac{n\pi}{2} - \frac{2}{n} \cos \frac{n\pi}{2} + \frac{1}{n} \cos n\pi \right] \sin \frac{n\pi x}{2}$

17. $\frac{12}{\pi} + 6 \sin 10t + \frac{12}{\pi} \sum_{n=2}^{\infty} \frac{(-1)^n}{1-n^2} \cos 10nt$

20. $\frac{1}{\pi} \sum_{n=1}^{\infty} \frac{1}{n} \left[\cos \frac{n\pi}{2} + 1 - 2(-1)^n \right] \sin \frac{n\pi x}{2}$

SECTION 6

23. (a) $2x + i3x^2$
(b) $-5 \sin 5x + i2 \sin x \cos x$

25. (b) $\frac{1+i}{2} e^{(1+i)x}$

(d) $\frac{x^2}{2} - i \ln x$

(e) $\frac{1}{2}[\sin^2 x + i(x - \sin x \cos x)]$

27. (a) $2 \sinh \frac{1}{2} \sum_{n=-\infty}^{\infty} \frac{(-1)^n(1 + i2n\pi)}{1 + 4n^2\pi^2} \exp(i2n\pi x)$

(e) 100

(f) $\frac{i20}{\pi}\left[\sum_{n=-\infty}^{-1} \frac{(-1)^n}{n} e^{in\pi x/4} + \sum_{n=1}^{\infty} \frac{(-1)^n}{n} e^{in\pi x/4}\right]$

(k) $\frac{2}{3} + \frac{1}{2\pi^2} \sum_{\substack{n=-\infty \\ n \neq 0}}^{\infty} \left\{\frac{9(1 + in\pi)}{n^2} e^{-in\pi} - \frac{9 + i3n\pi}{n^2} e^{-in\pi/3}\right\} \exp \frac{in\pi x}{3}$

28. $\frac{100}{e} + 100 \sum_{\substack{n=-\infty \\ n \neq 0}}^{\infty} \frac{(1 - i2n\pi)}{1 + 4n^2\pi^2}\left[\frac{1}{e} - 1\right] \exp(i2n\pi t)$

SECTION 7

30. (a) Piecewise continuous
(c) Not piecewise continuous
(d) Piecewise continuous
(f) Neither piecewise differentiable nor piecewise continuous

31. Yes

32. (a) Piecewise differentiable
(b) Neither piecewise differentiable nor piecewise continuous
(d) Not piecewise differentiable but piecewise continuous
(f) Neither piecewise differentiable nor piecewise continuous

33. Continuous on $[0, \frac{1}{120}]$
Piecewise differentiable on $[0, \frac{1}{120}]$
Therefore, by Definition 10, it is piecewise differentiable for all t.

34. (b)

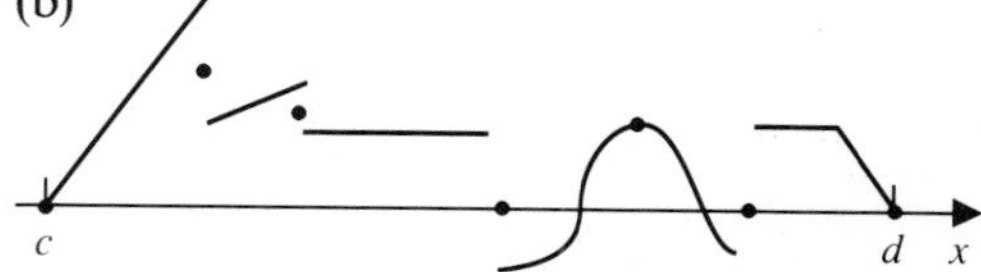

CHAPTER 4

SECTION 2

1. $\frac{400}{\pi} \sum_{n=1,3,5,\ldots}^{\infty} \frac{1}{n} \exp\left(\frac{-kn^2\pi^2 t}{L^2}\right) \sin \frac{n\pi x}{L}$

3. $\pi \sin 30 \sum_{n=1}^{\infty} \frac{(-1)^n n}{900 - n^2\pi^2} \exp\left(\frac{-n^2\pi^2 t}{9 \times 10^6 RC}\right) \sin\left(\frac{n\pi x}{3000}\right)$

5. $\frac{3200}{\pi^3} \sum_{n=1,3,5,\ldots}^{\infty} \frac{1}{n^3} \exp\left(1 - \frac{kn^2\pi^2}{400}\right) t \sin\left(\frac{n\pi x}{20}\right)$

8. $\dfrac{1}{1250\pi^2} \sum_{n=1,3,5,\ldots}^{\infty} \dfrac{1}{n^2} \sin\left(\dfrac{n\pi x}{2}\right) \sin 5000 n\pi t$

9. $\dfrac{80L}{\pi c} \sum_{n=1,3,5,\ldots}^{\infty} \dfrac{1}{n} \sin\left(\dfrac{n\pi x}{2L}\right) \sin\left(\dfrac{n\pi ct}{2L}\right)$

12. $\dfrac{4}{a\pi^2} \sum_{n=1,3,5,\ldots}^{\infty} \dfrac{(-1)^{(n-1)/2}}{n^2} \sin(0.01)n\pi \sin n\pi x \sin an\pi t$

SECTION 3

13. $\dfrac{1}{3} \sum_{n=1}^{\infty} \dfrac{(-1)^{n+1}}{n} \left(\dfrac{\rho}{c}\right)^{6n} \sin 6n\theta$

15. $\dfrac{40}{\pi} \sum_{n=1,3,5,\ldots}^{\infty} \left(\dfrac{r}{20}\right)^n \dfrac{\sin n\theta}{n}$

24. $\dfrac{200}{a^2\pi^4} \sum_{n=1}^{\infty} \dfrac{1+n^2\pi^2}{n^4} \left[\int_0^{100} \sigma h(\sigma) \sin \dfrac{n\pi\sigma}{100}\, d\sigma\right] \dfrac{\sin(n\pi\rho/100)}{\rho} \sin \dfrac{an\pi t}{100}$

25. $\dfrac{1}{\rho^2}(\rho^2 p_\rho)_\rho = \dfrac{1}{a^2} p_{tt}, \quad p(a, t) = 1, \quad \dfrac{\partial p}{\partial \rho}(b, t) = 0,$

$p(\rho, 0) = 0, \quad \dfrac{\partial p}{\partial t}(\rho, 0) = \rho$

SECTION 5

28. $\sum_{n=1,3,5,\ldots}^{\infty} \sin \dfrac{n\pi x}{L} \left[\dfrac{40}{n\pi} \cos \dfrac{an\pi t}{L} - \dfrac{20L}{an^2\pi^2} \sin \dfrac{an\pi t}{L}\right]$

30. $\dfrac{L^2}{6} + 2t + \dfrac{2L^2}{\pi^2} \sum_{n=1}^{\infty} [(-1)^n(1-L) - 1] \cos \dfrac{n\pi x}{L} \cos \dfrac{an\pi t}{L}$

33. $2 - 0.0019x + \dfrac{1}{5\pi} \sum_{n=1}^{\infty} \dfrac{(-1)^n - 20}{n} \sin \dfrac{n\pi x}{100} \exp\left(-\dfrac{n^2\pi^2 t}{10^6 RC}\right)$

35. $\Phi(x) = \dfrac{x}{2a^2}\left[L^2 - \dfrac{x^2}{3}\right]$

$z_{tt} = a^2 z_{xx} \quad z(0, t) = 0, \quad \dfrac{\partial z}{\partial x}(L, t) = 0$

$z(x, 0) = f(x) - \Phi(x) \quad \dfrac{\partial x}{\partial t}(x, 0) = 0$

37. $\Phi(x) = \dfrac{\sin x}{a^2} - x\left(1 + \dfrac{\cos L}{a^2}\right) \quad w_{tt} = a^2 w_{xx} \quad w(0, t) = 0,$

$w_x(L, t) = 0, \quad w(x, 0) = -\Phi(x) \quad w_t(x, 0) = f(x)$

SECTION 6

39. (a) $(e^{2x}y')' + 3e^{2x}y = 0$
(d) $(xy')' + [(x^2 - \upsilon^2)/x]y = 0$
(g) $(y')' - xy = 0$
(j) $(xe^{-x}y')' + ne^{-x}y = 0$

42. (a) $\lambda_n = \dfrac{n^2\pi^2}{9}$ $\lim_{n\to\infty} \lambda_n = \infty$

44. (a) $\lambda_n = -\dfrac{2}{3} - \dfrac{n^2\pi^2}{3L^2}$ $\qquad y_n = \sin \dfrac{n\pi x}{L}$

(c) $\lambda_n = -\dfrac{L^2 + n^2\pi^2}{L^2}$ $\qquad y_n = \sin \dfrac{n\pi x}{L}$

(e) Positive eigenvalues: $\lambda_n = \alpha_n^2$ where α_n is solution of

$$\tan \alpha_n = \frac{2\alpha_n}{1 - \alpha_n^2}$$

and corresponding eigenfunctions

$$y_n = -\alpha_n \cos \alpha_n x + \sin \alpha_n x \qquad n = \pm 1, \pm 2, \dots$$

Negative eigenvalues: Just one $\lambda_1^- = -\beta_1^2$ where β_1 is solution of

$$\tanh \beta_1 = \frac{2\beta_1}{1 + \beta_1^2}$$

and the corresponding eigenfunction is $y = \beta_1 \cosh \beta_1 x + \sinh \beta_1 x$.

SECTION 7

46.

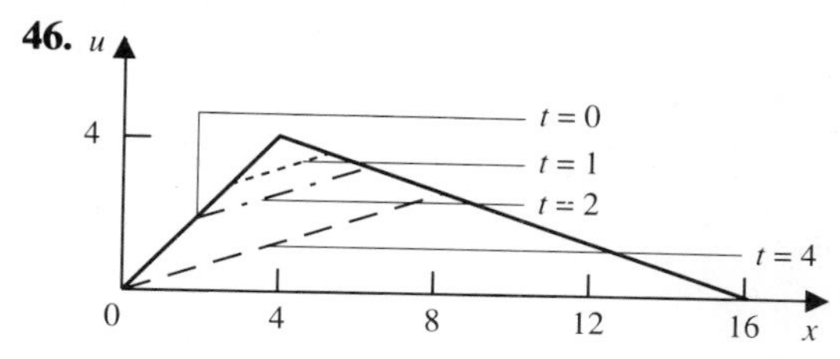

49.

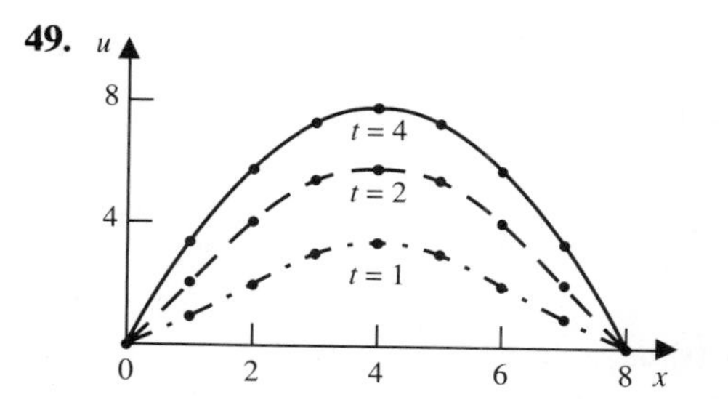

50.

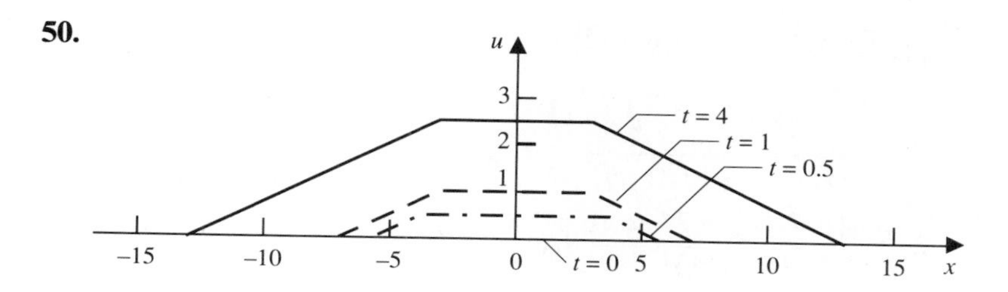

CHAPTER 5

SECTION 3

3. (b) $J_4(x) = x^4\left[1 - \frac{x^2}{20} + \frac{x^4}{960} - \frac{x^6}{80{,}640} + \cdots\right]$

4. (c) $J_{20}(x) = \frac{1}{20!}\left(\frac{x}{2}\right)^{20}\left[1 - \frac{1}{21}\left(\frac{x}{2}\right)^2 + \frac{1}{924}\left(\frac{x}{2}\right)^4 - \frac{1}{63{,}756}\left(\frac{x}{2}\right)^6 + \cdots\right]$

SECTION 4

9. (b) 120
(d) $\pm\infty$ (e) 99

10. (a) 2.98
(c) -0.588 (e) 2.29

SECTION 5

26. (a) $x^{10}J_{10}(x)$
(b) $-x^{-3/2}J_{3/2}(x)$
(c) $x^5J_3(x) + 2x^4J_4(x)$
(d) $-x^{2-\nu}J_\nu(x) - 2x^{1-\nu}J_{\nu-1}(x)$
(f) $-2J_1(x) + \int J_0(x)\,dx$

27. (b) $s^4J_2(s) - 2s^3J_3(s)$

SECTION 6

37. No. The second integral is $\to -\infty$.

SECTION 8

39. $\frac{200}{c}\sum_{j=1}^{\infty}\frac{J_0(\alpha_j x)}{\alpha_j J_1(\alpha_j c)}$

41. $-\frac{20}{c^2}\sum_{j=1}^{\infty}\frac{\alpha_j cJ_1(\alpha_j c) + 2J_0(\alpha_j c)}{[\alpha_j J_3(\alpha_j c)]^2}J_2(\alpha_j x)$

43. $2\sum_{j=1}^{\infty}\frac{5\alpha_j J_1(5\alpha_j) - 2J_2(5\alpha_j)}{[\alpha_j J_1(5\alpha_j)]^2}J_0(\alpha_j x)$

45. $\frac{2}{c^2}\sum_{j=1}^{\infty}\left\{\frac{1}{\alpha_j^4[J_4(\alpha_j c)]^2}\int_0^{\alpha_j c}s^3J_3(s)\,ds\right\}J_3(\alpha_j x)$

48. $400\sum_{j=1}^{\infty}\frac{\alpha_j J_2(10\alpha_j)}{(200\alpha_j^2 - 1)J_1(10\alpha_j)}J_1(\alpha_j x)$

50. $100\sum_{j=1}^{\infty}\frac{\int_0^{\alpha_j c}J_0(s)\,ds - \alpha_j cJ_0(\alpha_j c)}{(\alpha_j^2c^2 - 1)[J_1(\alpha_j c)]^2}J_1(\alpha_j x)$

51. $\frac{1}{2} + 2\sum_{j=1}^{\infty}\frac{\{\alpha_j J_1(\alpha_j) - 2J_2(\alpha_j)]}{[\alpha_j J_0(\alpha_j)]^2}J_0(\alpha_j x)$

SECTION 9

54. (a) $20 \sum_{j=1}^{\infty} \frac{\exp(-k\alpha_j^2 t)}{x_j J_1(x_j)} J_0(\alpha_j r)$

56. (a) $80 \sum_{j=1}^{\infty} \frac{\exp(-k\alpha_j^2 t)}{\alpha_j J_1(5\alpha_j)} J_0(\alpha_j r)$

(b) $t = \frac{8}{k}$ seconds r in centimeters

58. $6666.7 \sum_{j=1}^{\infty} \frac{\cosh(\alpha_j z)}{\alpha_j J_1(0.03\alpha_j) \cosh(0.15\alpha_j)} J_0(\alpha_j r)$ r in meters

64. $\frac{40}{a} \sum_{j=1}^{\infty} \frac{\sin a\alpha_j t}{\alpha_j^2 J_1(2.5\alpha_j)} J_0(\alpha_j r)$ r in meters

68. $\frac{2}{a} \sum_{j=1}^{\infty} \frac{\sin \sqrt{c^2\alpha_j^2 + w_c}\, t[J_0(\alpha_j r)]}{\sqrt{c^2\alpha_j^2 + wc}[\alpha_j J_1(\alpha_j a)]}$

CHAPTER 6

SECTION 1

1. $\frac{1}{\rho^2}\left\{\frac{\partial}{\partial\rho}\left(\rho^2 \frac{\partial T}{\partial\rho}\right) + \frac{\partial^2 T}{\partial\theta^2}\right\} = 0$

2. $\frac{\partial}{\partial\rho}\left[\rho^2 \frac{\partial T}{\partial\rho}\right] = 0$

3. (a) $u(a, \theta) = 100 \qquad 0 \le \theta \le \pi$

(c) $u(L, \theta) = 0 \qquad 0 \le \theta < \frac{\pi}{2}$

$u\left(\rho, \frac{\pi}{2}\right) = 100 \qquad 0 \le \rho < L$

(f) $u(a, \theta) - \frac{\partial u}{\partial\rho}(a, \theta) = 0$

SECTION 2

5. (a) $P_5(x) = \frac{1}{8}[63x^5 - 70x^3 + 15x]$
(b) $P_8(x) = \frac{1}{128}[6435x^8 - 12{,}012x^6 + 6930x^4 - 1260x^2 + 35]$

SECTIONS 3–5

23. (a) $100 + x$
(d) $\frac{1}{3}P_0(x) + \frac{2}{3}P_2(x)$
(g) $\frac{3}{2}P_1(x) - \frac{7}{8}P_3(x) + \frac{11}{16}P_5(x) - \cdots$
(i) $\frac{1}{4}P_0(x) + \frac{1}{2}P_1(x) + \frac{5}{16}P_2(x) + \cdots$
(k) $\frac{3}{2}P_0(x) - P_1(x) - \frac{5}{8}P_2(x) + \cdots$

25. (a) $\cos\theta$

(c) $\frac{1}{4} + \frac{11}{12}\cos\theta + \frac{5}{16}[3\cos^2\theta - 1] + \cdots$

26. $P_0\left(\frac{x}{4}\right) + P_1\left(\frac{x}{4}\right) - \frac{5}{4}P_2\left(\frac{x}{4}\right) + \cdots$

27. (a) (1) $\frac{1}{3}P_0(x) + \frac{2}{3}P_2(x)$

(2) $\frac{3}{4}P_1(x) + \frac{7}{24}P_3(x) - \frac{11}{192}P_5(x) + \cdots$

(d) (1) $0.4597P_0(x) + 0.5248P_2(x) - 0.2052P_4(x) + \cdots$

(2) $0.9036P_1(x) - 0.0633P_3(x) + 0.0032P_5(x) + \cdots$

SECTION 6

28. $\frac{1.4 \times 10^{10}}{\rho}$; $93°\,\mathrm{C}$

31. $\frac{6400}{\rho}\left[199.75P_0(\cos\theta) - \frac{6400}{2\rho}P_1(\cos\theta) - \frac{5}{16}\frac{6400^2}{\rho^2}P_2(\cos\theta) + \cdots\right]$

33. $\left(\frac{c}{\rho}\right)^2 P_1(\cos\theta) + \frac{7}{8}\left(\frac{c}{\rho}\right)^4 P_3(\cos\theta) + \frac{11}{28}\left(\frac{c}{\rho}\right)^6 P_5(\cos\theta) + \cdots$

CHAPTER 7

SECTION 2

1. (a) $A(\omega) = \frac{2}{\omega\pi}\sin 4\omega$; $B(\omega) = 0$

(b) $A(\omega) = \frac{1}{\pi\omega^2}[\cos\omega + \omega\sin\omega - 1]$;

$B(\omega) = \frac{1}{\pi\omega^2}[\sin\omega - \omega\cos\omega]$

(c) $A(\omega) = \frac{1}{\pi\omega^3}[4000\omega\cos 1000\omega + 2(10^6\cdot\omega^2 - 2)\sin 1000\omega]$;

$B(\omega) = 0$

(e) $A(\omega) = \frac{2}{\pi(4+\omega^2)}$; $B(\omega) = \frac{\omega}{\pi(4+\omega^2)}$

2. $A(\omega) = 0$; $B(\omega) = \frac{1}{\pi\omega}$

3. $A(\omega) = \frac{a}{\pi(a^2 - \omega^2)}$; $B(\omega) = 0$

8. $f(t) = \frac{1}{\pi}\int_0^\infty \frac{1}{\omega^2}[(1 - \cos\omega)\cos\omega t + (\omega - \sin\omega)\sin\omega t]\,d\omega$

11. $f(t) = \frac{1}{\pi}\int_0^\infty \left[\frac{\sin(1-\omega)\pi}{1-\omega} - \frac{\sin(1+\omega)\pi}{1+\omega}\right]\sin\omega t\,d\omega$

The term in brackets is replaced by 1 at $\omega = 1$.

13. $u(x, t) = \frac{1}{\pi} \int_0^\infty \exp(-k\alpha^2 t)\left[\frac{\sin\alpha}{\alpha}\cos\alpha x + \frac{1}{\alpha}(1 - \cos\alpha)\sin\alpha x\right] d\alpha$

16. $u(x, t) = \frac{1}{\pi} \int_0^\infty \frac{\cos\alpha x \cos c\alpha t}{1 + \alpha^2} d\alpha$

19. $u(x, t) = 2\frac{1}{\pi} \int_0^\infty \left[\frac{\sin(1 - \alpha)\pi}{1 - \alpha} - \frac{\sin(1 + \alpha)\pi}{1 + \alpha}\right] \cos c\alpha t \sin\alpha x \, d\alpha$

The term inside the brackets is replaced by 1 at $\alpha = 1$,

23. (a) $\varphi(x) = A(e^{-x} - 1)$

(b) $v_{xx} + \frac{1}{k} v_t = 0, \qquad v(0, t) = 0, \qquad v(x, 0) = f(x) + A(1 - e^{-x})$

SECTION 4

25. $F(\omega) = \frac{i}{2\pi\omega}[e^{-i\omega} - 1]$

27. $F(\omega) = \frac{1}{\pi\omega^2}[1 - \cos 2\omega]$

28. $F(\omega) = \frac{1}{\pi\omega^2}[1 - \cos 3\omega + 3\omega \sin 3\omega]$

32. $F(\omega) = \frac{1}{2\pi(b + i\omega)^2}$

SECTION 5

39. $u(x, t) = \frac{2}{\pi} \int_{-\infty}^\infty \frac{1}{\omega}[1 - \cos\pi\omega] \exp(-k\omega^2 t) \sin\omega x \, d\omega$

40. $u(x, t) = \int_{-\infty}^\infty F(\omega) \cos c\omega t e^{i\omega x} \, d\omega$

where $F(\omega) = \frac{1}{2\pi} \int_{-\infty}^\infty f(x) e^{-i\omega x} \, dx$

44. $u(y, t) = 2 \int_0^\infty F(v) \exp(-kv^2 t) \cos vy \, dv$

where $F(v) = \frac{1}{\pi} \int_0^\infty f(y) \cos vy \, dy$

CHAPTER 8

SECTION 1

5. (a) Odd in x, even in y
(c) Neither in x, odd in y

(e) Odd in x and y
(g) Neither in x nor y

6. (a) Odd in x and y
(b) Odd in x and y

7. (a) $\frac{4}{\pi}\sum_{n=1}^{\infty}\frac{(-1)^{n+1}}{n}\sin\frac{n\pi y}{4}$

9. (b) $\pi\sum_{m=1}^{\infty}\frac{(-1)^{m+1}}{m}\sin mx+\frac{8}{\pi}\sum_{m=1}^{\infty}\sum_{n=1,3,5,\ldots}^{\infty}\frac{(-1)^m}{n^2m}\sin mx\cos ny$

(c) $\frac{4}{\pi}\sum_{m=1}^{\infty}\sum_{n=1}^{\infty}\frac{(-1)^n}{nm^3}[(-1)^m(m^2\pi^2-2)+2]\sin mx\sin ny$

SECTION 2

10. (b) $u(x, y, t)=\frac{4000}{\pi^2}\sum_{m=1}^{\infty}\sum_{n=1,3,5,\ldots}^{\infty}\frac{(-1)^{m+1}}{mn}\sin\frac{m\pi x}{10}\sin\frac{n\pi y}{20}$
$\times\exp[-k(\lambda_m+\mu_n)t]$

where $\lambda_m=\frac{m^2\pi^2}{100}$, $\mu_n=\frac{n^2\pi^2}{400}$

(d) $\frac{16T}{\pi^2}\sum_{m=1,3,5,\ldots}^{\infty}\sum_{n=1,3,5,\ldots}^{\infty}\frac{1}{mn}\sin\frac{m\pi x}{8}\sin\frac{n\pi y}{16}\exp[-k(\lambda_m+\mu_n)t]$

where $\lambda_m=\frac{m^2\pi^2}{64}$, $\mu_n=\frac{n^2\pi^2}{256}$

(f) $-\frac{16}{\pi^2}\sum_{m=1}^{\infty}\sum_{n=1}^{\infty}\frac{1}{m^2n}\cos mx\sin ny\exp[-k(m^2+n^2)]t$

$+\sum_{n=1}^{\infty}\frac{1}{n}[1-3(-1)^n]\sin ny\exp[-kn^2t]$

12. $\frac{8}{c\pi^2}\sum_{m=1}^{\infty}\frac{(-1)^{m+1}}{m^2}\sin\frac{m\pi x}{2}\sin\frac{cm\pi t}{2}$

15. $\frac{2304}{\pi^2}\sum_{m=1}^{\infty}\sum_{n=1}^{\infty}\frac{(-1)^{m+n}}{mn}\sin\frac{m\pi x}{6}\sin\frac{n\pi y}{4}\cos c\pi\sqrt{\frac{m^2}{36}+\frac{n^2}{16}}\,t$

16. $\frac{2000}{\pi}\sum_{m=1,3,5,\ldots}^{\infty}\frac{1}{m\cosh m\pi}\sin\frac{m\pi x}{8}\cosh\frac{m\pi z}{8}$

SECTION 4

20. (a) $f_0=11.18$
(b) $f_{12}=20.62$ $f_{21}=14.14$ $f_{31}=18.02$
(c) $\frac{f_{12}}{f_0}=1.84$

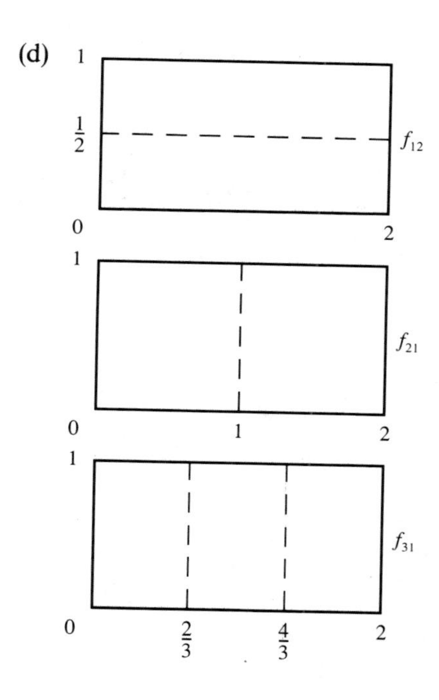

23. (a) $f_0 = 5.89$

(b) $f_{33} = 17.67, \quad f_{45} = 26.68 \quad f_{41} = 17.18$

(c) $\dfrac{f_{33}}{f_0} = \dfrac{17.67}{5.89} = 3, \qquad \dfrac{f_{45}}{f_0} = \dfrac{26.68}{5.89} = 4.53$

(d)

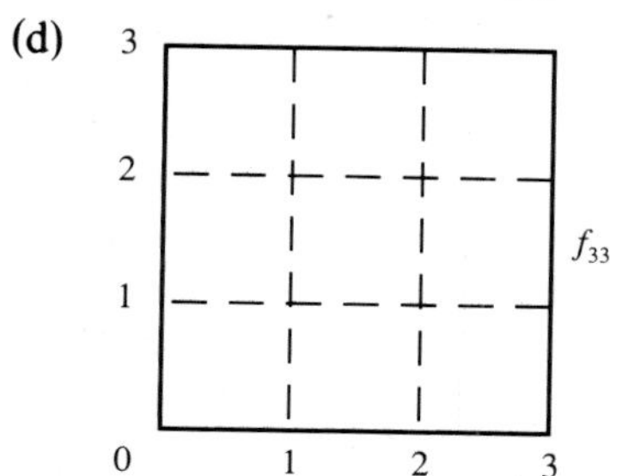

3
12/5
9/5
6/5
3/5
0
3/4
3/2
9/4
3
f_{45}

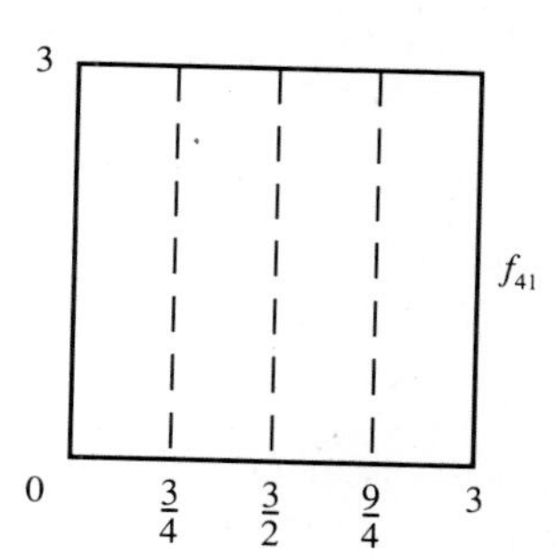

CHAPTER 9

SECTION 2

2. $\delta(x-1)+\delta(x-2)$

5. $f(x)=\begin{cases} x & x>0 \\ 0 & x\le 0 \end{cases}$

7. (a) $\dfrac{1}{r}\,\delta(r)$

SECTION 3

9. Use Equation (9.39).
11. Use Equation (9.29) and (9.39).

SECTION 4

13. Follow the same steps as for the Helmholtz operator.

SECTION 5

19. Replace ρ^{-n} for ρ^{n} in Equation (9.38).

SECTION 7

24. In R^2 a rotation is given by

$$\tilde{\mathbf{x}} = \begin{pmatrix} \cos(\theta), & \sin(\theta) \\ -\sin(\theta), & \cos(\theta) \end{pmatrix}\mathbf{x}$$

26. Seven image points are needed, one in each octant.

CHAPTER 10

SECTION 1

1. Introduce $T=t-\tau$ in Equation (10.15).
3. Integrate Equation (10.15) by parts n times using the initial conditions.

SECTIONS 2 and 3

5. Follow the steps in Example 7.

12. $t+\delta\left(t-\dfrac{ax^2}{2}\right)+c$

CHAPTER 11

SECTION 1

1. $x=0.557$, $y=-0.696$, $z=-0.722$, $w=-1.013$
3. $x=0.476$, $y=-0.190$, $z=0.175$, $w=-0.119$

SECTION 2

5. The coefficient matrix has a dominant diagonal.

7. The exact solution is

$$u = e^{-x}[3 + e(4 + 3e)x] + x - 2$$

SECTION 3

8. Use the analog of Equation (11.39).

10. A particular solution of the equation is

$$r^3\left(\frac{1}{18} + \frac{\cos 2\theta}{10}\right)$$

13. Use Equations (11.20) and (11.39).

SECTION 4

19. An approximation formula for $\partial^2 u/\partial x \partial t$ is

$$\left(\frac{\partial u^2}{\partial x \partial t}\right)_{ij} = \frac{u_{i+1,\,j+1} - u_{i+1,\,j-1} - u_{i-1,\,j+1} + u_{i-1,\,j-1}}{4\,\Delta x\,\Delta t}$$

Bibliography

There are many books dealing with boundary value problems and partial differential equations. We list here a few titles that can provide the reader a more advanced exposition on topics treated in this book.

ANDREWS, L. C. *Special Functions.* New York: Macmillan Publishing Co, 1985.

BERG, P. W. AND J. L. MCGREGOR. *Elementary Partial Differential Equations.* Oakland, Calif.: Holden–Day, 1966.

CARRIER, G. F. AND C. E. PEARSON. *Partial Differential Equations*, 2nd ed. New York: Academic Press, 1988.

GRIFFEL, D. H. *Applied Functional Analysis.* New York: Halsted Press, 1985.

HABERMAN, R. *Elementary Applied Partial Differential Equations.* Englewood Cliffs, N.J.: Prentice-Hall, 1983.

JOHNSON, L. W. AND R. D. RIESS. *Numerical Analysis.* Reading, Mass.: Addison–Wesley, 1982.

Index